AIR MONITORING METHODS
FOR INDUSTRIAL CONTAMINANTS

AIR MONITORING METHODS
FOR INDUSTRIAL CONTAMINANTS

David A. Halliday
Editor

Biomedical Publications
Davis, California

Copyright © 1983 by Biomedical Publications, Box 495, Davis, California 95617, U.S.A. All rights reserved. Printed in the United States of America. No part of this book may be reproduced in any form or by any means without permission from the publisher.

Library of Congress Catalog Card No. 83-073097
ISBN 0-931890-12-8

Library of Congress Cataloging in Publication Data

Halliday, David A., 1944-
    Air monitoring methods for industrial contaminants.

    Bibliography: p.
    Includes index.
    1. Air sampling apparatus. 2. Chemistry, Analytic.
3. Industrial hygiene. 4. Hazardous substances—
Environmental aspects. I. Title.
TD890.H35 1983   628.5'3'0287   83-73097
ISBN 0-931890-12-8

PREFACE

The need to monitor exposure to toxic chemicals in the workplace is as great as ever before. Industries also have a need to be profitable, however, and, with the current state of economic affairs, industrial hygiene monitoring is often viewed as a nonessential luxury. Unless absolutely required by governmental regulations, the quantitative assessment of employee exposure to toxic atmospheres is usually viewed critically by budget analysts and senior plant administrators.

In recent years, commercial suppliers have made available an ever increasing number of industrial instruments for air monitoring of toxic substances. Because they are often portable, direct-reading, and easy to operate, these instruments have largely supplanted manual chemical methods. It is the purpose of this volume to present a compilation of the known commercial devices for quantitatively determining each of 200 different chemical substances in the workplace environment to aid the industrial hygienist in selecting the most cost-effective device for his particular application. The data presented include the initial cost of the system, the measurement technique used, the duration of sampling attainable with the instrument, whether it is portable or fixed, its useful concentration span, the additional cost per test (if any), and the manufacturer. An appendix provides the addresses and telephone numbers of the manufacturers cited.

Also presented in detailed form is one NIOSH method for each chemical, to be used in lieu of or as a backup to an instrumental device. The NIOSH methods were taken from the first six volumes of the NIOSH Manual of Analytical Methods (1977-1980), several of which are now out of print. Where more than one method existed in the Manual for a chemical, the one involving the least operator and hardware sophistication was chosen. In many instances, methods for structurally similar chemicals differed only in the choice of a gas chromatographic column or extraction solvent, and so in the interests of conserving space the reader has been directed to the parent method, with appropriate guidance as to the necessary modifications.

It is hoped that in succeeding editions we will be able to add those suppliers who have been inadvertently omitted from this volume, bring the information up to date in terms of new techniques and other important chemicals, and correct any errors we may have committed.

GLOSSARY

ACGIH	American Conference of Governmental Industrial Hygienists, Cincinnati, Ohio
AIHA	American Industrial Hygiene Association, Akron, Ohio
BA	Beta (particle) absorption
Ceiling limit	An exposure concentration that should never be exceeded
CM	Colorimetry
Cont	Continuous sampling device
CS	Chemical sensor
DT	Detector tube
EC	Electrochemical sensor
GC	Gas chromatograph
GF	Gold film
Grab	Grab sample
IR	Infrared spectrometer
LS	Light scatter
NIOSH	National Institute for Occupational Safety and Health, Cincinnati, Ohio
OSHA	Occupational Safety and Health Administration, Washington, DC
Personal	Personal sampling monitor
PI	Photoionization detector
Simple asphyxiant	Designation for so-called inert gases or vapors that may not have significant effects other than reduction of the oxygen content of the atmosphere
Skin	Refers to the potential of a chemical substance for absorption by the cutaneous route
TLV	Threshold limit value; the airborne concentration of a substance to which a worker may be exposed over a working lifetime without adverse effects. Set by concensus and updated annually by ACGIH.
TWA	Time-weighted average of the exposure concentration, determined by integrating the air concentrations over the total time base of the period under study

CONTENTS

Acetaldehyde 1
Acetic acid 6
Acetic anhydride 11
Acetone 14
Acetonitrile 20
Acetylene tetrabromide 25
Acrolein 30
Acrylonitrile 33
Allyl alcohol 37
Allyl chloride 42
Allyl glycidyl ether 43
Ammonia 49
n-Amyl acetate 51
Aniline 51
Arsine 57

Benzene 62
Benzyl chloride 62
Bis(chloromethyl)ether 63
Bromoform 66
1,3-Butadiene 67
2-Butanone 67
2-Butoxyethanol 68
n-Butyl acetate 73
n-Butyl alcohol 73
sec-Butyl alcohol 78
tert-Butyl alcohol 79
n-Butylamine 79
n-Butyl glycidyl ether 84
n-Butyl mercaptan 85
p-tert-Butyltoluene 90

Carbon dioxide 91
Carbon disulfide 94
Carbon monoxide 100
Carbon tetrachloride 103
Chlorine 104
Chlorobenzene 107
Chlorobromomethane 107
Chloroform 108
1-Chloro-1-nitropropane 108
Chloroprene 114
Chromic acid 120
Cobalt dust or fume 123
Copper dust or fume 127

Cresol	131
Crotonaldehyde	136
Cumene	140
Cyanide	141
Cyclohexane	145
Cyclohexanol	146
Cyclohexanone	146
Cyclohexene	147
Diacetone alcohol	147
o-Dichlorobenzene	148
p-Dichlorobenzene	148
Dichlorodifluoromethane	149
1,1-Dichloroethane	155
1,2-Dichloroethylene	155
Dichloroethyl ether	156
Dichloromonofluoromethane	156
Dichlorotetrafluoroethane	157
Dichlorvos	157
Diethylamine	163
2-Diethylaminoethanol	163
Difluorodibromomethane	169
Diisobutyl ketone	174
Diisopropylamine	174
Dimethylacetamide	175
Dimethylformamide	180
Dimethyl sulfate	180
2,4-Dinitrotoluene	185
p-Dioxane	189
Epichlorohydrin	190
Ethanolamine	190
2-Ethoxyethanol	191
2-Ethoxyethyl acetate	191
Ethyl acetate	192
Ethyl acrylate	192
Ethyl alcohol	193
Ethylamine	193
Ethylbenzene	194
Ethyl bromide	194
Ethyl butyl ketone	195
Ethyl chloride	195
Ethylene chlorohydrin	196
Ethylenediamine	196
Ethylene dibromide	201
Ethylene dichloride	206
Ethyleneimine	206
Ethylene oxide	211
Ethyl ether	217

Ethyl formate 222
Ethyl silicate 222

Fluorotrichloromethane 228
Formaldehyde 228
Formic acid 234
Furfural 238
Furfuryl alcohol 243

Glycidol 243

n-Heptane 248
Hexachloroethane 249
n-Hexane 249
2-Hexanone 250
Hexone 250
Hydrazine 251
Hydrogen chloride 254
Hydrogen cyanide 258
Hydrogen fluoride 259
Hydrogen sulfide 263

Iron oxide fume 269
Isoamyl acetate 273
Isoamyl alcohol 274
Isobutyl acetate 274
Isobutyl alcohol 275
Isophorone 275
Isopropyl acetate 276
Isopropyl alcohol 276
Isopropylamine 277
Isopropyl ether 277

Liquified petroleum gas 278

Magnesium oxide fume 279
Mercury vapor 283
Mesityl oxide 291
Methyl acetate 291
Methyl acetylene 292
Methyl acrylate 296
Methylal 296
Methyl alcohol 297
Methylamine 302
Methyl n-amyl ketone 302
Methyl bromide 303
Methyl cellosolve 303
Methyl cellosolve acetate 304
Methyl chloride 304

Methyl chloroform ... 305
Methylcyclohexane ... 305
Methylcyclohexanol ... 306
o-Methylcyclohexanone ... 306
Methylene chloride ... 307
Methyl formate ... 307
5-Methyl-3-heptanone ... 308
Methyl iodide ... 308
Methylisobutylcarbinol ... 313
Methyl methacrylate ... 314
α-Methylstyrene ... 314
Monomethylaniline ... 315
Morpholine ... 315

Nickel ... 316
Nitric acid ... 319
Nitric oxide ... 324
Nitrobenzene ... 329
Nitroethane ... 334
Nitrogen dioxide ... 334
Nitroglycol ... 335
Nitromethane ... 340
Nitrotoluene ... 340

Octane ... 341
Ozone ... 341

n-Pentane ... 344
2-Pentanone ... 345
Petroleum distillate ... 345
Phenol ... 346
Phenyl ether ... 350
Phenylhydrazine ... 350
Phosgene ... 353
Phosphine ... 356
Propane ... 361
n-Propyl acetate ... 362
n-Propyl alcohol ... 362
Propylene dichloride ... 363
Propylene oxide ... 363
n-Propyl nitrate ... 364
Pyridine ... 364

Rhodium dust or fume ... 369

Silica dust ... 373
Stoddard solvent ... 376
Styrene ... 377
Sulfur dioxide ... 378

Sulfur hexafluoride 382
Sulfuric acid 386
Sulfuryl fluoride 389

1,1,2,2-Tetrachloro-1,2-difluoroethane 392
1,1,2,2-Tetrachloroethane 393
Tetrachloroethylene 393
Tetrahydrofuran 394
Toluene 394
Toluene diisocyanate 395
o-Toluidine 399
1,1,2-Trichloroethane 399
Trichloroethylene 400
1,2,3-Trichloropropane 400
1,1,2-Trichloro-1,2,2-trifluoroethane 401
Triethylamine 401
Trifluoromonobromomethane 402
Turpentine 402

Vanadium pentoxide fumes 403
Vinyl acetate 406
Vinyl chloride 411
Vinylidene chloride 418
Vinyltoluene 418

Xylene 419

Appendix 420

Index 423

ACETALDEHYDE

ACGIH TLV: 100 ppm (180 mg/m^3)
OSHA Standard: 200 ppm

CH_3CHO

Method	Sampling Duration	Sampling Location	Useful Range (ppm)	System Cost ($)	Test Cost ($)	Manufacturer
DT	Grab	Portable	100-1000	150	2.50	Nat'l Draeger
DT	Grab	Portable	40-10,000	165	1.70	Matheson
DT	Grab	Portable	10-750	180	2.25	Bendix
IR	Cont	Portable	1-400	4,374	0	Foxboro
PI	Cont	Fixed	2-200	4,950	0	HNU
GC	1 min	Portable	1-10^6	10,000	0	Microsensor

NIOSH METHOD NO. S345

Principle

A known volume of air is drawn through a midget bubbler containing a buffered solution of Girard T reagent. Acetaldehyde is derivatized with the Girard T reagent. The acetaldehyde-Girard T derivative is analyzed by high pressure liquid chromatography.

Range and Sensitivity

This method was validated over the range of 170-670 mg/m^3 at an atmospheric temperature of 21° C. and pressure of 756 mm Hg, using a 60 L sample. The upper limit of the range of the method is dependent on the concentration of Girard T reagent. While the collection efficiency has not been tested outside of the method range, at least a 2-fold molar excess of Girard T reagent should be maintained over the total amount of acetaldehyde and other carbonyl compounds sampled.

Under the instrumental conditions used in the validation study, a 50 μL injection of a 0.07 mg/mL standard solution of acetaldehyde in Girard T reagent solution resulted in a peak whose height was 40% of full scale. The HPLC detector had an attenuation setting of 0.16 absorbance unit full scale and the electronic integrator had an attenuation of 64.

The detection limit of the method is estimated to be less than 0.325 μg acetaldehyde, corresponding to a 50 μL aliquot of a 6.5 μg/mL standard. This corresponds to less than 0.1 mg/bubbler sample. The limit of detection may be extended by not diluting the collected sample before analysis and by attenuating the liquid chromatograph output less.

Interferences

When interfering compounds are known or suspected to be present in the air, such information, including their suspected identities, should be transmitted with the sample. Other volatile aldehydes or ketones such as formaldehyde, propionaldehyde, acrolein, or acetone may cause significant interference or compete with acetaldehyde for reaction with Girard T reagent. Chromatographic conditions can be adjusted to separate the various substances.

Precision and Accuracy

The coefficient of variation for the total analytical and sampling method in the range of 170-670 mg/m^3 was 0.053. This value corresponds to a 19 mg/m^3 standard deviation at the OSHA standard level. Statistical information can be found in reference 1. Details of the test procedures are found in reference 2.

In validation experiments, this method was found to be capable of coming within 25% of the "true value" on the average 95% of the time over the validation range. The concentrations obtained at 0.5, 1, and 2 times the OSHA environmental limit averaged 1.2% higher than the dynamically generated test concentrations (n = 18). The analytical method recovery was determined to be 1.004 for a collector loading of 11 mg. Storage stability studies on samples collected from a test atmosphere at a concentration of 374 mg/m^3 indicated that collected samples are stable for at least 7 days if protected from exposure to light. The mean of samples analyzed after 7 days were within 1% of the mean of samples analyzed immediately after collection. Collection efficiency of the midget bubbler containing Girard T reagent was determined to be at least 0.998 within the range tested; therefore, no correction for collection efficiency is necessary. Experiments performed in the validation study are described in reference 2.

Advantages and Disadvantages

The acetaldehyde-Girard T reagent derivative has adequate storage stability if protected from light. Collected samples are analyzed by a quick instrumental method. A disadvantage of the method is the awkwardness in using midget bubblers for collecting personal samples. If the worker's job performance requires much body movement, loss of the collection solution during sampling may occur.

The bubblers are more difficult to ship than absorption tubes or filters due to possible breakage and leakage of the bubblers during shipping. The Girard T reagent solution should be stored in annealed glassware in the dark. In the validation study, an interfering substance was formed when the Girard T reagent solution was stored in annealed glassware over a period of 20 days.

Apparatus

Glass midget bubblers with fritted glass stems.

Personal sampling pump. A calibrated personal sampling pump whose flow rate can be determined to an accuracy of 5%. The sampling pump is protected from splashover or solvent condensation by a trap. The trap is a midget bubbler or impinger with the stem broken off which is used to collect spillage. The trap is attached to the pump with a metal holder. The outlet of the trap is connected to the pump by flexible tubing. Each sampling pump must be calibrated with a representative bubbler and trap in the line to minimize errors associated with uncertainties in the volume sampled.

Thermometer.

Manometer.

High pressure liquid chromatograph (HPLC). Equipped with a variable wavelength uv detector set at 245 nm and a sample injection valve with a 50 µL external sample loop.

HPLC column. Packed with Zipax SCX (50 cm long × 2 mm I.D. stainless steel). This column can be obtained from Dupont.

Electronic integrator. Or some other suitable method for measuring peak areas.

Syringes. 2 mL Luer-lock.

Fractional distillation apparatus. For preparing high purity acetaldehyde.

Volumetric flasks. 500 mL and other convenient sizes for preparing standard solutions.
Pipets. Convenient sizes for preparing stock standard solutions and for measuring the collection medium.
China marker.
Teflon plugs. Or equivalent for sealing the inlet and outlet of the bubbler stem before shipping.

Reagents

All reagents used must be ACS reagent grade or better.
Acetaldehyde.
Citric acid.
Disodium hydrogen phosphate.
Girard T reagent. (Carboxymethyl)trimethylammonium chloride hydrazide, recrystallized from 95% ethanol.
Distilled, deionized water.
Ethanol, 95%.
Sodium dihydrogen phosphate monohydrate.
0.2 M Girard T Reagent. Dissolve 5.39 g citric acid, 6.63 g disodium hydrogen phosphate, and 16.77 g Girard T reagent in approximately 400 mL of distilled, deionized water. Transfer the solution to a 500 mL volumetric flask and bring to volume with distilled, deionized water. This solution must be used within 2 weeks and should be stored in annealed glassware in the dark.
HPLC eluant. 0.022 M Na_2HPO_4 and 0.019 M NaH_2PO_4 in 20% ethanol. Prepare a stock eluant of 31.2 g Na_2HPO_4 and 26.2 g $NaH_2PO_4.H_2O$ made up to 1 liter in distilled, deionized water. To prepare the HPLC eluant, combine 100 mL of stock and 200 mL of 95% ethanol. Dilute this solution to 1 liter with distilled, deionized water. Filter the solution through a 5 micron Teflon filter and degas prior to use. Slowly bubble helium through the eluant reservoir during use to prevent bacterial growth. This solution is also used for sample dilution prior to analysis.
Stock standard acetaldehyde-Girard T derivative. Prepare a standard containing 4.32 mg/mL acetaldehyde in 0.2 M Girard T reagent by weighing 216 mg of freshly distilled acetaldehyde into a 50 mL volumetric flask containing approximately 49 mL of Girard T reagent solution. Make up to volume with Girard T reagent solution. This solution must be used within 1 day.

Procedure

Cleaning of equipment. All glassware used for the laboratory analysis should be detergent washed and thoroughly rinsed with tap water and distilled water, and dried. All midget bubblers should be heated in an oxidizing atmosphere at a temperature of approximately 580° C. This procedure will remove organic contaminants.
Collection and shipping of samples. Pipet 15 mL of the buffered Girard T reagent into each midget bubbler. Mark the liquid level on the bubbler with a china marker. The outlet of the midget bubbler is attached to a trap which is used to protect the pump during personal sampling. The trap is a midget impinger or bubbler with the stem broken off which is used to collect spillage. The trap is attached to the pump with a metal holder. The outlet of the trap is connected to the pump by flexible tubing.

Air being sampled should not be passed through any hose or tubing before entering the midget bubbler. A sample size of 60 L is recommended. Sample at a flow rate between 0.1-0.5 L/min. Do not sample at a flow rate less than 0.1 L/min. Sampling at higher flow rates causes frothing of the collection medium.

Turn the pump on and begin sample collection. Terminate sampling at the predetermined time and record sample flow rate, collection time, and ambient temperature and pressure. If pressure reading is not available, record the elevation. Also record the type of sampling pump used. The inlet and outlet of the bubbler stem should be sealed by inserting Teflon plugs, or equivalent, in the inlet and outlet. Do not seal with rubber. The standard taper joint of the bubbler should be taped securely to prevent leakage during shipping. Care should be taken to minimize spillage or loss of sample by evaporation at all times. Wrap the bubblers with foil to protect them from light.

The bubblers should be shipped in a suitable container designed to prevent damage in transit. The samples should be shipped to the laboratory as soon as possible. With each batch or partial batch of ten samples, submit one bubbler containing 15 mL of the collection medium prepared from the same stock as that used for sample collection. This bubbler must be subjected to exactly the same handling as the samples except that no air is drawn through it. Label this bubbler as the blank.

Do not ship the material in the trap. If more than 1 mL of material is collected in the trap after sampling, the sample should be considered invalid. A sample of the bulk material should be submitted in a glass container with a Teflon-lined cap or equivalent. This sample must not be transported in the same container as the bubblers.

Analysis of samples. Sample analysis should be done in a room with a fairly stable temperature to prevent retention time fluctuations on the ion exchange column. Each sample and blank is analyzed separately. Remove the bubbler stem and tap the stem lightly against the flask to drain the contents into the bubbler flask. If necessary bring the liquid volume to the 15 mL mark with distilled water. Transfer a 5 mL aliquot to a 100 mL flask and bring to volume with the HPLC eluant.

HPLC conditions. The typical operating conditions for the high pressure liquid chromatograph are:

Column temperature: ambient
Flow rate: 0.75 mL/min
Mobile phase: 0.022 M Na_2HPO_4 and 0.019 M NaH_2PO_4 in 20% ethanol
Detector: UV at 245 nm

Under the above conditions, the Girard T reagent will elute in approximately 2 minutes and the acetaldehyde-Girard T derivative will elute in approximately 7 minutes.

Injection. The first step in the analysis is to inject the sample into the high pressure liquid chromatograph. The chromatograph is fitted with a sample injection valve and a 50 μL sample loop. Flush this loop thoroughly with the sample (500 μL), and inject the sample. Duplicate injections should compare within 3%. The area of the sample peak is measured by an electronic integrator or some other suitable form of area measurement, and results are read from a standard curve prepared as discussed below.

Calibration and Standards

A series of standards, varying in concentration over the range corresponding to approximately 0.1 to 3 times the OSHA standard for the substance under study, is prepared and analyzed under the same HPLC conditions and during the same time period as the unknown samples. Curves are established by plotting concentration in mg/15 mL versus peak area. Note: Since no internal standard is used in this method, standard solutions must be analyzed at the same time as the samples. This will minimize the effect of known day-to-day variations and variations during the same day of the UV detector response.

Prepare working standards by diluting appropriate aliquots from the stock standard solution with the buffered Girard T solution to the desired concentration. Prepare at least 5 working standards to cover the range of 2-64 mg/sample (130-4300 μg/mL). This range is based on a 60 L sample.

Transfer a 5 mL aliquot of each working standard to a 100 mL volumetric flask and bring to volume with HPLC eluant. Dilution and preparation of standards in this way ensures that the derivative/Girard T reagent ratio will be the same as in collected samples. Analyze the samples as described above. Prepare at least one blank and analyze it at the same time as the standards. Prepare a standard calibration curve by plotting concentration of acetaldehyde in mg/15 mL versus peak area.

Calculations

Read the weight, in mg, corresponding to each peak area from the standard curve. No volume correction is needed, because the standard curve is based on mg/15 mL Girard T reagent and the volume of sample injected is identical to the volume of the standards injected, and the final dilution with HPLC eluant is the same for both standards and samples. A correction for the blank must be made for each sample:

mg = mg sample - mg blank

where:

mg sample = mg found in sample bubbler
mg blank = mg found in blank bubbler

For personal sampling pumps with rotameters only, the following volume correction should be made:

Corrected volume = $ft(P_1T_2/P_2T_1)^{1/2}$

where:

f = flow rate sampled
t = sampling time
P_1 = pressure during calibration of sampling pump (mm Hg)
P_2 = pressure of air sampled (mm Hg)
T_1 = temperature during calibration of sampling pump (°K)
T_2 = temperature of air sampled (°K)

The concentration of acetaldehyde in the air sample can be expressed in mg/m^3:

mg/m^3 = mg x 1000 (L/m^3) ÷ corr. air volume sampled (L)

References

1. *Documentation of NIOSH Validation Tests,* National Institute for Occupational Safety and Health, Cincinnati, Ohio (DHEW-NIOSH Publication #77-185), 1977.
2. Backup Data Report No. S345 for Acetaldehyde, prepared under NIOSH Contract NO. 210-76-0123.

ACETIC ACID

ACGIH TLV: 10 ppm (25 mg/m³)
OSHA Standard: 10 ppm

CH_3COOH

Method	Sampling Duration	Sampling Location	Useful Range (ppm)	System Cost ($)	Test Cost ($)	Manufacturer
DT	Grab	Portable	5-80	150	2.60	Nat'l Draeger
DT	Grab	Portable	1-80	180	2.25	Bendix
IR	Cont	Portable	1-20	4374	0	Foxboro

NIOSH METHOD NO. S169

Principle
A known volume of air is drawn through a tube containing activated coconut charcoal to trap acetic acid vapors. Acetic acid is desorbed from the charcoal with formic acid, and the sample is analyzed by gas chromatography.

Range and Sensitivity
This method was validated over the range of 12.5-50 mg/m³ at an atmospheric temperature of 22° C. and atmospheric pressure of 767 mm Hg, using a 173 L sample. This sample size is based on the capacity of the charcoal to collect vapors of acetic acid in air. The method may be capable of measuring smaller amounts if the desorption efficiency is adequate. Desorption efficiency must be determined over the range used.

The upper limit of the range of the method depends on the adsorptive capacity of the charcoal. This capacity may vary with the concentrations of acetic acid and other substances in the air. Breakthrough is defined as the time that the effluent concentration from the collection tube (containing 100 mg of charcoal) reaches 5% of the concentration in the test gas mixture. In a breakthrough test, which was discontinued after 4.6 hours, the charcoal tube collected 10.4 mg of acetic acid without breakthrough for a 269 L air sample with a relative humidity of 90% and a temperature of 23° C. This is equivalent to approximately 2.5 times the OSHA standard for this sample size. It is likely that breakthrough would not occur with even larger amounts of acetic acid.

Interferences
When other compounds are known or suspected to be present in the air, such information, including their suspected identities, should be transmitted with the sample. Any compound that has the same retention time as acetic acid at the operating conditions described in this method is an interference. Retention time data on a single column cannot be considered proof of chemical identity.

Precision and Accuracy
The coefficient of variation for the total analytical and sampling method in the range of 12.5-50 mg/m³ was 0.058. This value corresponds to a 1.5 mg/m³ standard deviation at the OSHA standard level. Statistical information can be found in reference 2. Details of the test procedures

are found in reference 3.

On average the concentrations obtained in the laboratory validation study at 0.5X, 1X, and 2X the OSHA standard were 5.4% higher than the concentrations calculated for the reference method, which was based on syringe-drive introduction of the acetic acid into the sample generator. Any difference between the found concentration and the reference concentration (sometimes called the "true" concentration) may not represent a bias in the sampling and analytical method, but rather an uncertainty in the reference concentration. It was therefore assumed that there is no bias in the sampling and analytical method and that no correction for bias is needed. The coefficient of variation is a good measure of the accuracy of the method since the recoveries and storage stability were good. Storage stability studies on samples collected from a test atmosphere at a concentration of 25 mg/m^3 indicate that collected samples are stable for at least 7 days.

Advantages and Disadvantages

The sampling device is small, portable, and involves no liquids. Interferences are minimal, and most of those that occur can be eliminated by altering chromatographic conditions. The tubes are analyzed by means of a quick, instrumental method. One disadvantage of the method is that the amount of sample that can be taken is limited by the number of mg that the tube will hold before overloading. When the amount of acetic acid found on the backup section of the charcoal tube exceed 25% of that found on the front section, the probability of sample loss exists. The precision of the method is limited by the reproducibility of the pressure drop across the tubes. This drop will affect the flow rate and cause the volume to be imprecise, because the pump is usually calibrated for using one tube only.

Apparatus

Personal sampling pump. A calibrated personal sampling pump whose flow rate can be determined within 5% at the recommended flow rate.

Charcoal tubes. Glass tube with both ends flame sealed, 7 cm long with a 6 mm O.D. and a 4 mm I.D., containing two sections of 20/40 mesh activated charcoal separated by a 2 mm portion of urethane foam. The activated charcoal is prepared from coconut shells and is fired at 600° C. prior to packing. The adsorbing section contains 100 mg of charcoal, the backup section 50 mg. A 3 mm portion of urethane foam is placed between the outlet end of the tube and the backup section. A plug of silylated glass wool is placed in front of the adsorbing section. The pressure drop across the tube must be less than 25 mm of mercury at a flow rate of 1 L/min.

Gas chromatograph. Equipped with a flame ionization detector.

Column. 1 m long × 4 mm I.D. glass packed with 60/80 mesh Carbopack B/3% Carbowax 20M/0.5% H_3PO_4.

Electronic integrator. Or some other suitable method of determining peak areas.

Sample containers. Two mL glass sample containers with glass stoppers or Teflon caps.

Syringes. 10 microliter and other convenient sizes for preparing standards.

Pipets. Delivery type, 1.0 mL and other convenient sizes.

Volumetric flasks. 10 mL and other convenient sizes for preparing standard solutions.

Stopwatch.

Manometer.

Reagents

Formic acid, 88%. To prevent interference, the formic acid must contain less than 0.02% acetic acid. Obtain the lot analysis from the manufacturer.

Glacial acetic acid, reagent grade.

Nitrogen, purified.
Hydrogen, prepurified.
Air, filtered, compressed.

Procedure

Cleaning of equipment. All glassware used for the laboratory analysis should be detergent washed, thoroughly rinsed with tap water and distilled water, and dried.

Calibration of sampling pumps. Each personal sampling pump must be calibrated with a representative charcoal tube in the line to minimize errors associated with uncertainties in the volume sampled.

Collection and shipping of samples. Immediately before sampling, break the ends of the charcoal tube to provide an opening at least one-half the internal diameter of the tube. All tubes must be from the same manufacturer's lot.

The smaller section of charcoal is used as a backup and should be positioned nearer the sampling pump. The charcoal tube should be placed in a vertical direction during sampling to minimize channeling through the charcoal. Air being sampled should not be passed through any hose or tubing before entering the charcoal tube.

A sample size of 168 L is recommended. Sample at a flow rate of 1.0 L/min or less. Record the sampling time, flow rate, and type of sampling pump used. The temperature, pressure, and relative humidity of the atmosphere being sampled should be recorded. If pressure reading is not available, record the elevation.

The charcoal tube should be capped with plastic caps immediately after sampling. Under no circumstances should rubber caps be used. With each batch of ten samples, submit one tube from the same lot of tubes used for sample collection. This tube must be subjected to exactly the same handling as the samples except that no air is drawn through it. This tube should be labeled as the blank. A minimum of 18 extra charcoal tubes should be shipped for desorption efficiency determinations. Capped tubes should be packed tightly and padded before they are shipped to minimize tube breakage during shipping.

Analysis of samples. In preparation for analysis, each charcoal tube is scored with a file in front of the first section of charcoal and broken open. The glass wool is removed and discarded. The charcoal in the first (larger) section is transferred to a 2 mL stoppered sample container. The separating section of foam is removed and discarded; the second section is transferred to another sample container vial. These two sections are analyzed separately.

Desorption of samples. Prior to analysis, 1.0 mL of formic acid is pipetted into each sample container. Desorption should be done for 60 minutes. Tests indicated that this is adequate if the sample is agitated occasionally during this period. The sample vials should be capped as soon as the solvent is added to minimize volatilization. Care should be taken to avoid contacting formic acid and/or acetic acid with the skin. These reagents may cause severe burns.

GC conditions. The typical operating conditions for the gas chromatograph are:

60 mL/min (60 psig) nitrogen carrier gas flow
50 mL/min (24 psig) hydrogen gas flow to detector
500 mL/min (50 psig) air flow to detector
230° C. injector temperature
230° C. detector manifold temperature
Column temperature, program from 130-180° C. as described below

A retention time of approximately 3-4 minutes is to be expected for acetic acid under these conditions and using the column recommended.

Injection. The first step in the analysis is the injection of the sample into the gas chromatograph. To eliminate difficulties arising from blow back or evaporation of solvent within the syringe needle, one should employ the solvent flush injection technique. The 10 microliter syringe is first flushed with solvent several times to wet the barrel and plunger. Three microliters of solvent are drawn into the syringe to increase the accuracy and reproducibility of the injected sample volume. The needle is removed from the solvent, and the plunger is pulled back about 0.2 microliter to separate the solvent flush from the sample with a pocket of air to be used as a marker. The needle is then immersed in the sample, and a 5 microliter aliquot is withdrawn, taking into consideration the volume of the needle, since the sample in the needle will be completely injected. After the needle is removed from the sample and prior to injection, the plunger is pulled back 1.2 microliters to minimize evaporation of the sample from the tip of the needle. Observe that the sample occupies 4.9-5.0 microliters in the barrel of the syringe. Begin temperature programming at a rate of 10° C. per minute. The initial temperature should be 130° C. and the final temperature should be 180° C. Duplicate injections of each sample and standard should be made. No more than a 3% difference in area is to be expected. The area of the sample peak is measured by an electronic integrator or some other suitable form of area measurement, and results are read from a standard curve prepared as discussed below.

Determination of desorption efficiency. The desorption efficiency of a particular compound can vary from one laboratory to another and also from one batch of charcoal to another. Thus, it is necessary to determine the fraction of the specific compound that is removed in the desorption process for a particular batch of charcoal.

Activated charcoal equivalent to the amount in the first section of the sampling tube (100 mg) is measured into a 64 mm, 4 mm I.D. glass tube, flame sealed at one end. This charcoal must be from the same batch as that used in obtaining the samples and can be obtained from unused charcoal tubes. The open end is capped with Parafilm. A known amount of acetic acid is injected directly into the activated charcoal with a microliter syringe, and the tube is capped with more Parafilm. The density of acetic acid is 1.049 g/mL at 20° C. The amount injected is equivalent to that present in a 168 L air sample at the selected level.

Six tubes at each of three levels (0.5X, 1X, and 2X the OSHA standard) are prepared in this manner and allowed to stand for at least overnight to assure complete adsorption of the acetic acid onto the charcoal. These tubes are referred to as the samples. A parallel blank tube should be treated in the same manner except that no sample is added to it. The sample and blank tubes are desorbed and analyzed in exactly the same manner as the sampling tube described above.

Two or three standards are prepared by injecting the same volume of acetic acid into 1.0 mL of formic acid with the same syringe used in the preparation of the samples. These are analyzed with the samples. These standards are used to confirm the calibration of the gas chromatograph. The desorption efficiency (D.E.) equals the average weight in mg recovered from the tube divided by the weight in mg added to the tube, or:

D.E. = average weight recovered (mg) ÷ weight added (mg)

The desorption efficiency is dependent on the amount of acetic acid collected on the charcoal. Plot the desorption efficiency versus weight of acetic acid found. This curve is used to correct for adsorption losses.

Calibration and Standards

A series of standards, varying in concentration over the range corresponding to approximately 0.1 to 3 times the OSHA standard for the sample under study, is prepared and analyzed under the same GC conditions and during the same time period as the unknown samples. Curves are established by plotting concentration in mg/1.0 mL versus peak area. Note: Since no internal standard is used in this method, standard solutions must be analyzed at the same time that the sample analysis is done. This will minimize the effect of known day-to-day variations and variations during the same day of the FID response.

Prepare a stock standard solution containing 42 mg/mL acetic acid in formic acid. From the above stock solution, appropriate aliquots are withdrawn and dilutions are made in formic acid. Prepare at least 5 working standards to cover the range of 0.42-12.6 mg/1.0 mL. This range is based on a 168 L sample. Prepare a standard calibration curve by plotting concentration of acetic acid in mg/1.0 mL versus peak area.

Calculations

Read the weight, in mg, corresponding to each peak area from the standard curve. No volume corrections are needed because the standard curve is based on mg/1.0 mL formic acid and the volume of sample injected is identical to the volume of the standards injected. Corrections for the blank must be made for each sample:

mg = mg sample - mg blank

where:

mg sample = mg found in front section of sample tube
mg blank = mg found in front section of blank tube

A similar procedure is followed for the backup sections. Add the weights found in the front and backup sections to determine the total weight of the sample. Read the desorption efficiency from the curve for the amount found in the front section. Divide the total weight by this desorption efficiency to obtain the corrected mg/sample:

Corrected mg/sample = total weight ÷ D.E.

For personal sampling pumps with rotameters only, the following correction should be made:

Corrected volume = ft $(P_1T_2/P_2T_1)^{1/2}$

where:

f = flow rate sampled
t = sampling time
P_1 = pressure during calibration of sampling pump (mm Hg)
P_2 = pressure of air sampled (mm Hg)
T_1 = temperature during calibration of sampling pump (°K)
T_2 = temperature of air sampled (°K)

The concentration of acetic acid in the air sampled can be expressed in mg/m^3:

mg/m^3 = corrected mg × 1000 (L/m^3) ÷ corrected air volume sampled (L)

References
1. White, L.D. et al. A convenient optimized method for the analysis of selected solvent vapors in the industrial

atmosphere. Amer. Ind. Hyg. Assoc. J. 31: 225, 1970.
2. *Documentation of NIOSH Validation Tests,* NIOSH Contract CDC-99-74-49.
3. Backup Data Report for Acetic Acid, prepared under NIOSH Contract No. 210-76-0123.

ACETIC ANHYDRIDE

ACGIH TLV: 5 ppm (20 mg/m^3) ceiling $(CH_3CO)_2O$
OSHA Standard: 5 ppm

Method	Sampling Duration	Sampling Location	Useful Range (ppm)	System Cost ($)	Test Cost ($)	Manufacturer
DT	Grab	Portable	0.5-40	180	2.25	Bendix
IR	Cont	Portable	1-10	4374	0	Foxboro
GC	1 min	Portable	1-10^6	10,000	0	Microsensor

NIOSH METHOD NO. S170

Principle

Acetic anhydride is collected in a standard midget bubbler charged with an alkaline solution containing hydroxylamine. The resulting product is reacted with ferric chloride to form a purple complex. The absorption maximum at 540 nm is used to measure the acetic anhydride reaction product.

Range and Sensitivity

This method was validated over the range of 9.35-37.4 mg/m^3 at an atmospheric temperature and pressure of 22° C. and 761 mm Hg. The probable useful range of the method is 5-45 mg/m^3 based on the range of standards used to prepare the standard curve. For samples of high concentration where the absorbance is greater than the limits of the standard curve, the samples can be diluted with the collection medium prior to color development to extend the upper limit of the range. A concentration of 5 mg/m^3 of acetic anhydride can be determined in a 100 L air sample based on a difference of 0.05 absorbance unit from the blank using a 1 cm cell. Greater sensitivity could be obtained by use of a longer path length cell.

Interferences

Any material containing an organic carbonyl group will interfere. Among such materials are esters, acid chlorides, and aldehydes. A potential interfering substance that is most likely to coexist with acetic anhydride in air is ketene.

Precision and Accuracy

The coefficient of variation for the total analytical and sampling method in the range of 9.35-37.4 mg/m^3 was 0.060. This value corresponds to a 1.2 mg/m^3 standard deviation at the OSHA standard level. Statistical information and details of the validation and experimental test procedures can be found in reference 2.

A collection efficiency of 1.00 ± 0.01 was determined for the collection medium; thus no bias was introduced in the sample collection step. On the average the concentrations obtained at the OSHA standard level using the overall sampling and analytical method were 2.5% higher than the "true" concentrations for a limited number of laboratory experiments. Any difference between the "found" and "true" concentrations may not represent a bias in the sampling and analytical method, but rather a random variation from the experimentally determined "true" concentration. Therefore, no recovery correction should be applied to the final result.

Advantages and Disadvantages

The samples collected in bubblers are analyzed by means of a quick, instrumental method. The alkaline solution containing hydroxylamine can produce dermatitis if it comes in contact with the skin. The method is simple and requires approximately 1.5 hours to complete.

Apparatus

Glass standard midget bubbler. With stem which has a fritted glass end and contains the absorbing solution. The fritted end should have porosity approximately equal to that of Corning EC (170-220 micron maximum pore diameter).

Pump. Suitable for delivering at least 1 L/min for 100 minutes. The sampling pump is protected from splashover or water condensation by a 5 cm (6 mm I.D. and 8 mm O.D.) glass tube loosely packed with a plug of glass wool and inserted between the exit arm of the impinger and the pump.

Integrating volume meter. Such as a dry gas meter or wet test meter.

Thermometer.

Manometer.

Stopwatch.

Spectrophotometer. Capable of measuring the developed color at 540 nm.

Matched glass cells or cuvettes. 1 cm pathlength.

Assorted laboratory glassware. Pipets, volumetric flasks, and graduated cylinders of appropriate capacities.

Reagents

All reagents must be ACS reagent grade or better.

Distilled water.

Hydroxylamine hydrochloride solution. Prepare this solution by dissolving 200 g in distilled water in a 1 L volumetric flask. Dilute to mark with distilled water and transfer the solution to a light-protected container. Store the solution in a refrigerator and reject any solution when it is two weeks old.

Sodium hydroxide solution. Prepare this solution by dissolving 200 g in distilled water in a 1 L volumetric flask. Cool the solution to room temperature and dilute to volume.

Absorbing solution. Mix equal volumes of the hydroxylamine hydrochloride and sodium hydroxide solution. This mixture is stable for only two hours and should be prepared fresh just prior to use.

Ferric chloride solution. Dissolve 100 g of ferric chloride ($FeCl_3.6H_2O$) in a 1 L volumetric flask in a mixture of equal volumes of water and concentrated hydrochloric acid. Make to volume.

Acetic anhydride standard solution. Pipet 10 mL of acetic anhydride into a 100 mL volumetric flask and make to volume with acetone. The solution must be prepared within two hours of use.

Procedure

Cleaning of equipment. No specialized cleaning of glassware is required. However, since known interferences occur with esters and aldehydes, cleaning techniques should insure the absence of all organic materials.

Calibration of personal sampling pump. The pump should be calibrated using an integrating volume meter or other means.

Collection and shipping of samples. Pour 10 mL of the absorbing solution into each bubbler. Connect the bubbler with a 5 cm glass adsorption tube (6 mm I.D. and 8 mm O.D.) containing the glass wool plug, then to the personal sampling pump using short pieces of flexible tubing. The air being sampled should not pass through any tubing or other equipment before entering the bubbler. Turn on pump to begin sample collection. Care should be taken to measure the flow rate, the time and/or volume as accurately as possible. Record atmospheric pressure and temperature. If pressure reading is not available, record the elevation. The sample should be taken at a flow rate of 1 L/min. A sample size of 100 L is recommended.

After sampling, the bubbler stem can be removed and cleaned. Tap the stem gently against the inside wall of the bubbler bottle to recover as much of the sampling solution as possible. Wash the stem with 1-2 mL of the absorbing solution and add the wash to the bubbler. Then seal the bubbler with a hard, non-reactive stopper (preferably Teflon or glass). Do not seal with rubber. The stoppers on the bubblers should be tightly sealed to prevent leakage during shipping. Care should be taken to minimize spillage or loss by evaporation at all times. Whenever possible, hand delivery of the samples is recommended. Otherwise, special bubbler shipping cases designed by NIOSH should be used to ship the samples. A blank bubbler should be handled as the other samples (fill, seal, and transport) except that no air is sampled through this bubbler.

Analysis of samples. The sample in each bubbler is analyzed separately. Transfer the solution to a 50 mL volumetric flask. Rinse the bubbler twice with 1 mL of distilled water and add the rinses to the volumetric flask. Pipet 5 mL of the ferric chloride solution into the volumetric flask. Dilute to the mark with a solution containing equal volumes of sodium hydroxide solution, hydroxylamine hydrochloride solution, and ferric chloride solution. This solution should be made in an ice bath. The purple complex is formed rapidly. Read the absorbance at 540 nm in the spectrophotometer against a blank prepared from the alkaline hydroxylamine solution and ferric chloride in the same fashion as the samples.

Calibration and Standards

Preparation of the calibration curve. Pipet 10 mL of the absorbing solution into six 50 mL volumetric flasks. Carefully transfer O (blank), 10, 20, 40, 80, and 100 microliters of the standard solution into the flasks. Add to this mixture 5 mL of ferric chloride solution using a 5 mL pipet. Continue as described above.

Adjust the baseline of the spectrophotometer to zero with distilled water in both cells. Measure the absorbance of the blank and standards at 540 nm. Construct a calibration curve by plotting absorbance against equivalent mg of acetic anhydride.

Calculations

Substract the absorbance of the blank from the absorbance of each sample. Determine from the calibration curve the mg of acetic anhydride present in each sample. The concentration of the analyte in the air sampled can be expressed in mg/m^3:

$$mg/m^3 = (\text{mg acetic anhydride} \times 1000) \div (\text{air volume sampled in L})$$

Another method of expressing concentration is ppm:

$$ppm = (mg/m^3)(24.45)(760)(T + 273) \div (M.W.)(P)(298)$$

where:

P = pressure (mm Hg) of air sampled
T = temperature (°C.) of air sampled
24.45 = molar volume (L/mol) at 25° C. and 760 mm Hg
M.W. = molecular weight (g/mol) of analyte
760 = standard pressure (mm Hg)
298 = standard temperature (°K)

References

1. Diggle, W.M. and Gage, J.C. The determination of ketene and acetic anhydride in the atmosphere. Analyst 78: 473, 1953.
2. Documentation of NIOSH Validation Tests, NIOSH Contract No. CDC-99-74-45.

ACETONE

ACGIH TLV: 750 ppm (1780 mg/m^3)
OSHA Standard: 1000 ppm
NIOSH Recommendation: 250 ppm

CH_3COCH_3

Method	Sampling Duration	Sampling Location	Useful Range (ppm)	System Cost ($)	Test Cost ($)	Manufacturer
DT	Grab	Portable	100-12,000	150	2.60	Nat'l Draeger
DT	Grab	Portable	100-50,000	165	1.70	Matheson
DT	Grab	Portable	100-20,000	180	2.25	Bendix
DT	8-hr	Personal	62.5-10,000	850	3.10	Nat'l Draeger
IR	Cont	Portable	1-2000	4374	0	Foxboro
PI	Cont	Fixed	20-2000	4950	0	HNU
GC	1 min	Portable	1-10^6	10,000	0	Microsensor

NIOSH METHOD NO. P&CAM 127

Principle

A known volume of air is drawn through a charcoal tube to trap the organic vapors present. The charcoal in the tube is transferred to a small, graduated test tube and desorbed with carbon disulfide. An aliquot of the desorbed sample is injected into a gas chromatograph. The area of the resulting peak is determined and compared with areas obtained from the injection of standards.

Range and Sensitivity

The lower limit in mg/sample for the specific compound at 16 × 1 attenuation on a gas chromatograph fitted with a 10:1 splitter is shown in Table 1. This value can be lowered by reducing the attenuation or by eliminating the 10:1 splitter.

Interferences

When the amount of water in the air is so great that condensation actually occurs in the tube, organic vapors will not be trapped. Preliminary experiments indicate that high humidity severely decreases the breakthrough volume. When two or more solvents are known or suspected to be present in the air, such information (including their suspected identities), should be transmitted with the sample, since with differences in polarity, one may displace another from the charcoal.

It must be emphasized that any compound which has the same retention time as the specific compound under study at the operating conditions described in this method is an interference. Hence, retention time data on a single column, or even on a number of columns, cannot be considered as proof of chemical identity. For this reason it is important that a sample of the bulk solvent(s) be submitted at the same time so that identity(ies) can be established by other means. If the possibility of interference exists, separation conditions (column packing, temperatures, etc.) must be changed to circumvent the problem.

Precision and Accuracy

The mean relative standard deviation of the analytical method is 8%. The mean relative standard deviation of the analytical method plus field sampling using an approved personal sampling pump is 10%. Part of the error associated with the method is related to uncertainties in the sample volume collected. If a more powerful vacuum pump with associated gas-volume integrating equipment is used, sampling precision can be improved. The accuracy of the overall sampling and analytical method is 10% (NIOSH-unpublished data) when the personal sampling pump is calibrated with a charcoal tube in the line.

Advantages and Disadvantages

The sampling device is small, portable, and involves no liquids. Interferences are minimal, and most of those which do occur can be eliminated by altering chromatographic conditions. The tubes are analyzed by means of a quick, instrumental method. The method can also be used for the simultaneous analysis of two or more solvents suspected to be present in the same sample by simply changing gas chromatographic conditions from isothermal to a temperature-programmed mode of operation.

One disadvantage of the method is that the amount of sample which can be taken is limited by the number of mg that the tube will hold before overloading. When the sample value obtained for the backup section of the charcoal tube exceeds 25% of that found on the front section, the possibility of sample loss exists. During sample storage, the more volatile compounds will migrate throughout the tube until equilibrium is reached (33% of the sample on the backup section). Furthermore, the precision of the method is limited by the reproducibility of the pressure drop across the tubes. This drop will affect the flow rate and cause the volume to be imprecise, because the pump is usually calibrated for one tube only.

Apparatus

Approved and calibrated sampling pump for personal samples. For an area sample, any vacuum pump whose flow can be determined accurately at 1 L/min or less.

Charcoal tubes. Glass tube with both ends flame sealed, 7 cm long with 6 mm O.D. and a 4 mm I.D., containing 2 sections of 20/40 mesh activated charcoal separated by a 2 mm portion of urethane foam. The activated charcoal is prepared from coconut shells and is fired at 600° C. prior to packing. The absorbing section contains 100 mg of charcoal, the backup section 50 mg. A 3 mm portion of urethane foam is placed between the outlet end of the tube and the backup section. A plug of silylated glass wool is placed in front of the absorbing section. The pressure drop across the tube must be less than one inch of mercury at a flow rate of 1 L/min.

Gas chromatograph. Equipped with a flame ionization detector.

Column. 10 ft × 1/8 in with 10% FFAP stationary phase on 80/100 mesh, acid-washed DMCS Chromosorb W solid support. Other columns capable of performing the required separations may be used.

Mechanical or electronic integrator. Or a recorder and some method for determining peak area.

Microcentrifuge tubes, 25 mL, graduated.

Hamilton syringes. 10 µL, and convenient sizes for making standards.

Pipets. 0.5 mL delivery pipets or 1.0 mL type graduated in 0.1 mL increments.

Volumetric flasks. 10 mL or convenient sizes for making standard solutions.

Reagents

Spectroquality carbon disulfide (Matheson Coleman and Bell).
Sample of the specific compound under study, preferably chromatoquality grade.
Bureau of Mines Grade A helium.
Prepurified hydrogen.
Filtered compressed air.

Procedure

Cleaning of equipment. All glassware used for the laboratory analysis should be detergent washed and thoroughly rinsed with tap water and distilled water.

Calibration of personal pumps. Each personal pump must be calibrated with a representative charcoal tube in the line. This will minimize errors associated with uncertainties in the sample volume collected.

Collection and shipping of samples. Immediately before sampling, the ends of the tube should be broken to provide an opening at least one-half the internal diameter of the tube (2 mm). The small section of charcoal is used as a back-up and should be positioned nearest the sampling pump. The charcoal tube should be vertical during sampling to reduce channeling through the charcoal. Air being sampled should not be passed through any hose or tubing before entering the charcoal tube.

The flow, time, and/or volume must be measured as accurately as possible. The sample should be taken at a flow rate of 1 L/min or less to attain the total sample volume required. The minimum and maximum sample volumes that should be collected for each solvent are shown in Table 1. The minimum volume quoted must be collected if the desired sensitivity is to be achieved. The temperature and pressure of the atmosphere being sampled should be measured and recorded.

The charcoal tubes should be capped with the supplied plastic caps immediately after sampling. Under no circumstances should rubber caps be used. One tube should be handled in the same manner as the sample tube (break, seal, and transport), except that no air is sampled through this tube. This tube should be labeled as a blank. Capped tubes should be packed tightly before they are shipped to minimize tube breakage during shipping.

Samples of the suspected solvent(s) should be submitted to the laboratory for qualitative characterization. These liquid bulk samples should not be transported in the same container as the samples or blank tube. If possible, a bulk air sample (at least 50 L air drawn through tube) should be shipped for qualitative identification purposes.

Analysis of samples. In preparation for analysis, each charcoal tube is scored with a file in front of the first section of charcoal and broken open. The glass wool is removed and discarded. The charcoal in the first (larger) section is transferred to a small stoppered test tube. The separating section of foam is removed and discarded; the second section is transferred to another test tube. These two sections are analyzed separately.

Desorption of samples. Prior to analysis, one-half mL of carbon disulfide is pipetted into each test tube. (All work with carbon disulfide should be performed in a hood because of its high toxicity). Tests indicate that desorption is complete in 30 minutes if the sample is stirred occasionally during this period.

GC conditions. The typical operating conditions for the gas chromatograph are:

85 cc/min (70 psig) helium carrier gas flow
65 cc/min (24 psig) hydrogen gas flow to detector
500 cc/min (50 psig) air flow to detector
200° C. injector temperature
200° C. detector temperature
Isothermal oven or column temperature — refer to Table 1 for specific compounds

TABLE 1
Parameters Associated With P&CAB Analytical Method No. 127

Organic Solvent	Method Classification	Detection limit (mg/sample)	Sample Volume (liters) Minimum[a]	Maximum[b]	GC Column Temp.(°C)	Molecular Weight
Acetone	D	—	0.5	7.7	60	58.1
Benzene	A	0.01	0.5	55	90	78.1
Carbon tetrachloride	A	0.20	10	60	60	154.0
Chloroform	A	0.10	0.5	13	80	119
Dichloromethane	D	0.05	0.5	3.8	85	84.9
p-Dioxane	A	0.05	1	18	100	88.1
Ethylene dichloride	D	0.05	1	12	90	99.0
Methyl ethyl ketone	B	0.01	0.5	13	80	72.1
Styrene	D	0.10	1.5	34	150	104
Tetrachloroethylene	B	0.06	1	25	130	166
1,1,2-trichloroethane	B	0.05	10	97	150	133
1,1,1-trichloroethane (methyl chloroform)	B	0.05	0.5	13	150	133
Trichloroethylene	A	0.05	1	17	90	131
Toluene	B	0.01	0.5	22	120	92.1
Xylene	A	0.02	0.5	31	100	106

(a) Minimum volume, in liters, required to measure 0.1 times the OSHA standard
(b) These are breakthrough volumes calculated with data derived from a potential plot (rf 2) for activated coconut charcoal. Concentrations of vapor in air at 5 times the OSHA standard (rf 3) or 500 ppm, whichever is lower, 25°C, and 760 torr were assumed. These values will be as much as 50% lower for atmospheres of high humidity. The effects of multiple contaminants have not been investigated, but it is suspected that less volatile compounds may displace more volatile compounds.

Injection. The first step in the analysis is the injection of the sample into the gas chromatograph. To eliminate difficulties arising from blowback or distillation within the syringe needle, one should employ the solvent flush injection technique. The 10 μL syringe is first flushed with solvent several times to wet the barrel and plunger. Three microliters of solvent are drawn into the syringe to increase the accuracy and reproducibility of the injected sample volume. The needle is removed from the solvent, and the plunger is pulled back about 0.2 µL to separate the solvent flush from the sample with a pocket of air to be used as a marker. The needle is then immersed in the sample, and a 5 µL aliquot is withdrawn, taking into consideration the volume of the needle, since the sample in the needle will be completely injected. After the needle is removed from the sample and prior to injection, the plunger is pulled back a short distance to minimize evaporation of the sample from the tip of the needle. Duplicate injections of each sample and standard should be made. No more than a 3% difference in area is to be expected.

Measurement of area. The area of the sample peak is measured by an electronic integrator or some other suitable form of area measurement, and preliminary results are read from a standard curve prepared as discussed below.

Determination of desorption efficiency. The desorption efficiency of a particular compound can vary from one laboratory to another and also from one batch of charcoal to another. Thus, it is necessary to determine at least once the percentage of the specific compound that is removed in the desorption process for a given compound, provided the same batch of charcoal is used. NIOSH has found that the desorption efficiencies for the compounds in Table 1 are between 81% and 100% and vary with each batch of charcoal.

Procedure for determining desorption efficiency. Activated charcoal equivalent to the amount in the first section of the sampling tube (100 mg) is measured into a 5 cm, 4 mm I.D. glass tube, flame-sealed at one end (similar to commercially available culture tubes). This charcoal must be from the same batch as that used in obtaining the samples and can be obtained from unused charcoal tubes. The open end is capped with Parafilm. A known amount of the compound is injected directly into the activated charcoal with a microliter syringe, and the tube is capped with more Parafilm. The amount injected is usually equivalent to that present in a 10 L sample at a concentration equal to the federal standard.

At least five tubes are prepared in this manner and allowed to stand at least overnight to assure complete absorption of the specific compound onto the charcoal. These five tubes are referred to as the samples. A parallel blank tube should be treated in the same manner except that no sample is added to it. The sample and blank tubes are desorbed and analyzed in exactly the same manner as the sampling tube described earlier.

Two or three standards are prepared by injecting the same volume of compound into 0.5 mL of CS_2 with the same syringe used in the preparation of the sample. These are analyzed with the samples. The desorption efficiency equals the difference between the average peak area of the samples and the peak area of the blank divided by the average peak area of the standards, or:

D.E. = (area sample - area blank) ÷ area standard

Calibration and Standards

It is convenient to express concentration of standards in terms of mg/0.5 mL CS_2 because samples are desorbed in this amount of CS_2. To minimize error due to the volatility of carbon disulfide, one can inject 20 times the weight into 10 mL of CS_2. For example, to prepare a 0.3 mg/0.5 mL standard, one would inject 6.0 mg into exactly 10 mL of CS_2 in a glass-stoppered flask. The density of the specific compound is used to convert 6.0 mg into microliters for easy measurement with a microliter syringe. A series of standards, varying in concentration over the range of interest, is prepared and analyzed under the same GC conditions and during the same time period as the unknown samples. Curves are established by plotting concentration in mg/0.5 mL versus peak area.

NOTE: Since no internal standard is used in the method, standard solutions must be analyzed at the same time that the sample analysis is done. This will minimize the effect of known day-to-day variations and variations during the same day of the FID response.

Calculations

The weight, in mg, corresponding to each peak area is read from the standard curve for the particular compound. No volume corrections are needed, because the standard curve is based on mg/0.5 mL CS_2 and the volume of sample injected is identical to the volume of the standards injected. Corrections for the blank must be made for each sample:

Corrected mg = mg sample - mg blank

where:

mg sample = mg found in front section of sample tube
mg blank = mg found in front section of blank tube

A similar procedure is followed for the backup sections.

The corrected amounts present in the front and backup sections of the same sample tube are added to determine the total measured amount in the sample. This total weight is divided by the determined desorption efficiency to obtain the corrected mg per sample. The concentration of the analyte in the air sampled can be expressed in mg/m^3:

mg/m^3 = (corrected mg)(1000 L/m^3) ÷ (air sample volume in L)

Another method of expressing concentration is ppm (corrected to standard conditions of 25° C. and 760 mm Hg):

ppm = (mg/m^3)(24.45)(760)(T + 273) ÷ (M.W.)(P)(298)

where:

P = pressure (mm Hg) of air sampled
T = temperature (°C.) of air sampled
24.45 = molar volume (L/mol) at 25° C. and 760 mm Hg
M.W. = molecular weight
760 = standard pressure (mm Hg)
298 = standard temperature (°K)

References

1. White, L.D., D.G. Taylor, P.A. Mauer and R.E. Kupel. A convenient optimized method for the analysis of selected solvent vapors in the industrial atmosphere. Am. Ind. Hyg. Assoc. J. 31: 225, 1970.
2. Young, D.M. and A.D. Crowell. *Physical Adsorption of Gases,* pp. 137-146, Butterworths, London, 1962.
3. Federal Register, 37:202:22139-22142, October 18, 1972.
4. NIOSH Contract HSM-99-72-98, Scott Research Laboratories, Inc., "Collaborative Testing of Activated Charcoal Sampling Tubes for Seven Organic Solvents", pp. 4-22, 4-27, 1973.

ACETONITRILE

ACGIH TLV: 40 ppm (70 mg/m³) - Skin
OSHA Standard: 40 ppm

CH_3CN

Method	Sampling Duration	Sampling Location	Useful Range (ppm)	System Cost ($)	Test Cost ($)	Manufacturer
IR	Cont	Portable	5-80	4374	0	Foxboro
PI	Cont	Fixed	0.2-20	4950	0	HNU
GC	1 min	Portable	$1-10^6$	10,000	0	Microsensor

NIOSH METHOD NO. S165

Principle

A known volume of air is drawn through a large charcoal tube to trap the organic vapors present. The charcoal in the tube is transferred to a small, stoppered sample container, and the analyte is desorbed with benzene. An aliquot of the desorbed sample is injected into a gas chromatograph. The area of the resulting peak is determined and compared with areas obtained from the injection of standards.

Range and Sensitivity

This method was validated over the range of 31.4-140.2 mg/m³ at an atmospheric temperature and pressure of 22° C. and 762 mm Hg, using a 10 L sample. Under the conditions of sample size (10 L) the probable useful range of this method is 10-210 mg/m³ at a detector sensitivity that gives nearly full deflection on the strip chart recorder for a 2.1 mg sample. The method is capable of measuring much smaller amounts if the desorption efficiency is adequate. Desorption efficiency must be determined over the range used.

The upper limit of the range of the method is dependent on the adsorptive capacity of the charcoal tube. This capacity varies with the concentrations of acetonitrile and other substances in the air. The first section of the charcoal tube was found to hold 5.2 mg of acetonitrile when a test atmosphere containing 140 mg/m³ of acetonitrile in air was sampled at 0.196 L/min for 190 minutes; breakthrough was observed at this time, i.e., the concentration of acetonitrile in the effluent was 5% of that in the influent. (The charcoal tube consists of two sections of activated charcoal separated by a section of urethane foam.) If a particular atmosphere is suspected of containing a large amount of contaminant, a smaller sampling volume should be taken.

Interferences

When the amount of water in the air is so great that condensation actually occurs in the tube, organic vapors will not be trapped efficiently. Preliminary experiments using toluene indicate that high humidity severely decreases the breakthrough volume.

When two or more substances are known or suspected to be present in the air, such information, including their suspected identities, should be transmitted with the sample. Since benzene is used rather than carbon disulfide to desorb acetonitrile from the charcoal, it would not be possible to measure benzene in the sample with this desorbing solvent. If it is suspected that benzene is present, a separate sample should be collected for benzene analysis.

It must be emphasized that any compound which has the same retention time as the analyte at the operating conditions described in this method is an interference. Retention time data on a single column cannot be considered proof of chemical identity. If the possibility of interference exists, separation conditions (column packing, temperature, etc.) must be changed to circumvent the problem.

Precision and Accuracy

The coefficient of variation for the total analytical and sampling method in the range of 31.4-140.2 mg/m^3 was 0.072. This value corresponds to a 5.0 mg/m^3 standard deviation at the OSHA standard level. Statistical information and details of the validation and experimental test procedures can be found in reference 2.

On the average the concentrations obtained at the OSHA standard level using the overall sampling and analytical method were 5.3% higher than the "true" concentrations for a limited number of laboratory experiments. Any difference between the "found" and "true" concentrations may not represent a bias in the sampling and analytical method, but rather a random variation from the experimentally determined "true" concentration. Therefore, no recovery correction should be applied to the final result.

Advantages and Disadvantages

The sampling device is small, portable, and involves no liquids. Interferences are minimal, and most of those which do occur can be eliminated by altering chromatographic conditions. The tubes are analyzed by means of a quick, instrumental method. The method can also be used for the simultaneous analysis of two or more substances suspected to be present in the same sample by simply changing gas chromatographic conditions from isothermal to a temperature-programmed mode of operation.

One disadvantage of the method is that the amount of sample which can be taken is limited by the number of milligrams that the tube will hold before overloading. When the sample value obtained for the backup section of the charcoal tube exceeds 25% of that found on the front section, the possibility of sample loss exists. Furthermore, the precision of the method is limited by the reproducibility of the pressure drop across the tubes. This drop will affect the flow rate and cause the volume to be imprecise, because the pump is usually calibrated for one tube only.

Apparatus

Calibrated personal sample pump. Whose flow can be determined within 5% at the recommended flow rate.

Charcoal tubes. Glass tube with both ends flame sealed, 9 cm long with 8 mm O.D. and 6 mm I.D., containing 2 sections of 20/40 mesh activated charcoal separated by a 2 mm portion of urethane foam. The activated charcoal is prepared from coconut shells and is fired at 600° C. prior to packing. The adsorbing section contains 400 mg of charcoal, the backup section 200 mg. A 3 mm portion of urethane foam is placed between the outlet end of the tube and the backup section. A plug of silylated glass wool is placed in front of the adsorbing section. The pressure drop across the tube must be less than one inch of mercury at a flow rate of 1 L/min.

Gas chromatograph. Equipped with a flame ionization detector.

Column. 4 ft × 1/4 in stainless steel packed with 50/80 mesh Porapak, Type Q.

Electronic integrator. Or some other suitable method for measuring peak areas.

Volumetric flasks. 10 mL or convenient sizes for making standard solutions and for desorbing the samples. If an automatic sample injector is used, aliquots of the samples may be transferred to the associated vials.

Microliter syringes. 10 microliter for injection of samples into the gas chromatograph.

Pipets. 5 mL delivery pipets.

Reagents
Chromatographic quality benzene.
Acetonitrile, reagent grade.
Purified nitrogen.
Prepurified hydrogen.
Filtered compressed air.

Procedure
Cleaning of equipment. All glassware used for the laboratory analysis should be detergent washed and thoroughly rinsed with tap water and distilled water.

Calibration of personal pumps. Each personal pump must be calibrated with a representative charcoal tube in the line. This will minimize errors associated with uncertainties in the sample volume collected.

Collection and shipping of samples. Immediately before sampling, break the ends of the tube to provide an opening at least one-half the internal diameter of the tube. The smaller section of charcoal is used as a back-up and should be positioned nearest the sampling pump. The charcoal tube should be placed in a vertical direction during sampling to minimize channeling through the charcoal. Air being sampled should not be passed through any hose or tubing before entering the charcoal tube.

A sample size of 10 L is recommended. Sample at a flow of 0.20 L/min or less. The flow rate should be known with an accuracy of at least 5%. The temperature and pressure of the atmosphere being sampled should be recorded. If pressure reading is not available, record the elevation. The charcoal tubes should be capped with the supplied plastic caps immediately after sampling. Under no circumstances should rubber caps be used.

One tube should be handled in the same manner as the sample (break, seal, and transport), except that no air is sampled through this tube. This tube should be labeled as a blank. Capped tubes should be packed tightly and padded before they are shipped to minimize tube breakage during shipping. A sample of the bulk material should be submitted to the laboratory in a glass container with a Teflon-lined cap. This sample should not be transported in the same container as the charcoal tubes.

Analysis of samples. In preparation for analysis, each charcoal tube is scored with a file in front of the first section of charcoal and broken open. The glass wool is removed and discarded. The charcoal in the first (larger) section is transferred to a 10 mL stoppered sample container. The separating section of foam is removed and discarded; the second section is transferred to another stoppered container. These two sections are analyzed separately.

Desorption of samples. Prior to analysis, 5 mL of benzene is pipetted into each sample container. (All work with benzene should be performed in a hood because of its high toxicity.) Desorption should be done for 30 minutes. Tests indicate that this is adequate if the sample is agitated occasionally during this period. If an automated sample injector is used, transfer an appropriate amount of the sample to the automatic sample injector vials after desorption is complete. The sample vials should be capped as soon as the sample is added to minimize volatilization.

GC conditions. The typical operating conditions for the gas chromatograph are:

50 mL/min (60 psig) nitrogen carrier gas flow
65 mL/min (24 psig) hydrogen gas flow to detector
500 mL/min (50 psig) air flow to detector

270° C. injector temperature
285° C. detector temperature
180° C. column temperature

Injection. The first step in the analysis is the injection of the sample into the gas chromatograph. To eliminate difficulties arising from blow back or distillation within the syringe needle, one should employ the solvent flush injection technique. The 10 microliter syringe is first flushed with solvent several times to wet the barrel and plunger. Three microliters of solvent are drawn into the syringe to increase the accuracy and reproducibility of the injected sample volume. The needle is removed from the solvent, and the plunger is pulled back about 0.2 microliter to separate the solvent flush from the sample with a pocket of air to be used as a marker. The needle is then immersed in the sample, and a 5 microliter aliquot is withdrawn, taking into consideration the volume of the needle, since the sample in the needle will be completely injected. After the needle is removed from the sample and prior to injection, the plunger is pulled back 1.2 microliters to minimize evaporation of the sample from the tip of the needle. Observe that the sample occupies 4.9-5.0 microliters in the barrel of the syringe. Duplicate injections of each sample and standard should be made. No more than a 3% difference in area is to be expected. An automatic sample injector can be used if it is shown to give reproducibility at least as good as the solvent flush method.

Measurement of area. The area of the sample peak is measured by an electronic integrator or some other suitable form of area measurement, and preliminary results are read from a standard curve prepared as discussed below.

Determination of desorption efficiency. The desorption efficiency of a particular compound can vary from one laboratory to another. Thus, it is necessary to determine at least once the percentage of the specific compound that is removed in the desorption process, provided the same batch of charcoal is used. Activated charcoal equivalent to the amount in the first section of the sampling tube (400 mg) is measured into a 9 cm × 6 mm I.D. glass tube, flame sealed at one end. This charcoal must be from the same batch as that used in obtaining the samples and can be obtained from used charcoal tubes. The open end is capped with Parafilm.

A known amount of acetonitrile in benzene (157 mg/mL) is injected directly into the activated charcoal with a microliter syringe, and the tube is capped with more Parafilm. The amount injected is equivalent to that present in a 10 L air sample at the selected level. Six tubes at each of three levels (0.5X, 1X, and 2X of the OSHA standard) are prepared in this manner and allowed to stand for at least overnight to assure complete adsorption of the analyte onto the charcoal. These tubes are referred to as the samples. A parallel blank tube should be treated in the same manner except that no sample is added to it. The sample and blank tubes are desorbed and analyzed in exactly the same manner as the sampling tube described earlier.

Two or three standards are prepared by injecting the same volume of compound into 5.0 mL of benzene contained in a 10 mL volumetric flask. These are analyzed with the samples. The desorption efficiency (D.E.) equals the average weight in mg recovered from the tube divided by the weight in mg added to the tube or:

D.E. = (average weight recovered in mg) ÷ (weight added in mg)

The desorption efficiency is dependent on the amount of analyte collected on the charcoal. Plot the desorption efficiency versus weight of analyte found. This curve is used to correct for adsorption losses.

Calibration and Standards

It is convenient to express concentration of standards in terms of mg/5 mL benzene, because samples are desorbed in this amount of benzene. The density of the analyte is used to convert mg into microliters for easy measurement with a microliter syringe. A series of standards, varying in concentration over the range of interest, is prepared and analyzed under the same GC conditions and during the same time period as the unknown samples. Curves are established by plotting concentration in mg/5.0 mL versus peak area. Note: Since no internal standard is used in the method, standard solutions must be analyzed at the same time that the sample analysis is done. This will minimize the effect of known day-to-day variations and variations during the same day of the FID response.

Calculations

Read the weight, in mg, corresponding to each peak area from the standard curve. No volume corrections are needed, because the standard curve is based on mg/5.0 mL benzene and the volume of sample injected is identical to the volume of the standards injected. Corrections for the blank must be made for each sample:

mg = mg sample - mg blank

where:

mg sample = mg found in front section of sample tube
mg blank = mg found in front section of blank tube

A similar procedure is followed for the backup sections.

Add the weights found in the front and backup sections to get the total weight in the sample. Read the desorption efficiency from the curve for the amount found in the front section. Divide the total weight by this desorption efficiency to obtain the corrected mg/sample:

Corrected mg/sample = total weight ÷ D.E.

The concentration of the analyte in the air sampled can be expressed in mg/m^3:

mg/m^3 = (corrected mg)(1000 L/m^3) ÷ (air volume sampled in L)

Another method of expressing concentration is ppm:

ppm = (mg/m^3)(24.45)(760)(T + 273) ÷ (M.W.)(P)(298)

where:

P = pressure (mm Hg) of air sampled
T = temperature (°C.) of air sampled
24.45 = molar volume (L/mol) at 25° C. and 760 mm Hg
M.W. = molecular weight (g/mol) of analyte
760 = standard pressure (mm Hg)
298 = standard temperature (°K.)

References

1. White, L.D. et al. A convenient optimized method for the analysis of selected solvent vapors in the industrial atmosphere. Amer. Ind. Hyg. Assoc. J. 31: 225, 1970.
2. Documentation of NIOSH Validation Tests, NIOSH Contract No. CDC-99-74-45.
3. Final Report, NIOSH Contract HSM-99-71-31. Personal sampler pump for charcoal tubes, September 15, 1972.

ACETYLENE TETRABROMIDE

ACGIH TLV: 1 ppm (15 mg/m^3) $\qquad$ $CHBr_2CHBr_2$
OSHA Standard: 1 ppm

Method	Sampling Duration	Sampling Location	Useful Range (ppm)	System Cost ($)	Test Cost ($)	Manufacturer
IR	Cont	Portable	0.1-2	4374	0	Foxboro

NIOSH METHOD NO. S117

Principle

A known volume of air is drawn through a silica gel tube to trap the organic vapors present. The silica gel in the tube is transferred to a small, stoppered sample container, and the analyte is desorbed with tetrahydrofuran. An aliquot of the desorbed sample is injected into a gas chromatograph. The area of the resulting peak is determined and compared with areas obtained from the injection of standards.

Range and Sensitivity

This method was validated over the range of 7.04-28.1 mg/m^3 at an atmospheric temperature and pressure of 22° C. and 760 mm Hg, using a 100 L sample. Under the conditions of sample size (100 L) the probable useful range of this method is 1.5-40 mg/m^3 at a detector sensitivity that gives nearly full deflection of the strip chart recorder for a 4 mg sample. The method is capable of measuring much smaller amounts if the desorption efficiency is adequate. Desorption efficiency must be determined over the range used.

The upper limit of the range of the method is dependent on the adsorptive capacity of the silica gel tube. This capacity varies with the concentrations of the analyte and other substances in the air. The first section of the silica gel tube was found to hold at least 6.32 mg of analyte when a test atmosphere containing 28.1 mg/m^3 of analyte in air was sampled at 0.91 L/min for 247 minutes. (The silica gel tube consists of two sections of silica gel separated by a section of urethane foam.) If a particular atmosphere is suspected of containing a large amount of contaminant, a smaller sampling volume should be taken.

Interferences

Silica gel has a high affinity for water, so organic vapors will not be trapped efficiently in the presence of a high relative humidity. This effect may be important even though there is no visual evidence of condensed water in the silica gel tube.

When two or more compounds are known or suspected to be present in the air, such information, including their suspected indentities, should be transmitted with the sample. Since tetrahydrofuran is used rather than carbon disulfide to desorb the analyte from the silica gel, it is not possible to measure tetrahydrofuran in the sample with this desorbing solvent. If it is suspected that tetrahydrofuran is present, a separate sample should be collected for tetrahydrofuran analysis.

It must be emphasized that any compound which has the same retention time as the analyte at the operating conditions described in this method is an interference. Retention time data on a single column cannot be considered proof of chemical identity. If the possibility of interference exists, separation conditions (column packing, temperature, etc.) must be changed to circumvent the problem.

Precision and Accuracy

The coefficient of variation for the total analytical and sampling method in the range of 7.04-28.1 mg/m^3 was 0.096. This value corresponds to a 1.34 mg/m^3 standard deviation at the OSHA standard level. Statistical information and details of the validation and experimental test procedures can be found in reference 2.

On the average the concentrations obtained at the OSHA standard level using the overall sampling and analytical method were 3.5% lower than the "true" concentrations for a limited number of laboratory experiments. Any difference between the "found" and "true" concentrations may not represent a bias in the sampling and analytical method, but rather a random variation from the experimentally determined "true" concentration. Therefore, no recovery correction should be applied to the final result.

Advantages and Disadvantages

The sampling device is small, portable, and involves no liquids. Interferences are minimal, and most of those which do occur can be eliminated by altering chromatographic conditions. The tubes are analyzed by means of a quick instrumental method. The method can also be used for simultaneous analysis of two or more substances suspected to be present in the same sample by simply changing gas chromatographic conditions from isothermal to a temperature-programmed mode of operation.

One disadvantage of the method is that the amount of sample which can be taken is limited by the number of milligrams that the tube will hold before overloading. When the sample value obtained for the backup section of the silica gel tube exceeds 25% of that found on the front section, the possibility of sample loss exists. Furthermore, the precision of the method is limited by the reproducibility of the pressure drop across the tubes. This drop will affect the flow rate and cause the volume to be imprecise, because the pump is usually calibrated for one tube only.

Apparatus

Calibrated personal sampling pump. Whose flow can be determined within 5% at the recommended flow rate.

Silica gel tubes. Glass tube with both ends flame sealed, 7 cm long with a 6 mm O.D. and a 4 mm I.D., containing 2 sections of 20/40 mesh silica gel separated by a 2 mm portion of urethane foam. The adsorbing section contains approximately 150 mg of silica gel and the backup section, approximately 75 mg. A 3 mm portion of urethane foam is placed between the outlet end of the tube and the backup section. A plug of silylated glass wool is placed in front of the adsorbing section. The pressure drop across the tube must be less than one inch of mercury at a flow rate of 1 L/min.

Gas chromatograph. Equipped with a flame ionization detector.
Column. 3 ft × 1/4 in glass packed with 5% SE-30 on 80/100 mesh Gas Chrom W.
Electronic integrator. Or some other suitable method for measuring peak areas.
Sample containers. 2 mL with glass stoppers or Teflon-lined caps.
Syringes. 10 microliter, and other convenient sizes for making standards.
Pipets. 1.0 mL delivery pipets.
Volumetric flasks. 10 mL or convenient sizes for making standard solutions.

Reagents
Chromatographic quality tetrahydrofuran.
Acetylene tetrabromide, 98% solution, reagent grade.
Prepurified hydrogen.
Filtered compressed air.
Purified nitrogen.

Procedure
Cleaning of equipment. All glassware used for the laboratory analysis should be detergent washed and thoroughly rinsed with tap water and distilled water.
Calibration of personal pumps. Each personal pump must be calibrated with a representative silica gel tube in the line. This will minimize errors associated with uncertainties in the sample volume collected.
Collection and shipping of samples. Immediately before sampling, break the ends of the tube to provide an opening at least one-half the internal diameter of the tube (2 mm). The smaller section of silica gel is used as a back-up and should be positioned nearest the sampling pump. The silica gel tube should be placed in a vertical direction during sampling to minimize channeling through the silica gel. Air being sampled should not be passed through any hose or tubing before entering the silica gel tube.
A sample size of 100 L is recommended. Sample at a flow of 1.0 L/min or less. The flow rate should be known with an accuracy of at least 5%. The temperature and pressure of the atmosphere being sampled should be recorded. If pressure reading is not available, record the elevation. The silica gel tubes should be capped with the supplied plastic caps immediately after sampling. Under no circumstances should rubber caps be used. One tube should be handled in the same manner as the sample tube (break, seal, and transport), except that no air is sampled through this tube. This tube should be labeled as a blank.
Capped tubes should be packed tightly and padded before they are shipped to minimize tube breakage during shipping. A sample of the bulk material should be submitted to the laboratory in a glass container with a Teflon-lined cap. This sample should not be transported in the same container as the silica gel.
Analysis of samples. In preparation for analysis, each silica gel tube is scored with a file in front of the first section of silica gel and broken open. The glass wool is removed and discarded. The silica gel in the first (larger) section is transferred to a 2 mL stoppered sample container. The separating section of foam is removed and discarded; the second section is transferred to another stoppered container. These two sections are analyzed separately.
Desorption of samples. Prior to analysis, 1.0 mL of tetrahydrofuran is pipetted into each sample container. Desorption should be done for 30 minutes. Tests indicate that this is adequate if the sample is agitated occasionally during this period.
GC conditions. The typical operating conditions for the gas chromatograph are:

50 mL/min (60 psig) nitrogen carrier gas flow

65 mL/min (24 psig) hydrogen gas flow to detector
500 mL/min (50 psig) air flow to detector
220° C. injector temperature
270° C. detector temperature
105° C. column temperature

Injection. The first step in the analysis is the injection of the sample into the gas chromatograph. To eliminate difficulties arising from blow back or distillation within the syringe needle, one should employ the solvent flush injection technique. The 10 microliter syringe is first flushed with solvent several times to wet the barrel and plunger. Three microliters of solvent are drawn into the syringe to increase the accuracy and reproducibility of the injected sample volume. The needle is removed from the solvent, and the plunger is pulled back about 0.2 microliter to separate the solvent flush from the sample with a pocket of air to be used as a marker. The needle is then immersed in the sample, and a 5 microliter aliquot is withdrawn, taking into consideration the volume of the needle, since the sample in the needle will be completely injected. After the needle is removed from the sample and prior to injection, the plunger is pulled back 1.2 microliters to minimize evaporation of the sample from the tip of the needle. Observe that the sample occupies 4.9-5.0 microliters in the barrel of the syringe. Duplicate injections of each sample and standard should be made. No more than a 3% difference in area is to be expected.

Measurement of area. The area of the sample peak is measured by an electronic integrator or some other suitable form of area measurement, and preliminary results are read from a standard curve prepared as discussed below.

Determination of desorption efficiency. The desorption efficiency of a particular compound can vary from one batch of silica gel to another. Thus, it is necessary to determine at least once the percentage of the specific compound that is removed in the desorption process, provided the same batch of silica gel is used.

Silica gel equivalent to the amount in the first section of the sampling tube (approximately 150 mg) is measured into a 2.5 in, 4 mm I.D. glass tube, flame sealed at one end. This silica gel must be from the same batch as that used in obtaining the samples and can be obtained from unused silica gel tubes. The open end is capped with Parafilm. A known amount of hexane solution of acetylene tetrabromide (350 mg/mL) is injected directly into the silica gel with a microliter syringe, and the tube is capped with more Parafilm.

The amount injected is equivalent to that present in a 100 L air sample at the selected level. Six tubes at each of three levels (using 2, 4, and 8 microliters of hexane solution for the 0.5X, 1X, and 2X OSHA standard levels) are prepared in this manner and allowed to stand for at least overnight to assure complete adsorption of the analyte onto the silica gel. These tubes are referred to as the samples. A parallel blank tube should be treated in the same manner except that no sample is added to it. The sample and blank tubes are desorbed and analyzed in exactly the same manner as the sampling tube described earlier.

Two or three standards are prepared by injecting the same volume of compound into 1.0 mL of tetrahydrofuran with the same syringe used in the preparation of the samples. These are analyzed with the samples. The desorption efficiency (D.E.) is dependent on the amount of analyte collected on the silica gel. The desorption efficiency equals the average weight in mg recovered from the tube divided by the weight in the mg added to the tube, or:

D.E. = (average weight recovered in mg) ÷ (weight added in mg)

The desorption efficiency is dependent on the amount of analyte collected on the silica gel. Plot the desorption efficiency versus weight of analyte found. This curve is used to correct for adsorption losses.

Calibration and Standards

It is convenient to express concentration of standards in terms of mg/1.0 mL tetrahydrofuran, because samples are desorbed in this amount of tetrahydrofuran. The density of the analyte is used to convert mg into microliters for easy measurement with a microliter syringe. A series of standards, varying in concentration over the range of interest, are prepared and analyzed under the same GC conditions and during the same time period as the unknown samples. Curves are established by plotting concentration in mg/1.0 mL versus peak area. Note: Since no internal standard is used in the method, standard solutions must be analyzed at the same time that the same analysis is done. This will minimize the effect of known day-to-day variations and variations during the dame day of the FID response.

Calculations

Read the weight, in mg, corresponding to each peak area from the standard curve. No volume corrections are needed, because the standard curve is based on mg/1.0 mL tetrahydrofuran and the volume of sample injected is identical to the volume of the standards injected. Corrections for the blank must be made for each sample:

mg = mg sample - mg blank

where:

mg sample = mg found in front section of sample tube
mg blank = mg found in front section of blank tube

A similar procedure is followed for the backup sections. Add the weights found in the front and backup sections to get the total weight in the sample. Read the desorption efficiency from the curve for the amount found in the front section. Divide the total weight by this desorption efficiency to obtain the corrected mg/sample:

Corrected mg/sample = total weight ÷ D.E.

The concentration of the analyte in the air sampled can be expressed in mg/m³:

mg/m^3 = (corrected mg × 1000 L/m^3) ÷ (air volume sampled in L)

Another method of expressing concentration is ppm:

ppm = (mg/m^3)(24.45)(760)(T + 273) ÷ (M.W.)(P)(298)

where:

P = pressure (mm Hg) of air sampled
T = temperature (°C.) of air sampled
24.45 = molar volume (L/mol) at 25° C. and 760 mm Hg
M.W. = molecular weight (g/mol) of analyte
760 = standard pressure (mm Hg)
298 = standard temperature (°K)

References

1. White, L.D. et al. A convenient optimized method for the analysis of selected solvent vapors in the industrial atmosphere. Amer. Ind. Hyg. Assoc. J. 31: 225, 1970.
2. Documentation of NIOSH Validation Tests, NIOSH Contract No. CDC-99-74-45.

ACROLEIN

Synonym: acrylaldehyde $\qquad$ $CH_2=CHCHO$
ACGIH TLV: 0.1 ppm (0.25 mg/m^3)
OSHA Standard: 0.1 ppm

Method	Sampling Duration	Sampling Location	Useful Range (ppm)	System Cost ($)	Test Cost ($)	Manufacturer
DT	Grab	Portable	50-18,000	165	1.70	Matheson
DT	Grab	Portable	7-800	180	2.25	Bendix
IR	Cont	Portable	0.3-2	4374	0	Foxboro
GC	1 min	Portable	1-10^6	10,000	0	Microsensor

NIOSH METHOD NO. P&CAM 211

Principle
Air is drawn through two midget impingers containing 1% NaHSO$_3$. The reaction of acrolein with 4-hexylresorcinol in an alcoholic trichloroacetic acid solvent medium in the presence of mercuric chloride results in a blue colored product with strong absorption maximum at 605 nm.

Range and Sensitivity
The absorbance at 605 nm is linear for at least 1-30 μg of acrolein in the 10 mL portions of the mixed reagent. A concentration of 10 ppb of acrolein can be determined in a 50 L air sample based on a difference of 0.05 absorbance unit from the blank using a 1 cm cell. Greater sensitivity could be obtained by use of a longer pathlength cell.

Interferences
There is no interference from ordinary quantities of sulfur dioxide, nitrogen dioxide, ozone and most organic air pollutants. A slight interference occurs from dienes: 1.5% for 1,3-butadiene and 2% for 1,3-pentadiene. The red color produced by some other aldehydes and undetermined materials does not interfere in spectrophotometric measurements.

Precision and Accuracy
Known standards can be determined to within ±5% of the true value. No data are available on precision and accuracy of air samples.

Advantages and Disadvantages

The method is sensitive to the extent that 0.01 ppm of acrolein can be determined in a 50 L air sample. Both the solid trichloroacetic acid and the solution are corrosive to the skin. Mercuric chloride is highly toxic. This reagent and the former should be handled carefully.

Apparatus

Absorbers. All glass standard midget impingers with fritted glass inlets are acceptable. The fritted end should have a porosity approximately equal to that of Corning EC (170-220 micron maximum pore diameter). A train of 2 bubblers is needed.

Air pump. A pump capable of drawing at least 2 L of air/min for 20 minutes through the sampling train is required. A trap at the inlet to protect the pump from the corrosive reagent is recommended.

Air metering device. Either a limiting orifice of approximately 1 or 2 L/min capacity or a meter can be used. If a limiting orifice is used, regular and frequent calibration is required.

Spectrophotometer. This instrument should be capable of measuring the developed color at 605 nm. The absorption band is rather narrow and thus a lower absorptivity may be expected in a broad-band instrument.

Reagents

Purity of chemicals. All reagents must be analytical reagent grade. An analytical grade of distilled water must be used.

$HgCl_2$-4-hexylresorcinol. Dissolve 0.30 g $HgCl_2$ and 2.5 g 4-hexylresorcinol in 50 mL of 95% ethanol (stable at least 3 weeks in refrigerator).

TCAA. To a 1 lb bottle of trichloroacetic acid, add 23 mL of distilled water and 25 mL of 95% ethanol. Mix until all the TCAA has dissolved.

Collection medium. 1% sodium bisulfite in water.

Acrolein, purified. Freshly prepare a small quantity (less than 1 mL is sufficient) by distilling 10 mL of the purest grade of acrolein commercially available. Reject the first 2 mL of the distillate. (The acrolein should be stored in a refrigerator to retard polymerization.) The distillation should be done in a hood because the vapors are irritating to the eyes.

Acrolein, standard solution A (1 mg/mL). Weigh 0.1 g (approximately 0.12 mL) of freshly prepared, purified acrolein into a 100 mL volumetric flask and dilute to volume with 1% aqueous sodium bisulfite. This solution may be kept for as long as a month if properly refrigerated.

Acrolein, standard solution B (10 µg/mL). Dilute 1 mL of standard solution A to 100 mL with 1% sodium bisulfite. This solution may be kept for as long as a month if properly refrigerated.

Procedure

Cleaning of Equipment. No specialized cleaning of glassware is required. However, since known interferences occur with dienes, cleaning techniques should insure the absence of all organic materials.

Collection of Samples. Two midget impingers, each containing 10 mL of 1% $NaHSO_3$ are connected in series with Tygon tubing. These are followed by and connected to an empty impinger (for meter protection) and a dry test meter and a source of suction. During sampling, the impingers are immersed in an ice bath. Sampling rate of 2 L/min should be maintained. Sampling duration will depend on the concentration of aldehydes in the air. One hour sampling time at 2 L/min is adequate for ambient concentrations.

Storage of samples. After sampling is complete, the impingers are disconnected from the train, the inlet and outlet tubes are capped, and the impingers stored in an ice bath or at 0° C. in a refrigerator until analyses are performed. Cold storage is necessary only if the acrolein determination cannot be performed within 4 hr of sampling.

Shipping of samples. The color takes 2 hours to fully develop at room temperature. However, if a sample can be refrigerated and analyzed within 48 hr after collection, then samples can be shipped or carried to a central lab at some distance from the sampling area.

Analysis of samples. Each impinger is analyzed separately. To a 25 mL graduated tube, add an aliquot of the collected sample in bisulfite containing no more than 30 μg acrolein. Add 1% sodium bisulfite (if necessary) to a volume of 4.0 mL. Add 1.0 mL of the $HgCl_2$-4-hexylresorcinol reagent and mix. Add 5.0 mL of TCAA reagent and mix again. Insert in a boiling water bath for 5-6 min, remove, and set aside until tubes reach room temperature. Centrifuge samples at 1500 rpm for 5 min to clear slight turbidity. One hour after heating, read in a spectrophotometer at 605 nm against a bisulfite blank prepared in the same fashion as the samples. Determine the acrolein content of the sampling solution from a curve previously prepared from the standard acrolein solutions.

Calibration and Standards

Pipette 0, 0.5, 1.0, 2.0, and 3.0 mL of standard solution B into glass-stoppered test tubes. Dilute each standard to exactly 4 mL with 1% aqueous sodium bisulfite. Develop color as described above, in section on analysis of samples. Plot, on linear paper, absorbance values against micrograms of acrolein in the color-developed solution. The color after forming is stable for about 2 hours.

Calculations

The concentration of acrolein in the sampled atmosphere may be calculated by using the following equations:

$\mu g/m^3$ = (total μg of acrolein in test sample $\times$ 1000) $\div$ (air sample volume in L)

ppm = (total μg of acrolein in sample) $\div$ (2.3 $\times$ sample volume in L)

Correct the air sample volume to 25° C. and 760 Torr.

References

1. Rosenthaler, L. and G. Vegessi. Determination of acrolein. Z. Lebensm. Unters. Forsch. 99: 352, 1954.
2. Cohen, I.R. and A.P. Altshuller. A new spectrophotometric method for the determination of acrolein in combustion gases and atmosphere. Anal. Chem. 33: 726, 1961.
3. Altshuller, A.P. and S.P. McPherson. Spectrophotometric analysis of aldehydes in the Los Angeles atmosphere. J. Air Pollut. Control Assoc. 13: 109, 1963.
4. Intersociety Committee. *Methods of Air Sampling and Analysis,* p. 187, American Public Health Association, Washington, D.C., 1972.
5. Levaggi, D.A. and M. Feldstein. The determination of formaldehyde, acrolein, and low molecular weight aldehydes in industrial emissions on a single collected sample. J. Air Pollut. Control Assoc. 20: 312, 1970.
6. Tentative method of analysis for low molecular weight aliphatic aldehydes in the atmosphere. Health Lab. Sci. 9: 75, 1972.

ACRYLONITRILE

Synonym: vinyl cyanide $\qquad$ CH$_2$=CHCN
ACGIH TLV: 2 ppm (4.5 mg/m^3) - Skin, human carcinogen
OSHA Standard: 20 ppm
NIOSH Recommendation: 4 ppm

Method	Sampling Duration	Sampling Location	Useful Range (ppm)	System Cost ($)	Test Cost ($)	Manufacturer
DT	Grab	Portable	0.5-30	150	2.60	Nat'l Draeger
DT	Grab	Portable	1-35,000	165	1.70	Matheson
DT	Grab	Portable	0.25-360	180	2.25	Bendix
DT	8-hr	Personal	0.25-40	850	3.10	Nat'l Draeger
IR	Cont	Portable	0.4-4	4374	0	Foxboro
PI	Cont	Fixed	0.2-20	4950	0	HNU
GC	1 min	Portable	1-10^6	10,000	0	Microsensor

NIOSH METHOD NO. P&CAM 202

Principle
A known volume of air is drawn through a small tube containing a sorbent (Carbosieve B, a type of pyrolyzed Saran) on which the acrylonitrile is collected. The acrylonitrile is desorbed from the sorbent with methanol and the resulting solution is analyzed by gas chromatography with a flame ionization detector. The area under the acrylonitrile peak is compared with the areas obtained by the injection of standards.

Range and Sensitivity
The limit of detection of the sampling and analytical method is estimated to be 100 μg. With the procedure described, 800 μg of acrylonitrile should be present to obtain a peak twice the height of the methanol background. This corresponds to an acrylonitrile concentration of 40 mg/m^3 in a 20 L sample.

The largest amount of acrylonitrile that can be collected with a sorbent tube is a function of the relative humidity of the air from which it is sampled. These upper limits at specified relative humidities (R.H.) are: 6 mg when R.H. exceeds 95%, 15 mg when R.H. = 50%, and 22 mg when R.H. is less than 5%. Each value is the approximate number of mg of acrylonitrile that the front section of the tube will hold at a specified relative humidity before a significant amount of acrylonitrile is found on the backup section. (The sorbent tube contains two sections of Carbosieve B separated by a section of glass wool.) The maximum volume to be sampled when the acrylonitrile concentration is 225 mg/m^3 (100 ppm) is 25 L at high humidity or 95 L at low humidity.

Interferences
High humidity decreases the loading capacity of the sorbent tube. Any compound that has the same retention time as acrylonitrile at the operating conditions described in this method is an interference. If the possibility of interference exists, separation conditions (column packing,

temperature, etc.) must be changed to eliminate the problem. Samples should be analyzed by an independent method when overlapping gas chromatographic peaks cannot be resolved.

Precision and Accuracy

The coefficient of variation for the analytical procedure has been found to be 0.05 for 4 mg samples of acrylonitrile. The acrylonitrile was transferred directly with a syringe to sorbent tubes and then analyzed according to the procedure described below. For the analytical method plus simulated field testing, the coefficient of variation over a range of concentrations from 150 to 200 mg/m^3 in a 10 L sample was approximately 0.10. Acrylonitrile in dry air was sampled from a laboratory generator through sorbent tubes. Under laboratory conditions, measurements resulting from application of the total sampling and analytical method were within 10% of expected values at acrylonitrile concentrations of 150 to 200 mg/m^3 in a 10 L sample.

Advantages and Disadvantages

The sampling device is small and portable, and involves no liquids. The tubes are analyzed by means of gas-liquid chromatography, rapid instrumental method. Field testing has not been carried out. It is expected that any interferences that may be encountered can be eliminated by altering chromatographic conditions. One disadvantage of the method is that the amount of sample that can be taken is limited by the amount that the tube will hold at ambient temperatures before overloading. High relative humidity reduces the loading capacity. When the sample value obtained for the backup section of the sorbent tube exceeds 20% of that found on the front section, the possibility of sample loss exists.

Also, sampling from an atmosphere where the temperature was 43 to 46° C. (109 to 115° F.) has demonstrated about a 10% loss in collection efficiency. The precision of the method is limited by the reproducibility of the pressure drop across the tubes. This drop affects the flow rate and causes the volume to be imprecise because the personal sampling pump is usually calibrated for one tube only. The pressure drop across a typical sorbent tube is approximately 1 in of Hg (12 to 13 in of water) at a sampling rate of 1 L/min.

Apparatus

Personal sampling pump. The pump should be calibrated with a representative sorbent tube in the sampling line. A dry or wet test meter or a glass rotameter that will determine the appropriate flow rate (1 L/min) to within 5% may be used for the calibration.

Sorbent tubes. The glass tubes have both ends flame sealed. Each is 8 cm long with a 7 mm O.D. and a 5 mm I.D. They contain two sections of 45/60 mesh Carbosieve B, separated by a 2 mm portion of glass wool. Carbosieve B is a pyrolyzed Saran manufactured and distributed by Supelco, Inc., Bellafonte, Pennsylvania. The sorbing section contains 150 mg of sorbent, the backup section 50 mg. A 3 mm plug of glass wool is placed between the outlet end of the tube and the backup section. A 3 mm plug of glass wool is also placed in front of the sorbing section.

Gas chromatograph. Equipped with a flame ionization detector.

Stainless steel column. 9 ft × 0.125 in with 50/80 mesh Porapak N as the stationary phase.

Recorder. With an appropriate means of determining peak area.

Vials with plastic caps (2 dram).

Syringe. 10 μL for injection into gas chromatograph.

Volumetric pipettes of appropriate sizes.

Reagents

Methanol, reagent grade.

Purified nitrogen.

Prepurified hydrogen.
Filtered compressed air.
Acrylonitrile, at least 99% pure.
Acrylonitrile standard solution (10 mg/mL). Dilute 1.000 g of acrylonitrile to 100 mL with methanol. This standard is stable for at least 2 weeks at ambient temperatures.

Procedure

Cleaning of equipment. All equipment used for the laboratory analysis should be washed with detergent and thoroughly rinsed with tap water and distilled water.

Collection and shipping of samples. Immediately before sampling, break the end of the tube to provide an opening at least one-half the internal diameter of the tube (2 mm). The smaller section of sorbent is used as a backup and should be positioned nearest the sampling pump. The sorbent tube should be placed in a vertical position with the larger section of sorbent pointing up during sampling to minimize channeling of acrylonitrile through the sorbent tubes.

Air being sampled should not be passed through any hose or tubing before entering the sorbent tube. The flow rate and time (or volume) must be measured as accurately as possible. The sample should be taken at a flow rate of 1 L/min or less to attain the total sample volume required. The minimum and maximum sample volumes that should be collected for acrylonitrile are 10 and 95 L, respectively. This maximum is for air of low humidity. The minimum volume quoted must be collected if the desired sensitivity is to be achieved.

At the time personal samples are taken, relatively large volumes of air also should be sampled through other sorbent tubes. These bulk air samples will be used by the analyst to identify possible interferences before the personal samples are analyzed. The temperature and pressure of the atmosphere being sampled should be measured and recorded if either differs greatly from standard conditions (760 mm Hg and 25° C.). The sorbent tubes should be capped immediately after sampling. Under no circumstances should rubber caps be used.

One tube should be handled in the same manner as the sample tube (break, seal, and transport), except that no air is sampled through this tube. This tube should be labeled as a blank. Capped tubes should be packed tightly before they are shipped to minimize tube breakage during shipping. A sample of the bulk material should be submitted to the laboratory in a glass container with a Teflon-lined cap. This sample should not be transported in the same container as the sorbent tubes.

Analysis of samples. In preparation for analysis, each sorbent tube is scored with a file in front of the first section of sorbent and broken open. The glass wool and the sorbent in the first (larger) section are transferred to a small stoppered vial. The separating section of glass wool and the second sorbent section are transferred to another vial. The contents of each vial are analyzed separately.

Desorption of samples. Prior to analysis, 2 mL of methanol is added to the vial containing the first sorbent section, and 1 mL to the vial containing the backup layer. Desorption is complete after 3 to 4 min if the sample is agitated frequently. No loss of sample or desorbing solvent has been noted if the vial is stoppered with a plastic cap during the desorption process.

Gas chromatographic conditions. Typical operation conditions for the gas chromatograph are:

Nitrogen carrier gas flow, 40 mL/min
Hydrogen gas flow to detector, 40 mL/min
Air flow to detector, 300 mL/min
Injection temperature, 200° C.
Column temperature, 170° C.
Detector temperature, 170° C. (In general, the detector temperature should be higher than the

column temperature. However, the gas chromatograph used in the development of this method required a common temperature for both.)

Injection. The first step in the analysis is the injection of the sample into the gas chromatograph. In order to inject as little methanol as possible and, therefore, to maximize resolution of methanol and acrylonitrile, the following injection procedure should be carried out. A 10 μL syringe is first flushed with methanol and dried. The tip of the needle is then immersed in the sample, the plunger is pushed in and pulled out several times until there are no air bubbles between the plunger and sample liquid, and 2 μL of sample is withdrawn. The needle is then withdrawn from the solution and the plunger is pulled back until a small volume of air (approximately 0.5 μL) is evident below the sample layer. The sample volume is then read from the calibrated syringe. The sample is then injected, after which the plunger is again pulled back until the volume of sample left in the syringe can be read from the syringe graduations. The net volume injected can then be computed. Duplicate injections of each sample and standard should be made. No more than a 5% difference in area under the sample peak is to be expected.

Measurement of area. The area under the sample peak is measured and the results are read from a standard curve as explained below.

Determination of desorption efficiency. Only small differences in desorption efficiency are expected from physical differences between lots of Carbosieve B. This commercial sorbent is reported to be clean and pure and is stored in sealed ampoules after its manufacture. Desorption efficiency is usually near 100% for Carbosieve B Lots K51 and K73.

Calibration and Standards

It is convenient to express concentrations of standards in terms of mg/2.0 mL methanol because the sorbing layer of each sample is desorbed in this amount of methanol. A series of standards, varying in concentration over the range of interest, is prepared from the 10 mg/mL standard and analyzed under the same gas chromatographic conditions and during the same time period as the unknown samples. When acrylonitrile concentrations in the air sampled are near the OSHA standard, 45 mg/m^3 (20 ppm), and 20 L samples are taken, an appropriate set of standards may be prepared by diluting to 100 mL with methanol 1, 2, 3, 4, and 5 mL aliquots of the 10 mg/mL standard. Injection of 2 μL aliquots should be carried out as described above. Calibration curves are established by plotting concentration in mg/2.0 mL versus peak area. Since no internal standard is used in the method, standard solutions must be analyzed at the same time that the sample is analyzed. This will minimize the effect of known day-to-day variations of the response of the flame ionization detector.

Calculations

Read the weight, in mg, corresponding to each peak area from the standard curve. No volume corrections are needed, because the standard curve is based on mg/2.0 mL and the volume of sample injected is identical to the volume of the standards injected. Corrections for the blank must be made for each sample:

corrected mg = mg sample - mg blank

where:

mg sample = mg found in front section of sample tube
mg blank = mg found in front section of blank tube

A similar procedure is followed for the backup sections with one exception. Since only 1 mL is used in extracting the backup layer, the amount in milligrams of acrylonitrile as read from the standard curve should be divided by 2.

Add the corrected amounts present in the front and backup sections of the same sample tube to determine the total amount of acrylonitrile in the sample. The concentration of acrylonitrile in air may be expressed in mg/m³:

mg/m^3 = (corrected weight in mg × 1000 L/m^3) ÷ (volume of air sample in L)

The concentration may also be expressed in terms of parts per million (ppm) by volume:

ppm = (mg/m^3)(24.45)(760)(T + 273) ÷ (M.W.)(P)(298)

where:

24.45 = molar volume (L/mol) at 25° C. and 760 mm Hg
M.W. = molecular weight
P = pressure (mm Hg) of air sampled
T = temperature (°C.) of air sampled

References

1. Barrett, W.J., H.K. Dillon, and R.H. James. Sampling and analysis of four organic compounds using solid sorbents. Southern Research Institute, Birmingham, Alabama, Final Report for Contract No. HSM 99-73-63 to the National Institute for Occupational Safety and Health, Division of Laboratories and Criteria Development, Physical and Chemical Analysis Branch, Cincinnati, Ohio, pp. 53-77, 105-114, 1974.
2. White, L.D., D.G. Taylor, P.A. Mauer, and R.E. Kupel. A convenient optimized method for the analysis of selected solvent vapors in the industrial atmosphere. Amer. Ind. Hyg. Assoc. J. 31: 225, 1970.

ALLYL ALCOHOL

ACGIH TLV: 2 ppm (5 mg/m^3) - Skin $CH_2=CHCH_2OH$
OSHA Standard: 2 ppm

Method	Sampling Duration	Sampling Location	Useful Range (ppm)	System Cost ($)	Test Cost ($)	Manufacturer
IR	Cont	Portable	0.5-5	4374	0	Foxboro
PI	Cont	Fixed	0.2-20	4950	0	HNU

NIOSH METHOD NO. S52

Principle

A known volume of air is drawn through a charcoal tube to trap the organic vapors present. The charcoal in the tube is transferred to a small, stoppered sample container and the analyte is desorbed with carbon disulfide containing 5% 2-propanol. An aliquot of the desorbed sample is injected into a gas chromatograph. The area of the resulting peak is determined and compared with areas obtained from the injection of standards.

Range and Sensitivity

This method was validated over the range of 1.8-8.4 mg/m^3 at an atmospheric temperature and pressure of 25° C. and 747 mm Hg, using a 10 L sample. Under the conditions of sample size (10 L) the probable range of this method is 0.5-15 mg/m^3 at a detector sensitivity that gives nearly full deflection on the strip chart recorder for a 0.1 mg sample. The method is capable of measuring much smaller amounts if the desorption efficiency is adequate. Desorption efficiency must be determined over the range used.

The upper limit of the range of the method is dependent on the adsorptive capacity of the charcoal tube. This capacity varies with the concentrations of the analyte and other substances in the air. The first section of the charcoal tube was found to hold at least 0.38 mg of the analyte when a test atmosphere of 8 mg/m^3 of the analyte in dry air was sampled at 0.2 L/min for 240 minutes. Breakthrough did not occur at this time, i.e., no allyl alcohol was found on the backup section of the charcoal tubes. (The charcoal tube consists of two sections of activated charcoal separated by a section of urethane foam.) If a particular atmosphere is suspected of containing a large amount of contaminant, a smaller sample volume should be taken.

Interferences

When the amount of water in the air is so great that condensation actually occurs in the tube, organic vapors will not be trapped efficiently. Preliminary experiments with toluene indicate that high humidity severely decreases the breakthrough volume. When two or more compounds are known or suspected to be present in the air, such information, including their suspected identities, should be transmitted with the sample.

It must be emphasized that any compound which has the same retention time as the specific compound under study at the operating conditions described in this method is an interference. Retention time data on a single column cannot be considered as proof of chemical identity. If the possibility of interference exists, separation conditions (column packing, temperature, etc.) must be changed to circumvent the problem.

Precision and Accuracy

The coefficient of variation for the total analytical and sampling method in the range of 1.8-8.4 mg/m^3 was 0.111. This value corresponds to a standard deviation of 0.52 mg/m^3 at this OSHA standard level. Statistical information and details of the validation and experimental test procedures can be found in reference 2. The average values obtained using the overall sampling and analytical method were 1.2% lower than the "true" value at the OSHA standard level. The above data are based on validation experiments using the internal standard method (reference 2).

Advantages and Disadvantages

The sampling device is small, portable, and involves no liquids. Interferences are minimal, and most of those which do occur can be eliminated by altering chromatographic conditions. The tubes are analyzed by means of a quick, instrumental method. The method can also be used for the simultaneous analysis of two or more compounds suspected to be present in the same sample by simply changing gas chromatographic conditions from isothermal to a temperature-programmed mode of operation.

One disadvantage of the method is that the amount of sample which can be taken is limited by the number of milligrams that the tube will hold before overloading. When the sample value obtained for the backup section of the charcoal tube exceeds 25% of that found on the front section, the possibility of sample loss exists. Furthermore, the precision of the method is limited by the reproducibility of the pressure drop across the tubes. This drop will affect the flow rate and cause the volumes to be imprecise, because the pump is usually calibrated for one tube only.

Apparatus

Calibrated personal sampling pump. Whose flow can be determined accurately (5%) at the recommended flow rate.

Charcoal tubes. Glass tube with both ends flame sealed, 7 cm long with 6 mm O.D. and 4 mm I.D., containing 2 sections of 20/40 mesh activated charcoal separated by a 2 mm portion of urethane foam. The activated charcoal is prepared from coconut shells and is fired at 600° C. prior to packing. The absorbing section contains 100 mg charcoal and the backup section 50 mg. A 3 mm portion of the urethane foam is placed between the outlet end of the tube and the backup section. A plug of silylated glass wool is placed in front of the absorbing section. The pressure drop across the tube must be less than one inch of mercury at a flow rate of 1 L/min.

Gas chromatograph. Equipped with a flame ionization detector.

Column. 10 ft × 1/8 in stainless steel packed with 10% FFAP on 80/100 Chromosorb W-AW.

Electronic integrator. Or some other suitable method for determining peak size areas.

Glass sample containers. 2 mL, with glass stoppers or Teflon-lined caps. If an automatic sample injector is used, the sample injector vials can be used.

Syringes. 10 μL, and other convenient sizes for making standards.

Pipets. 1.0 mL delivery type.

Volumetric flasks. 10 mL or convenient sizes for making standard solutions.

Reagents

Eluent. Carbon disulfide (chromatographic grade) containing 5% 2-propanol (reagent grade).

Allyl alcohol, reagent grade.

Internal standard. n-Dodecane (99%) or other suitable standard.

n-Heptane, reagent grade.

Purified nitrogen.

Prepurified hydrogen.

Filtered compressed air.

Procedure

Cleaning of equipment. All glassware for the laboratory analysis should be detergent washed and thoroughly rinsed with tap water and distilled water.

Calibration of personal pumps. Each personal pump must be calibrated with a representative charcoal tube in the line. This will minimize errors associated with uncertainties in the sample volume collected.

Collection and shipping of samples. Immediately before sampling, break the ends and the tube to provide an opening at least one-half the internal diameter of the tube (2 mm). The smaller section of charcoal is used as a backup and should be positioned nearest the sampling pump. The charcoal tube should be placed in a vertical direction during sampling to minimize channeling through the charcoal. Air being sampled should not be passed through any hose or tubing before entering the charcoal tube.

A maximum size of 10 L is recommended. Sample at a flow of 0.20 L/min. The flow rate should be known with an accuracy of at least 5%. The temperature and pressure of the atmosphere being sampled should be recorded. If the pressure reading is not available the elevation should be recorded. The charcoal tubes should be capped with the supplied plastic caps immediately after sampling. Under no circumstances should rubber caps be used.

One tube should be handled in the same manner as the sample tube (break, seal, and transport), except that no air is sampled through this tube. This tube should be labeled as a blank. Capped tubes should be packed tightly and padded before they are shipped to minimize tube breakage during shipping. A sample of the suspected compound should be submitted to the laboratory in glass containers with Teflon-lined caps. These liquid bulk samples should not be transported in the same container as the charcoal tubes.

Analysis of samples. In preparation for analysis, each charcoal tube is scored with a file in front of the first section of charcoal and broken open. The glass wool is removed and discarded. The charcoal in the first (larger) section is transferred to a 2 mL stoppered sample container or automatic sample injector vial. The separating section of foam is removed and discarded; the second section is transferred to another sample container or vial. These two sections are analyzed separately.

Desorption of samples. Prior to analysis, 1.0 mL of the eluent is pipetted into each sample container. For the internal standard method a 0.005% solution of internal standard in the eluent is used. (All work with carbon disulfide should be performed in a hood because of its high toxicity.) Desorption should be done for 30 minutes. Tests indicate that this is adequate if the sample is agitated occasionally during this period. The sample vials should be capped as soon as the solvent is added to minimize volatilization.

GC conditions. The typical operating conditions for the gas chromatograph are:

30 mL/min (80 psig) nitrogen carrier gas flow
30 mL/min (50 psig) hydrogen gas flow to detector
300 mL/min (50 psig) air flow to detector
200° C. injector temperature
300° C. detector temperature
80° C. column temperature

Injection. The first step in the analysis is the injection of the sample into the gas chromatograph. To eliminate difficulties arising from blow back or distillation within the syringe needle, one should employ the solvent flush injection technique. The 10 microliter syringe is first flushed with solvent several times to wet the barrel and plunger. Three microliters of solvent are drawn into the syringe to increase the accuracy and reproducibility of the injected sample volume. The needle is removed from the solvent, and the plunger is pulled back about 0.2 microliter to separate the solvent flush from the sample with a pocket of air to be used as a marker. The needle is then immersed in the sample, and a 5 microliter aliquot is withdrawn, taking into consideration the volume of the needle, since the sample in the needle will be completely injected. After the needle is removed from the sample and prior to injection, the plunger is pulled back 1.2 microliters to minimize evaporation of the sample from the tip of the needle. Observe that the sample occupies 4.9-5.0 microliters in the barrel of the syringe. Duplicate injections of each sample and standard should be made. No more than a 3% difference in area is to be expected. An automatic sample injector can be used if it is shown to give reproducibility at least as good as the solvent flush technique. In this case 2 microliter injections are satisfactory.

Measurement of area. The area of the sample peak is measured by an electronic integrator or some other suitable form of area measurement, and preliminary results are read from a standard curve prepared as discussed below.

Determination of desorption efficiency. The desorption efficiency of a particular compound can vary from one laboratory to another and also from one batch of charcoal to another. Thus, it is necessary to determine at least once the percentage of the specific compound that is removed in the desorption process, provided that the same batch of charcoal is used.

Activated charcoal equivalent to the amount in the first section of the sampling tube (100 mg) is measured into a 2.0 mL sample container. This charcoal must be from the same batch as that used in obtaining the samples and can be obtained from unused charcoal tubes. A 12 mg/mL stock solution of the analyte in n-heptane is prepared. A known amount of this solution is injected directly into the activated charcoal with a 10 μL syringe, and the container is capped. The amount injected is equivalent to that present in a a 10 L sample at the selected level. It is not practical to inject the neat liquid directly because the amounts to be added would be too small to measure accurately.

At least six tubes at each of three levels (0.5X, 1X, and 2X the OSHA standard) are prepared in this manner and allowed to stand for at least overnight to assure complete adsorption of the analyte onto the charcoal. These six tubes are referred to as the samples. A parallel blank tube should be treated in the same manner except that no sample is added to it. The sample and blank tubes are desorbed and analyze in exactly the same manner as the sampling tube described earlier. The weight of analyte found in each tube is determined from the standard curve. Desorption efficiency is determined by the following equation:

D.E = average weight (mg) recovered ÷ weight (mg) added

The desorption efficiency is dependent on the amount of analyte collected on the charcoal. Plot the desorption efficiency versus the weight of analyte found. This curve is used to correct for adsorption losses.

Calibration and Standards

It is convenient to express concentration of standards in terms of mg/mL of eluent. To minimize error due to volatility of the eluent, one can add 10 times the weight to 10 mL of the eluent. (For the internal standard method use eluent containing 0.005% of the internal standard.) A series of standards, varying in concentration over the range of interest, is prepared and analyzed under the same GC conditions and during the same time period as the unknown samples. Curves are established by plotting concentrations in mg/mL versus peak area. In the case of the internal standard method plot the concentration versus the ratio of peak area of analyte to peak area of internal standard. Note: Whether the absolute area or internal standard method is used standard solutions should be analyzed at the same time that the sample analysis is done. This will minimize the effect of variations of FID response.

Calculations

Read the weights, in mg, corresponding to each peak area (area ratio in case of the internal standard method) from the standard curve. No volume corrections are needed, because the standard curve is based on mg/mL eluent and the volume of sample injected is identical to the volume of the standards injected. Corrections for the blank must be made for each sample:

mg = mg sample - mg blank

where:

mg sample = mg found in front section of sample tube
mg blank = mg found in front section of blank tube

A similar procedure is followed for the backup sections.

Add the weights present in the front and backup sections of the same sample tube to determine the total weight in the sample. Read the desorption efficiency from the curve for the amount of analyte found in the front section. Divide the total weight by this desorption efficiency to obtain the corrected mg/sample:

Corrected mg/sample = total weight ÷ D.E.

The concentration of analyte in the air sampled can be expressed in mg/m^3, which is numerically equal to μg/L of air:

mg/m^3 = (corrected mg × 1000 L/m^3) ÷ (air volume sampled in L)

Another method of expressing concentration is ppm:

ppm = (mg/m^3)(24.45)(760)(T + 273) ÷ (M.W.)(P)(298)

where:

P = pressure (mm Hg) of air sampled
T = temperature (°C.) of air sampled
24.45 = molar volume (L/mol) at 25° C. and 760 mm Hg
M.W. = molecular weight (g/mol) of analyte
760 = standard pressure (mm Hg)
298 = standard temperature (°K)

References

1. White L.D., et al. A convenient optimized method for the analysis of selected solvent vapors in the industrial atmosphere. Amer. Ind. Hyg. Assoc. J. 31: 225, 1970.
2. Documentation of NIOSH Validation Tests, Contract No. CDC-99-74-45.
3. Final Report, NIOSH Contract No. HSM-99-71-31. Personal sampler pump for charcoal tubes, September 15, 1972.

ALLYL CHLORIDE

Synonym: 3-chloropropene $\quad\quad\quad\quad\quad\quad\quad\quad\quad\quad\quad\quad\quad\quad\quad$ CH$_2$=CHCH$_2$Cl
ACGIH TLV: 1 ppm (3 mg/m^3)
OSHA Standard: 1 ppm
NIOSH Recommendation: 3 ppm ceiling

Method	Sampling Duration	Sampling Location	Useful Range (ppm)	System Cost ($)	Test Cost ($)	Manufacturer
IR	Cont	Portable	0.4-4	4374	0	Foxboro
PI	Cont	Fixed	0.2-20	4950	0	HNU

NIOSH METHOD NO. S165

(Use as described for acetonitrile, with 160° C. column temperature.)

ALLYL GLYCIDYL ETHER

Synonyms: AGE; allyl 2,3-epoxypropyl ether $\quad CH_2=CHCH_2OCH_2CH(O)CH_2$
ACGIH TLV: 5 ppm (22 mg/m^3) - Skin
OSHA Standard: 10 ppm ceiling

Method	Sampling Duration	Sampling Location	Useful Range (ppm)	System Cost ($)	Test Cost ($)	Manufacturer
IR	Cont	Portable	0.2-20	4374	0	Foxboro

NIOSH METHOD NO. S346

Principle

A known volume of air is drawn through a Tenax-GC tube to trap the organic vapors present. The sampling tube consists of a front adsorbing section and a backup section. The Tenax-GC in each tube is transferred to a vial and the allyl glycidyl ether is desorbed with diethyl ether and analyzed by gas chromatography.

Range and Sensitivity

This method was validated over the range of 19-87 mg/m^3 at an atmospheric temperature of 17° C. and atmospheric pressure of 752 mm Hg using a 3 L sample volume. This sample volume is less than two-thirds of the 5% breakthrough capacity determined at 90% relative humidity when sampling a test atmosphere at 2X OSHA standard. This method is capable of measuring much smaller amounts if the desorption efficiency is adequate. Desorption efficiency must be determined over the range used. The upper limit of the range of the method is dependent on the adsorptive capacity of the Tenax-GC resin tube. This capacity can vary with the concentration of analyte and other substances in the air.

Interferences

When two or more compounds are known or suspected to be present in the air, such information, including their suspected identities, should be transmitted with the sample. It must be emphasized that any compound which has the same retention time as the analyte at the operating conditions described in this method is an interference. Retention time data on a single column cannot be considered as proof of chemical identity. If the possibility of interference exists, separation conditions (column packing, temperature, etc.) must be changed to circumvent the problem.

Precision and Accuracy

The coefficient of variation for the total analytical and sampling method in the range of 19-87 mg/m^3 was 0.057. This value corresponds to a 2.6 mg/m^3 standard deviation at the OSHA standard level. Statistical information and details of the validation and experimental test procedures can be found in references 1 and 2. On the average, the concentrations obtained at the OSHA standard level using the overall sampling and analytical method were 0.5% lower than the "true" concentrations for a limited number of laboratory experiments. Any difference between the "found" and "true" concentrations may not represent a bias in the sampling and analytical method, but rather a random variation from the experimentally determined "true" concentration. Therefore, no recovery correction should be applied to the final result. The data are based on validation experiments using the internal standard method.

Advantages and Disadvantages

The sampling device is small, portable, and involves no liquids. Interferences are minimal, and most of those which do occur can be eliminated by altering chromatographic conditions. The sorbent tubes are analyzed by means of a quick, instrumental method.

One disadvantage of the method is that the amount of sample which can be taken is limited by the number of milligrams that the tube will hold before overloading. When an atmosphere at 90% relative humidity containing 92 mg/m^3 of allyl glycidyl ether was sampled at 0.8 L/min, 5% breakthrough was observed after 15 minutes (capacity = 12 L or 1.1 mg). The sample size recommended is two-thirds of the 5% breakthrough capacity at 90% R.H. for a test atmosphere at 2X OSHA standard to minimize the probability of overloading the sampling tube and allow sampling with the 0.2 L/min sampling pump.

The precision of the method is affected by the reproducibility of the pressure drop across the tubes. This drop will affect the flow rate and cause the volume to be imprecise, because the pump is usually calibrated for one tube only.

Apparatus

Calibrated personal sampling pump. Whose flow can be determined within 5% at the recommended flow rate of 0.2 L/min.

Resin tubes. Glass tube with both ends flame-sealed, 10 cm long with 8 mm O.D. and 6 mm I.D., containing 2 sections of 35/60 mesh Tenax GC resin. The adsorbing section contains 100 mg of resin, the backup section 50 mg. A small wad of silylated glass wool is placed between the front adsorbing section and the backup section; a plug of silylated glass wool is also placed in front of the adsorbing section and at the end of the backup section. Since the pressure drop across the tube must be less than 25 mm of mercury at a flow rate of 1 L/min, it is necessary to avoid overpacking with glass wool.

Gas chromatograph. Equipped with a flame ionization detector.

Column. 20 ft × 1/8 in stainless steel packed with 10% FFAP stationary phase on 100/200 mesh Supelcoport.

Electronic integrator. Or some other suitable method for measuring peak areas.

Sample containers. With Teflon-lined caps, 5 mL.

Syringes. 10 and 500 microliter and other convenient sizes for making standards and for taking sample aliquots for dilution.

Pipettes. 2 mL, delivery type.

Volumetric flasks. 1 and 10 mL or convenient sizes for making standard solutions and dilution of samples.

Reagents

Diethyl ether, anhydrous.

Allyl glycidyl ether, 99%.

Isoamyl alcohol. Or other suitable internal standard. The appropriate solution of the internal standard is prepared in ether.

Hexane. This solvent is used to prepare the allyl glycidyl ether solutions for the desorption efficiency determination.

Nitrogen, purified.

Hydrogen, prepurified.

Air, filtered compressed.

Procedure

Cleaning of equipment. All glassware used for the laboratory analysis should be detergent washed and thoroughly rinsed with tap water and distilled water.

Calibration of personal pumps. Each personal pump must be calibrated with a representative sampling tube series in the line; the tube is described above. This will minimize errors associated with uncertainties in the sample volume collected.

Collection and shipping of samples. Immediately before sampling, break the two ends of the resin tube to provide an opening at least one-half the internal diameter of the tube (3 mm). The section containing 50 mg of resin is used as a backup and should be positioned nearest the sampling pump. The resin tube series should be placed in a vertical direction during sampling to minimize channeling through the resin.

Air being sampled should not be passed through any hose or tubing before entering the resin tube. A sample size of 3 L is recommended. Sample at a flow of 0.2 L/min for 15 minutes. The flow rate should be known with an accuracy of at least 5%. The temperature and pressure of the atmosphere being sampled should be recorded. If pressure reading is not available, record the elevation.

The resin tube should be labeled appropriately and capped with the supplied plastic caps. Under no circumstances should rubber caps be used. With each batch of 10 samples, submit one resin tube which has been handled in the same manner as the sample tubes (break, seal, and transport), except that no air is sampled through this tube. This tube should be labeled as a blank. Capped resin tubes should be packed tightly and padded before they are shipped to minimize tube breakage during shipping.

Analysis of samples. In preparation for analysis, each resin tube is scored with a file in front of the first section of resin and broken open. The glass wool is removed and discarded. The resin in the front 100 mg section is transferred to a 5 mL screw-capped sample container. The separating section of glass wool is removed and discarded; the second 50 mg section is transferred to another stoppered container. These two sections are analyzed separately.

Desorption of sample. Prior to analysis, 2 mL of ether is pipetted into each sample container. Desorption should be done for 30 minutes. Tests indicate that this is adequate if the sample is agitated occasionally during this period. The sample vials should be capped as soon as solvent is added to minimize volatilization. For the internal standard method, desorb using 2.0 mL of internal standard solution in ether.

GC conditions. The typical operating conditions for the gas chromatograph are:

30 mL/min (60 psig) nitrogen carrier gas flow
30 mL/min (25 psig) hydrogen gas flow to detector
300 mL/min (60 psig) air flow to detector
200° C. injector temperature
280° C. detector temperature
150° C. column temperature

A retention time of approximately 10.0 minutes is to be expected for the analyte using these conditions and the column recommended. The internal standard elutes between ether and the allyl glycidyl ether.

Injection. A 2 microliter aliquot of the sample solution is injected into the gas chromatograph. The solvent flush method or other suitable alternative such as an automatic sample injector can be used provided that duplicate injections of a solution agree well. No more than a 3% difference in area is to be expected.

Measurement of area. The area of the sample peak is measured by an electronic integrator or some other suitable form of area measurement, and preliminary results are read from a standard curve prepared as discussed below.

Determination of desorption efficiency. The desorption efficiency of a particular compound can vary from one laboratory to another and also from one batch of Tenax-GC to another. Thus, it is necessary to determine the percentage of the specific compound that is removed in the desorption process for the particular batch of resin used for sample collection and over the concentration range of interest. The desorption efficiency must be at least 75% at the equivalent 1X OSHA standard level.

The desorption efficiency must be determined over the sample concentration range of interest. In order to determine the sample concentration range which should be tested, the samples are analyzed first and then the analytical samples are prepared based on the relative amount of allyl glycidyl ether found in the samples. The desorption efficiency must be determined at least in duplicate for each concentration level of allyl glycidyl ether found in the sample.

The analytical samples are prepared as follows: Tenax-GC, equivalent to the amount in the front section (100 mg), is measured into a 5 mL screw capped vial. This resin must be from the same batch as that used in obtaining the samples. A known amount of a solution of allyl glycidyl ether in hexane (spiking solution) is injected directly into the resin by means of a microliter syringe. Adjust the concentration by means of a microliter syringe. Adjust the concentration of the spiking solution such that no more than a 10 μL aliquot is used to prepare the analytical samples.

For the validation studies conducted to determine the precision and accuracy of this method, six analytical samples at each of the three concentration levels (0.5X, 1X and 2X of the OSHA standard) were prepared by adding an amount of allyl glycidyl ether equivalent to that present in a 3 L sample at the selected level. A stock solution containing 67.34 milligrams of allyl glycidyl ether per milliliter of hexane was prepared. One, 2 and 4 microliter aliquots of the solution were added to the Tenax-GC resin tubes to produce 0.5, 1 and 2X the OSHA standard level. The analytical samples were allowed to stand at least overnight to assure complete adsorption of the analyte onto the resin. A parallel blank tube was treated in the same manner except that no sample was added to it.

The procedure described can be used to prepare the analytical samples which are analyzed to determine desorption efficiency over the concentration range of interest. Desorption and analysis experiments are done on the analytical samples. Calibration standards are prepared by adding the appropriate volume of spiking solution to 2.0 mL of ether with the same syringe used in the preparation of the samples. Standards should be prepared at the same time that the sample analysis is done and should be analyzed with the samples.

If the internal standard method is used, prepare calibration standards by using 2.0 mL of ether containing a known amount of the internal standard. The desorption efficiency (D.E.) equals the average weight in μg recovered from the tube divided by the weight in μg added to the tube, or:

D.E. = average weight recovered ÷ weight added

The desorption efficiency may be dependent on the amount of allyl glycidyl ether collected on the resin. Plot the desorption efficiency versus weight of allyl glycidyl ether found. This curve is used to correct for adsorption losses.

Calibration and Standards

Solutions of allyl glycidyl ether should be prepared over the appropriate concentration range in diethyl ether. The solutions should be prepared by addition of allyl glycidyl ether to 2.0 mL of diethyl ether in 5 mL vials. The allyl glycidyl ether may either be added neat using a microliter syringe or one may use the hexane solution described earlier. The concentration of standards can be expressed in terms of μg of allyl glycidyl ether per 2 mL of ether.

A series of standards, varying in concentration over the range of interest, is prepared as described above the analyzed under the same GC conditions and during the same time period as the unknown samples. Curves are established by plotting peak area (ordinate) against sample concentration in $\mu g/2.0$ mL.

For the internal standard method, use ether containing a predetermined amount of the internal standard. The internal standard concentration used was approximately 70% of the concentration at 2X the standard. The area ratio of the analyte to that of the internal standard is plotted against the analyte concentration in $\mu g/2.0$ mL. Note: Whether the external standard or internal standard method is used, standard solutions should be analyzed at the same time the sample analysis is done. This will minimize the effect of variations in FID response.

Calculations

Read the weight, in μg, corresponding to each peak area from the standard curve. No volume corrections are needed, because the standard curve is based on μg per 2.0 mL ether and the volume sample injected is identical to the volume of the standards injected. Corrections for the blank must be made for each sample:

$\mu g = \mu g$ sample - μg blank

where:

μg sample $= \mu g$ found in front (100 mg) sample section
μg blank $= \mu g$ found in front (100 mg) blank section

A similar procedure is followed for the backup (50 mg) section. Read the desorption efficiency from the curve for the amount found in the front section of the tube. Divide the total weight by this desorption efficiency to obtain the corrected μg/sample:

Corrected μg/sample = weight (front section) ÷ D.E.

Add the amounts present in the front and backup sections for the same sample to determine the total weight in the sample. Determine the volume of air sampled at ambient conditions in L based on the appropriate information, such as flow rate in L/min multiplied by sampling time. If a pump using a rotameter for flow rate control was used for sample collection, a pressure and temperature correction must be made for the indicated flow rate. The expression for this correction is:

Corrected volume = ft $(P_1T_2/P_2T_1)^{1/2}$

where:

    f = flow rate sampled
    t = sampling time
    P_1 = pressure during calibration of sampling pump (mm Hg)
    P_2 = pressure of air samples (mm Hg)
    T_1 = temperature during calibration of sampling pump (°K)
    T_2 = temperature of air sampled (°K)

The concentration of the analyte in the air sampled can be pressed in mg/m^3 which is numerically equal to $\mu g/L$:

mg/m^3 = corrected μg ÷ air volume sampled in L

Another method of expressing concentration is ppm (corrected to standard conditions of 25° C. and 760 mm Hg):

ppm = $(mg/m^3)(24.45)(760)(T + 273)$ ÷ $(M.W.)(P)(298)$

where:

    P = pressure (mm Hg) of air sampled
    T = temperature (°C) of air sampled
    24.45 = molar volume (L/mol) at 25° C. and 760 mm Hg
    M.W. = molecular weight
    760 = standard pressure (mm Hg)
    298 = standard temperature (°K)

References

1. Memoranda, Kenneth A. Busch, Chief, Statistical Services, DLCD, to Deputy Director, DLCD, dated 1/16/75, 11/8/74, subject: Statistical Protocol for Analysis of Data from Contract CDC-99-74-45.
2. Backup Data Report for Allyl Glycidyl Ether, No. S346, prepared under NIOSH Contract No. 210-76-0123.
3. Final Report, NIOSH Contract HSM-99-71-31, Personal Sampler Pump for Charcoal Tubes, September 15, 1972.

AMMONIA

ACGIH TLV: 25 ppm (18 mg/m^3)　　　　　　　　　　　　　　　　　　　　　　　　　NH$_3$
OSHA Standard: 50 ppm
NIOSH Recommendation: 50 ppm ceiling

Method	Sampling Duration	Sampling Location	Useful Range (ppm)	System Cost ($)	Test Cost ($)	Manufacturer
DT	Grab	Portable	5-700	150	2.70	Nat'l Draeger
DT	Grab	Portable	5-200,000	165	1.70	Matheson
DT	Grab	Portable	1-320,000	180	2.25	Bendix
DT	4-hr	Personal	5-100	850	3.10	Nat'l Draeger
CS	Grab	Portable	0.5-50	2465	0.80	MDA
IR	Cont	Portable	1-100	4374	0	Foxboro
CS	Cont	Fixed	2.5-50	4950	Variable	MDA
PI	Cont	Fixed	2-200	4950	0	HNU
CM	Cont	Port/Fixed	0.01-50	4995	Variable	CEA Inst

NIOSH METHOD NO. P&CAM 205

Principle

Ammonia is collected in a dilute sulfuric acid solution in a midget impinger to form ammonium sulfate. Nessler reagent is used to produce a yellow-brown complex. The ammonia concentration is determined by reading the absorbance of the yellow-brown solution at 440 nm and comparing it with a standard curve. Absorption peak may shift with concentration.

Range and Sensitivity

The Nessler reagent is said to be sensitive to as little as 0.002 mg of ammonia. The range of application of this method is 0.10 mg to 0.80 mg of ammonia in a 10 L air sample, about one-third to twice the present limit of 35 mg/m^3 of air.

Interferences

Ammonium salts will react with Nessler reagent to give a false high reading. These can be removed by filtration of the air before its passage into the midget impinger. Other interferences have not been reported.

Precision and Accuracy

The precision and accuracy of this method have not been reported.

Advantages and Disadvantages

The method is sensitive, but ammonium salts can interfere. The method does not distinguish between free and combined ammonia.

Apparatus

Battery-operated personal air sampling pump. Capable of drawing 1 L of air per minute through

10 mL of absorbing solution.
Impingers. Standard midget, all glass, calibrated.
Beckman Model B spectrophotometer, or equivalent.
Spectrophotometer cells, 1 cm.
Volumetric flasks. 50 mL, 100 mL, 200 mL, 1000 mL, glass-stoppered.
Pipettes. 0.5 mL, 1.0 mL, 5 mL, 10 mL, 20 mL.
Graduated cylinders. 25 mL, 50 mL, 100 mL.

Reagents

All reagents should be analytical reagent grade.
Double-distilled water free of ammonia.
Potassium iodide.
Mercuric chloride.
Potassium hydroxide.
Ammonium sulfate.

Nessler reagent. Dissolve 35 g of mercuric chloride in 500 mL of hot water. Filter and allow to cool. Dissolve 62.5 g of potassium iodide in 260 mL of cold water. Gradually add the mercuric chloride solution to 250 mL of the iodide solution until a slight permanent red precipitate is formed. Dissolve the precipitate with the remaining iodide solution and again add mercuric chloride slowly until a red precipitate remains. Dissolve 150 g of potassium hydroxide in 250 mL of distilled water. Add this solution to the potassium iodide-mercuric chloride solution and make up to 1 L with distilled water. Stir thoroughly and allow to stand a day or so, and decant the clear liquid. Warning: the Nessler reagent should be handled with caution because of its toxicity and corrosive properties.

Standard ammonium sulfate solution. Dissolve 77.6 mg of ammonium sulfate in distilled water and make up to 1 L. One mL of this solution contains 20 µg of ammonia. Discard solution after a week.

Absorbing solution. Dilute 2.8 mL of concentrated sulfuric acid (18 M) to 1 L with distilled water to form 0.1 N sulfuric acid.

Procedure

Sampling. Add 10 mL of absorbing solution to each midget impinger including one for the blank. Attach the bubbler to a personal air sampling pump and draw air through the bubbler at a rate of 1 L/min for 10-15 min. Record the volume of air sampled.

Color development. Dilute the sample to 50 mL with distilled water in a volumetric flask and shake well. Take 1 mL of this solution and make up to 50 mL with distilled water in another volumetric flask. Shake. Add 2 mL of Nessler reagent to the latter flask and determine absorbance after 10 min at 440 nm in a spectrophotometer using a 1 cm cell. Treat the blank in the same manner as the sample.

Calibration and Standards

Dilute samples of 5 mL, 10 mL, 20 mL, 30 mL, and 40 mL of the standard ammonium sulfate solution and process in the same manner as described above. Determine light absorbance in a spectrophotometer at 440 nm. Prepare a standard curve of absorbance versus micrograms of ammonia.

Calculations

$$\mu g\ NH_3/m^3 = (w/v) \times 1000$$

where:

w = μg of ammonia from the standard curve minus blank value
v = volume of air in L, corrected to 25° C. and 760 Torr

References
1. Goldman, F.H. and M.B. Jacobs. *Chemical Methods of Industrial Hygiene,* Interscience Publishers, New York, 1953.
2. Elkins, H.B. *The Chemistry of Industrial Toxicology,* 2nd ed., John Wiley and Sons, New York, 1959.

n-AMYL ACETATE

ACGIH TLV: 100 ppm (530 mg/m^3) $\hfill$ CH$_3$COOC$_5$H$_{11}$
OSHA Standard: 100 ppm

Method	Sampling Duration	Sampling Location	Useful Range (ppm)	System Cost ($)	Test Cost ($)	Manufacturer
DT	Grab	Portable	100-9,000	180	2.25	Bendix
IR	Cont	Portable	1-200	4374	0	Foxboro
PI	Cont	Fixed	0.2-20	4950	0	HNU

NIOSH METHOD NO. P&CAM 127

(Use as described for acetone, with 90° C. column temperature)

ANILINE

ACGIH TLV: 2 ppm (10 mg/m^3) - Skin $\hfill$ C$_6$H$_5$NH$_2$
OSHA Standard: 5 ppm

Method	Sampling Duration	Sampling Location	Useful Range (ppm)	System Cost ($)	Test Cost ($)	Manufacturer
DT	Grab	Portable	1-20	150	2.70	Nat'l Draeger
DT	Grab	Portable	1-30	165	2.30	Matheson
DT	Grab	Portable	1.25-60	180	2.25	Bendix
IR	Cont	Portable	1-10	4374	0	Foxboro
PI	Cont	Fixed	0.2-20	4950	0	HNU

NIOSH METHOD NO. P&CAM 168

Principle

A known volume of air is drawn through a tube containing silica gel to trap the aromatic amines present. The silica gel in the tube is transferred to a glass-stoppered tube and treated with ethanol. An aliquot of the desorbed aromatic amines in ethanol is injected into a gas chromatograph. Peak areas are determined and compared with calibration curves obtained from the injection of standards.

Range and Sensitivity

The lower limit of this method using a flame ionization detector is 0.01 mg/sample of any one compound when the analyte is desorbed with 5 mL ethanol and a 10 μL aliquot is injected into the gas chromatograph. Sensitivity for p-nitroaniline is 5 times less.

The upper limit is at least 14 mg/sample. This is a minimum amount of aniline which the large section (700 mg) and the center section (150 mg silica gel) of the sampling tube will retain before 2% of the sample is found on the third (150 mg) section after 8 hours of sampling (200 mL/min high humidity air containing 150 mg/m^3 aniline). The corresponding upper limit for sampling in the reverse direction through the tube is 3.5 mg/sample. The less volatile substituted anilines have higher upper limits than aniline.

Interferences

The most common possible sampling interference is water vapor. The sampling tube has been designed so that 96 L of high humidity air can be sampled over an 8 hour period at 200 mL/min without displacement of the collected aromatic amines by water vapor.

Any compound which has nearly the same retention time as one of these aromatic amines at the gas chromatograph analytical conditions described in this method is an interference. This type of interference often can be overcome by changing the operating conditions of the gas chromatograph or selecting another column. Retention time data on a single column, or even on a number of columns, cannot be considered as conclusive proof of chemical identity in all cases. For this reason it is important that whenever practical sample of the bulk compound or mixture be submitted at the same time as the sample tube (but shipped separately) so that chemical identification can be made by other means.

Precision and Accuracy

The accuracy of the method depends upon collection efficiencies and desorption efficiencies. If a negligible amount of aromatic amine is detected on the backup section, the collection efficiency of the tube must be essentially 100%. Desorption efficiencies for the range of 1-8 mg have been found to be 100% within experimental error of 5%. Precision of the analysis is quite dependent upon the precision and sensitivity of the technique used to quantitate gas chromatographic peaks of samples and standards. Electronic digital integrators with baseline correction capabilities can be used to maximize analytical precision, particularly at lower concentrations.

Precision of preparing chemical standards can be 1%. The precision of the overall method, 2s = ±9%, has been determined from 8 consecutive identical samples taken with a personal sampling pump. Analytical precision can be improved by reducing or eliminating the error associated with syringe injection into the gas chromatograph. This is best accomplished by the addition of internal standard to the ethanol used to both prepare standards and elute samples from the silica gel. Therefore, it is recommended that about 0.1% solution of n-heptanol in ethanol be used. For isothermal analyses at temperatures above 120° C., n-octanol may be preferable.

Advantages and Disadvantages

The sampling method uses a small, portable device involving no liquids. Taking the effect of humidity into account, a sample of up to 8 hours can be taken for an average work day concentration or a 15 minute sample can be taken to test for excursion concentrations. Desorption of the collected sample is simple and is accomplished with a solvent with low toxicity. The analysis is accomplished by a quick instrumental method. Most analytical interferences which occur can be eliminated by altering gas chromatographic conditions. Several aromatic amines can be collected and analyzed simultaneously; this is useful where the composition of the aromatic amine vapors may not be known.

A major disadvantage of the method is the limitation on its precision due to the use of personal sampling pumps currently available. After initial adjustment of flow any change in the pumping rate will affect the volume of air actually sampled. Furthermore, if the pump used is calibrated for one tube only, as is often the case, the precision of the volume of air sampled will be limited by the reproducibility of the pressure drop across the tubes.

Apparatus

Personal sampling pump. Whose flow can be set and maintained at 1 L/min for 15 min or 200 mL/min for 8 hours.

Pyrex glass sampling tubes. With the dimensions shown in Figure 1 packed with three sections of 45/60 mesh silica gel. The weights of the sections in order are 700 mg, 150 mg, and 150 mg. The silica gel should be the equivalent of silica gel D-08, chromatographic grade, activated and fines free, 45/60 mesh, from Applied Science Laboratories, Inc., State College, PA. Plugs of 100 mesh stainless steel screen are used to contain the silica gel sections. These plugs of negligible pressure drop are prepared from 11 mm diameter discs pushed into an 8 mm I.D. tube with a 7 mm O.D. rod. Pieces of Pyrex tubing 7 mm O.D. by 12 mm long and located between the sorbent sections greatly reduce migration of the sample throughout the tube prior to analysis. The ends of each tube should be flame sealed after packing to prevent contamination before sampling. The pressure drop of such tubes should not exceed 13 in of water at 1 L/min or 2.5 in of water at 200 mL/min air flows.

Gas chromatograph. Equipped with flame ionization detector. Linear temperature programming capability is desirable but not essential.

Column. 4 ft × 1/8 in O.D., stainless steel, packed with silicone OV-25 liquid phase, 10% on 80/100 mesh Supelcoport (or equivalent support). A column (2 ft × 1/8 in O.D.) packed with Chromosorb 103 (80/100 mesh) can be used for all amines except p-nitroaniline.

Recorder. And some method for determining peak height or area.

Glass-stoppered tubes or flasks, 2 mL and 10 mL.

Syringes, 10 µL.

Pipettes and volumetric flasks. For preparation of standard solutions.

Reagents

Ethanol, 95%.
n-Heptanol and n-octanol, reagent grade.
Reagent grade aromatic amines standards.
Bureau of Mines Grade A helium.
Prepurified hydrogen.
Filtered compressed air.

Procedure

Cleaning of equipment. All glassware used for the laboratory analysis is detergent washed followed by tap and distilled water rinses.

Calibration of personal pumps. Each pump must be calibrated with a representative tube in the line to minimize errors associated with uncertainties in the sample volume collected.

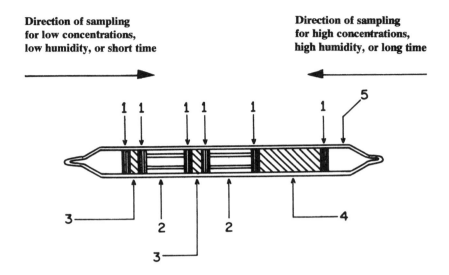

FIGURE 1. Silica gel sampling tube for aromatic amines. (1) 100 mesh stainless steel screen plugs; (2) 12-mm glass tube separator; (3) 150-mg silica gel section, 45/60 mesh; (4) 700-mg silica gel section, 45/60 mesh; (5) 8-mm I. D. glass tube.

Collection and shipping of samples. Immediately before sampling, the ends of the tube should be broken so as to provide an opening at least one half the internal diameter of the tube. The desired sampling direction is chosen, the initial section marked with a permanent marker, and tubing from the sampling pump attached to the other end. For low concentrations, low humidity, and short-term samples, the smaller (150 mg) section is used for the initial section. For expected high concentrations, high humidity, and long-term samples, the largest (700 mg) section is used for the initial section. See Table 1 footnotes. Sampled air should not pass through any hose or tubing before entering the sampling tube.

The atmosphere is sampled at the desired flow rate for the desired period of time. Recommended sampling volumes based on sensitivity and breakthrough studies are given in Table 1. The flow rate and sampling time or the volume of sampled air must be measure as accurately as possible. The temperature, pressure, and humidity of the atmosphere being sampled is measured and recorded. The sampling tubes are capped with the supplied plastic caps immediatlely after sampling. Under no circumstances should rubber caps be used.

One tube should be handled in the same manner as the sample tube (break, seal, and transport), except that no air is sampled through this tube. This tube is labeled as a blank. Capped tubes should be packed tightly before shipping to minimize tube breakage during shipping. Samples of the bulk liquids or solids from which the aromatic amine vapors arise should be submitted to the laboratory also, but not in the same container as the air samples or blank tubes. Tubes after sampling should be tightly capped and not subjected to extremes of high temperature or low pressure, if avoidable.

Analysis of samples. By removing and discarding the stainless steel plugs and glass spacers the silica gel sections are transferred to separate glass-stoppered tubes or flasks (2 mL for the smaller sections and 10 mL for the larger section). Note or mark which tubes contain the initial section, the backup section, and the third section. These are analyzed separately.

Desorption of samples. Prior to analysis, ethanol containing 0.1% n-heptanol internal standard is pipetted into each flask—5 mL for the larger section and 1 mL for all other sections. Tests indicate that desorption is complete in 30 min if the sample is stirred or shaken occasionally.

Gas chromatograph conditions. Typical operating conditions for the gas chromatograph are:

Carrier flow, 25 mL He/min
Injection port, 150° C.
Flame ionization detector, 250° C.
Gases to detector: 50 mL/min H_2, 470 mL/min air
Oven temperature program: 100° C. for 4 min, then increase at 8° C./min to 225° C.

An aliquot of the sample is injected into the gas chromatograph. With an internal standard in the eluent, direct injection of up to 10 μL with a 10 μL syringe is acceptably precise. At least duplicate injections of the same sample or standard are recommended. The areas of the sample peak and the internal standard peak are measured by an electronic integrator or some other suitable method of area measurement. The ratio of these areas is calculated and used to determine sample concentration in the eluent by using a standard curve prepared as discussed below.

TABLE I

Aromatic amines for which the method has been tested

	OSHA standard[a]		Recommended sample volumes (liters)	
	(ppm)	(mg/m^3)	Minimum[b]	Maximum[c]
Aniline	5	19	5	150
N,N-Dimethylaniline	5	25	5	190
o-Toluidine	5	22	5	300
2,4-Xylidine	5	25	5	430
o-Anisidine	0.1	0.5	250	35,800
p-Anisidine	0.1	0.5	250	33,000
p-Nitroaniline	1	6	80	12,100

[a]Taken or calculated from Federal Register, 37: #202, 22139 (18 October 1972).

[b]Based on analytical sensitivity and 0.1 OSHA standard detection when using the large section as the initial section. When the smaller section is used as the initial section, divide these minimum volumes by 5.

[c]Based on breakthrough studies at high humidities and a concentration of 5 times the OSHA standard when using the large section as the initial section. When the smaller section is used as the initial section, divide these maximum volumes by 5.

Calibration and Standards

For accuracy in the preparation of standards, it is recommended that one standard be prepared in a relatively large volume and at a high concentration. Aliquots of this standard can be diluted to prepare other standards. The solvent and diluent used must be the same ethanol/n-heptanol mixture used for the elution of the samples. For example, to prepare a 100 mL standard corresponding to a 96 L sample of air containing 95 mg/m³ of aniline (density = 1.022 g/mL) desorbed with 5 mL:

$$(95 \text{ mg/m}^3)(0.096 \text{ m}^3)(100 \text{ mL}) \div (5 \text{ mL})(1.022 \text{ mg/}\mu\text{L}) = 178 \; \mu\text{L}$$

of aniline is added to a 100 mL volumetric flask and diluted to the mark. The resulting concentration is:

$$(178 \; \mu\text{L})(1.022 \text{ mg/}\mu\text{L}) \div (100 \text{ mL}) = 1.82 \text{ mg/mL}$$

If 2 mL of this solution is diluted to 10 mL in a volumetric flask with the ethanol/n-heptanol mixture, the resulting concentration is:

$$(1.82 \text{ mg/mL})(2 \text{ mL}) \div (10 \text{ mL}) = 0.365 \text{ mg/mL}$$

When μL pipettes are used instead of μL syringes, it is better to prepare standards using a round number of μL (e.g., 200 μL aniline instead of 178 μL). For solids, the amount of compound used for the first standard should be weighed on an analytical balance. A series of standards is prepared varying in concentration over the range of interest. The standards prepared as above should be analyzed under the same GC conditions and during the same time period as the unknown samples. This will minimize the effect of day-to-day variations of the FID response.

A standard curve is prepared for each compound by plotting ratios of peak areas of the compound to internal standard against the concentration of the compound. From the resulting curve the concentration of an eluted sample is determined. This concentration (in mg/mL) is then converted to total sample weight by multiplying by the amount of ethanol used for that section (1 or 5 mL).

Calculations

Corrections for the blank must be made for each sample:

Correct mg = mg sample - mg blank

where:

mg sample = mg found in front section of sample tube
mg blank = mg found in front section of blank tube

A similar procedure is followed for the backup sections. Add the corrected amounts present in the front and backup sections of the same sample tube to determine the total measured amount (w) in the sample. Convert the volume of air sampled to standard conditions of 25° C. and 760 mm Hg:

$$V_s = (V)(P)(298) \div (760)(T + 273)$$

where:

V_s = volume of air in L at 25° C. and 760 mm Hg
V = volume of air in L as calculated (sampling time × correct flow rate)

P = barometric pressure in mm Hg
T = temperature of air in °C.

The concentration of the organic solvent in the air sampled can be expressed in mg/m³, which is numerically equal to µg/L of air:

$$mg/m^3 = \mu g/L = (w \text{ in mg})(1000 \, \mu g/mg) \div V_s$$

Another method of expressing concentration is ppm, defined as µL of compound vapor per L of air:

$$ppm = \mu L \text{ of vapor} \div V_s$$

$$ppm = (w \text{ in } \mu g)(24.45 \text{ L/mol}) \div (V_s)(M.W.)$$

where:

24.45 L/mol = molar volume at 25° C. and 760 mm Hg
M.W. = molecular weight of the compound (g/mol)

Reference

1. Campbell, E.E., G.O. Wood, and R.G. Anderson. Los Alamos Scientific Laboratory Progress Reports LA-5104-PR, LA-5164-PR, LA-5308-PR, LA-5389-PR, LA-5484-PR, and LA-5634-PR, Los Alamos, New Mexico, November 1972, January 1973, June 1973, August 1973, December 1973, and June 1974.

ARSINE

ACGIH TLV: 0.05 ppm (0.2 mg/m³) AsH_3
OSHA Standard: 0.05 ppm
NIOSH Recommendation: 0.0005 ppm (0.002 mg/m³) ceiling

Method	Sampling Duration	Sampling Location	Useful Range (ppm)	System Cost ($)	Test Cost ($)	Manufacturer
DT	Grab	Portable	0.05-60	150	2.70	Nat'l Draeger
DT	Grab	Portable	5-160	165	1.70	Matheson
DT	Grab	Portable	0.05-5	180	2.25	Bendix
CS	Grab	Portable	0.01-0.10	2465	0.80	MDA
PI	Cont	Portable	0.1-600	4295	0	Airco
IR	Cont	Portable	0.02-2	4374	0	Foxboro
CM	Cont	Fixed	0.002-0.2	6500	0	CEA
PI	Cont	Portable	0.01-600	6916	0	Airco
GC	1 min	Fixed	0.002-0.75	43,000	0	Airco

NIOSH METHOD NO. P&CAM 265

Principle

Arsine is collected in a charcoal tube containing activated coconut shell charcoal. The arsine as arsenic is desorbed from the charcoal with 0.01 M HNO_3. Nickel is added to the solution to enhance the analytical signal. The resulting solution is analyzed for arsenic by heated graphite atomization and atomic absorption spectroscopy at 193.7 nm.

Range and Sensitivity

This method has been evaluated for a 15 L sample containing between 12 and 110 ng. This corresponds to concentrations of 0.8 to 7.33 mg/m³ in air. Because of instrument sensitivity, this method probably cannot be extended to lower concentrations. The upper limit of the range of this method depends upon the adsorptive capacity of the charcoal tube. This capacity depends upon the sampling rate and concentrations of arsine and other contaminants in the air. The first section of the charcoal tube was found to hold at least 0.021 mg when a test atmosphere containing 0.405 mg/m³ of arsine was sampled at 0.22 L/min for 240 minutes. At that time, the concentration of arsine in the effluent was less than 5% of the influent. (The charcoal tube contains two sections.) When a test atmosphere containing 2 μg/m³ of arsine was sampled at a rate of 1.0 L/min for 60 minutes, no analyte was detected on the second section. Because of this information, the sampling rate should be limited to 0.2 L/min when the concentration of arsine exceeds 10 μg/m³. If the concentration of arsine exceeds 0.09 mg/m³, either dilute the sample as necessary or use NIOSH method S220. The slope of the calibration curve is 0.066 absorbance units/μg arsine. This is presumed to be the sensitivity of the method. The detection limit was not determined.

Interferences

Other arsenic compounds which exist as vapors and aerosols will interfere with the analysis. It is expected that interference from particulate arsenic compounds may be eliminated by placing a filter cassette before the charcoal tube. At a sampling rate of 1.0 L/min, aerosols are inefficiently collected by the charcoal tube. When two or more compounds are known or suspected to be present in the air, such information, including their suspected identities, should be transmitted with the sample. When the amount of water in the air is so great that condensation actually occurs in the tube, arsine vapors may not be trapped efficiently.

Precision and Accuracy

The coefficient of variation for the total analytical and sampling method in the range of 1 to 10 μg/m³ was 0.089. This value corresponds to a 0.23 μg/m³ standard deviation at the recommended ceiling of 2.0 μg/m³. Statistical information and details of the evaluation and experimental test procedures can be found in reference 2.

On the average, the concentrations obtained at the NIOSH recommended standard level, using the overall sampling and analytical method, were about 10-15% lower than the "true" concentrations for a limited number of laboratory experiments. Because either the "true" concentration or the concentration as measured by this method could be in error, no correction for the bias can be made.

Advantages and Disadvantages

The sampling device is small, portable, and involves no liquids. Interferences are minimal. The tubes are analyzed by means of an instrumental method. One disadvantage of the method is that the amount of sample which can be taken is limited by the number of milligrams that the tube will hold before overloading. When the sample value obtained for the backup section of the charcoal tube exceeds 25% of that found on the front section, the possibility of sample loss exists. The precision of the method is limited by the reproducibility of the pressure drop across the tubes. This drop will affect the flow rate and cause the volume to be imprecise, because the pump is usually calibrated for one tube only.

Apparatus

Calibrated personal sampling pump. Whose flow can be determined within 5% at the recommended flow rate.

Charcoal tubes. Glass tube with both ends flame-sealed, 7 cm long with 6 mm O.D. and 4 mm I.D., containing 2 sections of 20/40 mesh activated charcoal separated by a 2 mm portion of urethane foam. The activated charcoal is prepared from coconut shells and is fired at 600° C. prior to packing. The absorbing section contains 100 mg of charcoal, the backup section 50 mg. A 3 mm portion of urethane foam is placed between the outlet end of the tube and the backup section. A plug of silylated glass wool is placed in front of the adsorbing section. The pressure drop across the tube must be less than one inch of mercury at a flow rate of 1 L/min.

Atomic absorption spectrophotometer. With a direct readout (or recorder output) proportional to absorbance units, graphite furnace accessory and background corrector accessory. A background corrector is necessary in order to avoid false positive signals from molecular scatterings at 193.7 nm wavelength.

Arsenic electrodeless discharge lamp. This lamp is recommended because it is a brighter source and enhances sensitivity.

Electronic integrator. Or some other suitable method for measuring peak areas.

Volumetric flasks. 50 to 1000 mL volumetric flasks for making standards.

Volumetric pipets. Assorted sizes for making standards.

Eppendorf automatic pipettor. With disposable plastic tips for accurately injecting 50 μL sample aliquots into the graphite furnace tube.

Assorted glassware.

Ultrasonic agitator.

Centrifuge.

Centrifuge tubes, stoppered.

Reagents

All chemicals must be ACS reagent grade or better.

Nitric acid, 0.01 M.

Nitric acid, 0.1 M.

Ultrapure water. Distilled water was passed through a cation and anion exchange column and finally through a Millipore cellulose nitrate filter or equivalent.

Sodium hydroxide, 10 N.

Nickel solution. 1000 μg/mL as $Ni(NO_3)_2$ atomic absorption standard or equivalent.

Arsenic standards:

100 μg/mL As stock solution: 0.1322 g of dried certified reagent As_2O_3 is dissolved in 10 mL of 10 N NaOH and diluted to 1 L with ultrapure water.

1.00 μg/mL As standard solution: Pipet 5 mL of 100 μg/mL As solution into a 500 mL volumetric flask and dilute to the mark with 0.01 M HNO_3. Prepare daily.

0.01, 0.02, 0.05, 0.10, and 0.20 μg/mL As standard solutions: Using a 10 mL pipet, transfer 10 mL of 1.00 μg/mL As standard into 1000, 500, 200, 100 and 50 mL volumetric flasks, respectively, and dilute to the mark with 0.01 M HNO_3. Prepare daily.

Aqua regia, 5.0%. Mix by volume, a mixture of 3 or 4 parts hydrochloric acid to one part nitric acid. This mixture is diluted by a factor of 20 with ultrapure water to make a 5% aqua regia solution.

Procedure

Cleaning of equipment. All glassware must be cleaned with a detergent solution followed by both tap water and distilled water rinses. Then the glassware is agitated in 5% aqua regia with an ultrasonic bath for 15 minutes, and finally rinsed 3 times with distilled water and twice with distilled deionized water prior to use.

Calibration of personal pumps. Each personal pump must be calibrated with a representative charcoal tube in line. This will minimize errors associated with uncertainties in the sample volume collected.

Collection and shipping of samples. Immediately before sampling, break the ends of the tube to provide an opening at least one-half the internal diameter of the tube (2 mm). The smaller section of charcoal is used as a backup and should be positioned nearest the sampling pump. The charcoal tube should be placed in a vertical position during sampling to minimize channeling through the charcoal. Air being sampled should not be passed through any hose or tubing before entering the charcoal tube. Sample at a rate of 1.0 L/min for 15 min to evaluate compliance with the NIOSH recommended standard of 2.0 $\mu g/m^3$. The flow rate should be known with an accuracy of at least 5%.

The temperature and pressure of the atmosphere being sampled should be recorded. If pressure reading is not available, record the elevation. The charcoal tubes should be capped with the supplied plastic caps immediately after sampling. Under no circumstances should rubber caps be used. One tube should be handled in the same manner as the sample tubes (break, seal, and transport), except that no air is sampled through this tube. This tube should be labeled as a blank. Capped charcoal tubes should be packed tightly and padded before they are shipped to minimize tube breakage during shipping.

Analysis of samples. Each charcoal tube is scored with a file and broken open. The charcoal contents of each of the two sections are transferred to individual 10 mL capacity glass stoppered centrifuge tubes and analyzed separately. One mL of 0.01 M HNO_3 is pipetted into each centrifuge tube. The tube and contents are then agitated in an ultrasonic bath for 30 minutes. After centrifuging the supernatant liquid is analyzed.

Desorption efficiency tests would be experimentally difficult to perform because a small amount of arsine gas must be added to the charcoal. Therefore the experimentally determined value of 0.9 can be used when the charcoal used to collect the sample was from SKC lot 106 and when the samples are in the range of 1.0 to 15 $\mu g/m^3$ for 15 L samples. When these conditions are not met, desorption efficiency must be experimentally determined.

The instrument is adjusted to read the absorption of arsenic at 193.7 nm. Follow manufacturer's recommendations for source power, background correction, and furnace alignment as well as furnace parameters such as inert gas flow and time and temperature conditions. During validation, the following heated graphite atomizer program was followed:

dry: 40 sec at 110° C.
char: 15 sec at 1200° C.
atomize: 7 sec at 2540° C.

A 50 μL aliquot of sample is injected into the furnace followed by a 50 μL aliquot of 1000 $\mu g/mL$ Ni as $Ni(NO_3)_2$. The absorbance is then measured. Sample loadings greater than 500 ng are diluted to a lower range. A minimum of one duplicate analysis per sample should be done. To obtain reliable results, samples must be frequently alternated with standards which give responses close to that of the sample. The experimental protocol recommended is as follows: inject a standard solution in duplicate, inject a sample in duplicate, and reinject standard in duplicate, etc.

Calibration and Standards

To each centrifuge tube designated to contain a standard, add 100 mg SKC #106 activated charcoal. Add 1 mL of appropriate standard solution to each tube, stopper and agitate in an ultrasonic bath for 30 min as with samples. Repeat this procedure for each new sample batch. Prepare a calibration curve from graduated standards by plotting absorbance against As concentration to minimize the effect of variations in absorbance readings.

Calculations

Determine the weight in μg corresponding to the absorbance area of the sample by using calibration curves. Convert μg arsenic to μg arsine using molecular weight ratio, $78/75 = 1.04$. Corrections for the blank must be made for each sample:

$\mu g = \mu g$ sample - μg blank

where:

μg sample $= \mu g$ found in front section of the sample tube
μg blank $= \mu g$ found in front section of blank tube

Add the amount present in the front and backup sections of the sample tube to determine the total weight in the sample. Divide the total weight in μg found by the correction factor as discussed earlier; use 0.90 for samples in the range 1.0 to 10 $\mu g/m^3$. For personal sampling pumps with rotameters only, the following volume correction should be made:

Actual volume sampled $= ft(P_1T_2/P_2T_1)^{1/2}$

where:

f = flow rate indicated by rotameter
t = sampling time
P_1 = pressure during calibration of sampling pump (mm Hg)
P_2 = pressure of air sampled (mm Hg)
T_1 = temperature during calibration of sampling pump (°K)
T_2 = temperature of air sampled (°K)

The concentration of the analyte in the air sampled can be expressed in mg/m^3, which is numerically equal to $\mu g/L$:

mg/m^3 = corrected $\mu g \div$ air volume sampled in L

References

1. Denyszyn, R., P. Grohse, and D. Wagoner. Report on Evaluation and Refinement of Personal Sampling Method for Arsine, Research Triangle Park, North Carolina, 1977.
2. National Institute for Occupational Safety and Health, Method S229 for Arsine. In: *NIOSH Manual of Analytical Methods,* 2nd Edition, Cincinnati, Ohio, 1977.

BENZENE

C_6H_6

ACGIH TLV: 10 ppm (30 mg/m³) - Suspected carcinogen
OSHA Standard: 10 ppm
NIOSH Recommendation: 1 ppm

Method	Sampling Duration	Sampling Location	Useful Range (ppm)	System Cost ($)	Test Cost ($)	Manufacturer
DT	Grab	Portable	0.05-420	150	2.70	Nat'l Draeger
DT	Grab	Portable	1-200	165	2.30	Matheson
DT	Grab	Portable	0.25-120	180	2.25	Bendix
DT	4-hr	Personal	10-200	850	3.10	Nat'l Draeger
IR	Cont	Portable	1-40	4374	0	Foxboro
PI	Cont	Fixed	0.2-20	4950	0	HNU
GC	1 min	Portable	1-10^6	10,000	0	Microsensor

NIOSH METHOD NO. P&CAM 127

(See acetone)

BENZYL CHLORIDE

$C_6H_5CH_2Cl$

ACGIH TLV: 1 ppm (5 mg/m³)
OSHA Standard: 1 ppm
NIOSH Recommendation: 1 ppm ceiling (15 min)

Method	Sampling Duration	Sampling Location	Useful Range (ppm)	System Cost ($)	Test Cost ($)	Manufacturer
DT	Grab	Portable	0.5-25	180	2.25	Bendix
IR	Cont	Portable	1-10	4374	0	Foxboro
PI	Cont	Fixed	0.2-20	4950	0	HNU

NIOSH METHOD NO. P&CAM 127

(Use as described for acetone, with 160° C. column temperature).

BIS(CHLOROMETHYL) ETHER

Synonym: bis-CME $\qquad$ CH_2ClOCH_2Cl
ACGIH TLV: 0.001 ppm (0.005 mg/m^3) - Human carcinogen
OSHA Standard: 0.001 ppm

Method	Sampling Duration	Sampling Location	Useful Range (ppm)	System Cost ($)	Test Cost ($)	Manufacturer
IR	Cont	Portable	0.4-20	4374	0	Foxboro
PI	Cont	Fixed	0.2-20	4950	0	HNU

NIOSH METHOD NO. P&CAM 213

Principle

Bis-CME is concentrated from air by adsorption on Chromosorb 101 in a short tube. It is desorbed by heating and purging with helium through a gas chromatography column where it is separated from most interferences. The concentration of bis-CME is measured from the spectroscopic signal at m/e 79 and 81.

Range and Sensitivity

The range of the mass spectroscopic signal for the conditions listed corresponds to 0.5 to 10 ppb. A concentration of 0.5 ppb of bis-CME can be determined in a 5 L sample. Greater sensitivity can be obtained by using a larger sample.

Interferences

Interferences resulting from materials having retention times similar to bis-CME or simply background such as ions C_6H_7, $C_5C_1{}^{13}H_6$, C_4H_3Si and Br giving rise to m/e 79 can be encountered. However, the DuPont 21-491 mass spectrometer has sufficient resolution to resolve at least three of these ions at m/e 79, particularly the $C_5C_1{}^{13}H_6$ and C_6H_7 ions that are normally encountered in the plant samples. These are completely resolved from the bis-CME $C_2H_4Cl^{35}O$ ion.

Precision and Accuracy

The precision of this method has been determined to be 10% of relative standard deviation when different sampling tubes were spiked with 133 ng (corresponding to 1.4 ppb in 20 L of air) at intervals of approximately one hour. These data were obtained using 4 in × 1/4 in stainless steel columns packed with 60/80 mesh Chromosorb 101. The accuracy of the analysis is approximately 10% of the amount reported as determined from repeated analysis of several standards.

Advantages and Disadvantages

The gas chromatography-mass spectrometry technique interfaced with a Watson-Biemann helium separator is extremely sensitive and specific for the analysis of bis-CME. The gas chromatographic separation yields a retention time that is characteristic for bis-CME, but it is not highly specific for positive assignment of the signal as bis-CME. The mass spectrometer in combination with the gas chromatograph provides the high degree of specificity. The most intense ion in the mass spectrum of bis-CME is formed at m/e 79, with its chlorine-37 isotope at m/e 81. The chlorine isotopic abundances require that the intensity ratio of peaks at m/e 79:81 be approximately 3:1. The mass spectrometer is set to continuously monitor these ions as the sample elutes through the gas chromatograph. It is absolutely necessary that the retention time, observed masses at m/e 79 and 81 and the intensity ratio be correct in order to assign the signal to bis-CME.

Bis-CME, an impurity in chloromethyl methyl ether, has been reported to be a pulmonary carcinogen. Extreme safety precautions should be exercised in the preparation and disposal of liquid and gas standards and the analysis of air samples. Bis-CME may be destroyed in a methanol-caustic solution. Sampling must be carried out away from the mass spectrometer. GC-MS is not a good approach for continuous monitoring.

Apparatus

Sampling tubes. The sampling tubes are prepared by packing a 2.5 in × 0.25 in stainless steel tubing with 1.75 in of 60/80 mesh Chromosorb 101 with glass wool in the ends. These are conditioned overnight at 200° C. with helium or nitrogen flow set at 10-30 mL/min. The conditioned sampling tubes are then cooled under flow and capped immediately upon removal. An identification number is scribed on one of the nuts. The stainless steel tubing should be rinsed internally with water, acetone and methylene chloride before packing with Chromosorb 101. Longer sampling tubes may be prepared with a proportional amount of Chromosorb 101.

Gas chromatography column. A 4 ft × 1/8 in stainless steel separator column is rinsed internally with water, acetone and methylene chloride and air dried. The cleaned tubing is then packed with Chromosorb 101. The column is conditioned overnight at 200° C. and 10-30 mL/min helium or nitrogen flow.

Watson-Biemann helium separator.

Gas chromatograph. Hewlett-Packard 5750 gas chromatograph or equivalent. A gas chromatograph employing a single column oven and a temperature programmer is adequate.

Mass spectrometer. A mass spectrometer with a resolution of 1000-2000 and equipped with a repetitive scan attachment can be used in conjuction with a gas chromatograph. A CEC (now DuPont Analytical Instruments Div.) Model 21-491 modified by adding a follower amplifier with a long time constant (0.03 sec) to the output of the electron multiplier to increase the gain and signal to noise ratio has been found satisfactory for this purpose.

Syringe. 1 mL, gas tight (Hamilton Co., Inc.).

Syringe. 10 µL (Hamilton Co., Inc.).

Reagents

All reagents must be analytical reagent grade.

Bis(chloromethyl) ether (Eastman Kodak).

Acetone.

Dichloromethane.

n-Pentane.

Methanol.

Sodium hydroxide.

Plastic film gas bag, 13 L. Saran brand (trademark of the Dow Chemical Company abroad) plastic film gas bag has proven to be satisfactory for this purpose (Anspec Co., Inc.). Polyethylene film bags are not suitable.

Chromosorb 101 (Johns-Manville).

Procedure

Cleaning of equipment. None required.

Collection of samples. Spot samples may be obtained by attaching the sampling tube to a manual syringe pump. The sample is drawn into the end of the sampling tube with the marked nut. Continuous sampling of air may be accomplished by attaching a vacuum pump to a throttling valve and manometer. Each sample tube is calibrated so that the pressure drop is determined. Then the flow is established by adjusting the pressure drop with the throttling valve.

For larger sample size, it is important to realize that larger flows of air over a long period of time may cause elution of bis-CME through the sampling tube. It has been shown that a total of 25 L of air can be pumped through a 4 in × 1/4 in packing of Chromosorb 101 over a period of 24 hours without any elution of the bis-CME. This corresponds to a flow of 17 mL/min maximum. A maximum air volume of 5 L at a rate of 80 mL/min has been found to be safe for the 2.5 in × 1/4 in sampling tube. Bis-CME has been found to be stable in the sampling tubes for at least 10 days when kept tightly closed and at room temperature.

Analysis of samples. The mass spectrometer should be set to sweep peaks at m/e 79 to 81 with the visicorder speed at 0.8 in/sec for recording the resulting signal. Hold the temperature of the Watson-Biemann helium separator at 160° C. Set the helium flow at 30 mL/min. Balance the helium flow so that there is no vacuum applied to the outlet of the gas chromatographic column. Attach the sampling tube to the gas chromatograph via an adapter to the septum nut. The inscribed sample tube nut should be closest to the inlet of the analytical column.

A 1/4 in Swagelok Tee fitting is attached on the other end of the sampling tube to accomodate the carrier gas and a silicone rubber spectrum so that standards may be injected. After thoroughly checking for leaks, wrap the sampling tube with heat tape and heat to 150° C. for 5 min to assure complete desorption of the organic components onto the analytical column which is at ambient temperature. Temperature program the analytical column to 130° C. at 30° C./min and hold to yield a retention time for bis-CME of 8 to 9 min. Turn on recorder 30 sec. before elution of bis-CME. The analytical column should be cooled to ambient temperature before changing sampling tubes.

Calibration and Standards

Preparation of gas standard. Rinse out a 13 L Saran bag with prepurified nitrogen. Introduce 5 L of prepurified nitrogen through a dry test meter into the Saran bag and cap it with a cork. Place plastic tape on a Saran bag and inject 1 μL of bis-CME through the tape into the bag. The tape should prevent any cracking of the bag at the injection point. Tape the injection point at once upon removing the syringe needle from the bag. Clean the syringe immediately with methanol-caustic solution and then water and acetone.

Knead the bag for approximately 2 min. Carefully connect the bag with the meter to introduce an additional 5 L of prepurified nitrogen (the end of the Saran tubing inside the bag should be kept tight with a finger from the outside in order to avoid any loss of bis-CME while connecting the bag with the meter). Knead the bag for additional 2 min to assure complete mixing. With the 10 L bag, this procedure can prepare a ppm level standard (e.g., 1 μL of liquid bis-CME yields a 28.3 ppm by volume standard). This standard should not be kept for more than 3 days.

Preparation of liquid standard. Bis-CME standard (0.132 μg/μL) is prepared by dissolving 1 μL of bis-CME liquid in 10 mL n-pentane. This standard solution may be kept for as long as desired if evaporation is avoided.

Calibration. Inject a 1 mL portion of the 28.3 ppm standard bis-CME (0.132 μg) through a standard septum onto a sample tube attached to the inlet of the analytical column. This practice may be done routinely after two or three samples to ensure accuracy. Also 1 μL of the liquid standard (0.132 μg/μL) could be used for the calibration. Repeat steps as described under Analysis of Samples.

Calculations

The total ng of bis-CME in the air sampled is determined from the ratio of the peak height of m/e 79 of the sample and the standard multiplied by 132. The concentration in ppb is calculated as follows:

$$ppb = (ng\ bis\text{-}CME)(24.45) \div (V)(M.W.)$$

where:

ng bis-CME = total ng concentration as determined above
V = volume of air in L sampled at 25° C and 760 Torr
24.45 = molar volume of an ideal gas at 25° C. and 760 Torr
M.W. = molecular weight of bis-CME (115)

References

1. Van Dunren, B.L., A. Sinak, B.M. Goldschmidt, C. Katz, and S. Melchionne. J. Nat. Cancer Inst. 43: 481, 1969.
2. Laskin, S., M. Kuschner, R.T. Drew, V.P. Cauppiello, and N. Nelson. Technical Report, New York University Medical Center, Department of Environmental Medicine, August, 1971.
3. Leong, K.J., H.N. MacFarland, and W.H. Reese. Arch. Environ. Health 22: 663, 1971.
4. Watson, J.T. and K. Biemann. Anal. Chem. 36: 1135, 1964.
5. DuPont de Nemour, E.I., Analytical Instrument Division, Monrovia, California.
6. Shadoff, L.A., G.J. Kallos, and J.S. Woods. Analysis for bis-chloromethyl ether in air. Anal. Chem. 45: 2341, 1973.
7. Federal Register. Vol. 39, No. 20, pp 3773-3776, January 29, 1974.

BROMOFORM

Synonym: tribromomethane $CHBr_3$
ACGIH TLV: 0.5 ppm (5 mg/m^3) - Skin
OSHA Standard: 0.5 ppm

Method	Sampling Duration	Sampling Location	Useful Range (ppm)	System Cost ($)	Test Cost ($)	Manufacturer
DT	Grab	Portable	0.5-50	180	2.25	Bendix
IR	Cont	Portable	0.1-2	4374	0	Foxboro
PI	Cont	Fixed	0.2-20	4950	0	HNU

NIOSH METHOD NO. P&CAM 127

(Use as described for acetone, with 130° C. column temperature).

1,3-BUTADIENE

Synonyms: vinylethylene; divinyl
ACGIH TLV: 1000 ppm (2200 mg/m^3)
OSHA Standard: 1000 ppm

$CH_2=CHCH=CH_2$

Method	Sampling Duration	Sampling Location	Useful Range (ppm)	System Cost ($)	Test Cost ($)	Manufacturer
DT	Grab	Portable	30-26,000	160	1.70	Matheson
DT	Grab	Portable	50-800	180	2.25	Bendix
IR	Cont	Portable	20-2000	4374	0	Foxboro
PI	Cont	Fixed	20-2000	4950	0	HNU
GC	1 min	Portable	1-10^6	10,000	0	Microsensor

NIOSH METHOD NO. P&CAM 127

(Use as described for acetone, with 52° C. column temperature).

2-BUTANONE

Synonyms: methyl ethyl ketone; MEK
ACGIH TLV: 200 ppm (590 mg/m^3)
OSHA Standard: 200 ppm

$CH_3COC_2H_5$

Method	Sampling Duration	Sampling Location	Useful Range (ppm)	System Cost ($)	Test Cost ($)	Manufacturer
DT	Grab	Portable	100-50,000	160	1.70	Matheson
DT	Grab	Portable	200-6000	180	2.25	Bendix
IR	Cont	Portable	1-400	4374	0	Foxboro
PI	Cont	Fixed	0.2-20	4950	0	HNU
GC	1 min	Portable	1-10^6	10,000	0	Microsensor

NIOSH METHOD NO. P&CAM 127

(See acetone)

2-BUTOXYETHANOL

Synonym: butyl cellosolve
ACGIH TLV: 25 ppm (120 mg/m³) - Skin
OSHA Standard: 50 ppm

$C_4H_9OCH_2CH_2OH$

Method	Sampling Duration	Sampling Location	Useful Range (ppm)	System Cost ($)	Test Cost ($)	Manufacturer
IR	Cont	Portable	1-100	4374	0	Foxboro
PI	Cont	Fixed	0.2-20	4950	0	HNU

NIOSH METHOD NO. S76

Principle

A known volume of air is drawn through a charcoal tube to trap the organic vapors present. The charcoal in the tube is transferred to a small, stoppered sample container, and the analyte is desorbed with 5% methanol in methylene chloride. An aliquot of the desorbed sample is injected into a gas chromatograph. The area of the resulting peak is determined and compared with areas obtained from the injection of standards.

Range and Sensitivity

This method was validated over the range of 124-490 mg/m³ at an atmospheric temperature and pressure of 21° C. and 766 mm Hg, using a 10 L sample. Under the conditions of sample size (10 L) the probable useful range of this method is 25-720 mg/m³ at a detector sensitivity that gives nearly full deflection on the strip chart recorder for a 7 mg sample. This method is capable of measuring much smaller amounts if the desorption efficiency is adequate. Desorption efficiency must be determined over the range used.

The upper limit of the range of the method is dependent on the adsorptive capacity of the charcoal tube. This capacity varies with the concentrations of 2-butoxyethanol and other substances in the air. The first section of the charcoal tube was found to hold at least 21 mg of 2-butoxyethanol when a test atmosphere containing 480 mg/m³ of 2-butoxyethanol in air was sampled at 0.183 L/min for 240 minutes; at that time the concentration of analyte in the effluent was less than 5% of that in the influent. (The charcoal tube consists of two sections of activated charcoal separated by a section of urethane foam.) If a particular atmosphere is suspected of containing a large amount of contaminant, a smaller sampling volume should be taken.

Interferences

When the amount of water in the air is so great that condensation actually occurs in the tube, organic vapors will not be trapped efficiency. Preliminary experiments using toluene indicate that high humidity severely decreases the breakthrough volume.

When two or more compounds are known or suspected to be present in the air, such information, including their suspected identities, should be transmitted with the sample. Since 5% methanol in methylene chloride is used rather than carbon disulfide to desorb the 2-butoxyethanol from the charcoal, it would not be possible to measure methanol or methylene chloride in the sample with this desorbing solvent. If it is suspected that methanol or methylene chloride is present, a separate sample should be collected for methanol or methylene chloride analysis.

It must be emphasized that any compound which has the same retention time as the analyte at the operating conditions described in this method is an interference. Retention time data on a single column cannot be considered proof of chemical identity. If the possibility of interference exists, separation conditions (column packing, temperature, etc.) must be changed to circumvent the problem.

Precision and Accuracy

The coefficient of variation for the total analytical and sampling method in the range of 124-490 mg/m^3 was 0.060. This value corresponds to a 14 mg/m^3 standard deviation at the OSHA standard level. Statistical information and details of the validation and experimental test procedures can be found in reference 2. On the average the values obtained using the overall sampling and analytical method were 8% lower than the "true" values at the OSHA standard level.

Advantages and Disadvantages

The sampling device is small, portable, and involves no liquids. Interferences are minimal, and most of those which do occur can be eliminated by altering chromatographic conditions. The tubes are analyzed by means of a quick, instrumental method. The method can also be used for the simultaneous analysis of two or more substances suspected to be present in the same sample by simply changing gas chromatographic conditions from isothermal to a temperature-programmed mode of operation.

One disadvantage of the method is that the amount of sample which can be taken is limited by the number of mg that the tube will hold before overloading. When the sample value obtained for the backup section of the charcoal tube exceeds 25% of that found on the front section, the possibility of sample loss exists. Furthermore, the precision of the method is limited by the reproducibility of the pressure drop across the tubes. This drop will affect the flow rate and cause the volume to be imprecise, because the pump is usually calibrated for one tube only.

Apparatus

Calibrated personal sampling pump. Whose flow can be determined within 5% at the recommended flow rate.

Charcoal tubes. Glass tube with both ends flame sealed, 7 cm long with 6 mm O.D. and 4 mm I.D., containing 2 sections of 20/40 mesh activated charcoal separated by a 2 mm portion of urethane foam. The activated charcoal is prepared from coconut shells and is fired at 600° C. prior to packing. The adsorbing section contains 100 mg of charcoal, the backup section 50 mg. A 3 mm portion of urethane foam is placed between the outlet end of the tube and the backup section. A plug of silylated glass wool is placed in front of the adsorbing section. The pressure drop across the tube must be less than one inch of mercury at a flow rate of 1 L/min.

Gas chromatograph. Equipped with a flame ionization detector.

Column. 10 ft × 1/8 in stainless steel packed with 10% FFAP on 80/100 mesh, acid washed DMCS Chromosorb W.

Electronic integrator. Or some other suitable method for measuring peak area.

Sample containers. 1 mL, with glass stoppers or Teflon-lined caps.

Syringes. 10 μL, and other convenient sizes for making standards.

Pipets. 0.5 mL delivery pipets or 1.0 mL type graduated in 0.1 mL increments.

Volumetric flasks. 10 mL or convenient sizes for making standard solutions.

Reagents

Eluent. 5% methanol in methylene chloride (both reagent grade).
2-Butoxyethanol, reagent grade.
Purified nitrogen.
Prepurified hydrogen.
Filtered compressed air.

Procedure

Cleaning of equipment. All glassware used for the laboratory analysis should be detergent washed and thoroughly rinsed with tap water and distilled water.

Calibration of personal pumps. Each personal pump must be calibrated with a representative charcoal tube in the line. This will minimize errors associated with uncertainties in the sample volume collected.

Collection and shipping of samples. Immediately before sampling, break the ends of the tube to provide an opening at least one-half the internal diameter of the tube (2 mm). The smaller section of charcoal is used as a backup and should be positioned nearest the sampling pump. The charcoal tube should be placed in a vertical direction during sampling to minimize channeling through the charcoal.

Air being sampled should not be passed through any hose or tubing before entering the charcoal tube. A maximum sample size of 10 L is recommended. Sample at a flow of 0.20 L/min or less. The flow rate should be known with an accuracy of at least 5%. The temperature and pressure of the atmosphere being sampled should be recorded. If pressure reading is not available, record the elevation. The charcoal tubes should be capped with the supplied plastic caps immediately after sampling. Under no circumstances should rubber caps be used.

One tube should be handled in the same manner as the sample tube (break, seal, and transport), except that no air is sampled through this tube. This tube should be labeled as a blank. Capped tubes should be packed tightly and padded before they are shipped to minimize tube breakage during shipping. A sample of the bulk material should be submitted to the laboratory in a glass container with a Teflon-lined cap. This sample should not be transported in the same container as the charcoal tubes.

Analysis of samples. In preparation for analysis, each charcoal tube is scored with a file in front of the first section of charcoal and broken open. The glass wool is removed and discarded. The charcoal in the first (larger) section is transferred to a 1 mL stoppered sample container. The separating section of foam is removed and discarded; the second section is transferred to another stoppered container. These two sections are analyzed separately.

Desorption of samples. Prior to analysis, 0.5 mL of 5% methanol in methylene chloride is pipetted into each sample container. Desorption should be done for 30 minutes. Tests indicate that this is adequate if the sample is agitated occasionally during this period.

GC conditions. The typical operating conditions for the gas chromatograph are:

50 mL/min (60 psig) nitrogen carrier gas flow
65 mL/min (24 psig) hydrogen gas flow to detector
500 mL/min (50 psig) air flow to detector
215° C. injector temperature
263° C. detector temperature
145° C. column temperature

Injection. The first step in the analysis is the injection of the sample into the gas chromatograph. To eliminate difficulties arising from blow back or distillation within the syringe needle, one should employ the solvent flush injection technique. The 10 μL syringe is first flushed with solvent several times to wet the barrel and plunger. Three μL of solvent are drawn into the syringe to increase the accuracy and reproducibility of the injected sample volume. The needle is removed from the solvent, and the plunger is pulled back about 0.2 μL to separate the solvent flush from the sample with a pocket of air to be used as a marker. The needle is then immersed in the sample, and a 5 μL aliquot is withdrawn, taking into consideration the volume of the needle, since the sample in the needle will be completely injected. After the needle is removed from the sample and prior to injection, the plunger is pulled back 1.2 μL to minimize evaporation of the sample from the tip of the needle. Observe that the sample occupies 4.9-5.0 μL in the barrel of the syringe. Duplicate injections of each sample and standard should be made. No more than a 3% difference in area is to be expected.

Measurement of area. The area of the sample peak is measured by an electronic integrator or some other suitable form of area measurement, and preliminary results are read from a standard curve prepared as discussed below.

Determination of desorption efficiency. The desorption efficiency of a particular compound can vary from one laboratory to another and also from one batch of charcoal to another. Thus, it is necessary to determine at least once the percentage of the specific compound that is removed in the desorption process, provided the same batch of charcoal is used.

Activated charcoal equivalent to the amount in the first section of the sampling tube (100 mg) is measured into a 2.5 in, 4 mm I.D. glass tube, flame sealed at one end. This charcoal must be from the same batch as that used in obtaining the samples and can be obtained from unused charcoal tubes. The open end is capped with Parafilm. A known amount of the analyte is injected directly into the activated charcoal with a μL syringe, and the tube is capped with more Parafilm.

The amount injected is equivalent to that present in a 10 L air sample at the selected level. Six tubes at each of three levels (0.5X, 1X, and 2X of the OSHA standard) are prepared to assure complete adsorption of the analyte onto the charcoal. These tubes are referred to as the samples. A parallel blank tube should be treated in the same manner except that no sample is added to it. The sample and blank tubes are desorbed and analyzed in exactly the same manner as the sampling tube described above.

Two or three standards are prepared by injecting the same volume of compound into 0.5 mL of the eluent with the same syringe used in a the preparation of the samples. These are analyzed with the samples. The desorption efficiency (D.E.) equals the average weight in mg recovered from the tube divided by the weight in mg added to the tube or:

D.E. = average weight (mg) recovered ÷ weight (mg) added

The desorption efficiency is dependent on the amount of analyte collected on the charcoal. Plot the desorption efficiency versus weight of analyte found. This curve is used to correct for adsorption losses.

Calibration and Standards

It is convenient to express concentration of standards in terms of mg/0.5 mL of the eluent, because samples are desorbed in this amount of eluent. The density of the analyte is used to convert mg into μL for easy measurement with a μL syringe. A series of standards, varying in concentration over the range of interest, is prepared and analyzed under the same GC conditions and during the same time period as the unknown samples. Curves are established by plotting concentration in mg/0.5 mL versus peak area. Note: since no internal standard is used in the method, standard solutions must be analyzed at the same time that the sample analysis is done. This will minimize the effect of known day-to-day variations and variations during the same day of the FID response.

Calculations

Read the weight, in mg, corresponding to each peak area from the standard curve. No volume corrections are needed, because the standard curve is based on mg/0.5 mL of the eluent and the volume of sample injected is identical to the volume of the standards injected. Corrections for the blank must be made for each sample:

mg = mg sample - mg blank

where:

mg sample = mg found in front section of sample tube
mg blank = mg found in front section of blank tube

A similar procedure is followed for the backup sections. Add the weights found in the front and backup sections to get the total weight in the sample. Read the desorption efficiency from the curve for the amount found in the front section. Divide the total weight by this desorption efficiency to obtain the corrected mg/sample:

Corrected mg/sample = total weight/D.E.

The concentration of the analyte in the air sampled can be expressed in mg/m^3:

mg/m^3 = (corrected mg)(1000 L/m^3) ÷ (air volume sampled in L)

Another method of expressing concentration is ppm:

ppm = (mg/m^3)(24.45)(760)(T + 273) ÷ (M.W.)(P)(298)

where:

P = pressure (mm Hg) of air sampled
T = temperature (°C.) of air sampled
24.45 = molar volume (L/mol) at 25° C. and 760 mm Hg
M.W. = molecular weight (g/mol) of analyte
760 = standard pressure (mm Hg)
298 = standard temperature (°K)

References

1. White, L.D. et al. A convenient optimized method for the analysis of selected solvent vapors in the industrial atmosphere. Amer. Ind. Hyg. Assoc. J. 31: 225, 1970.
2. Documentation of NIOSH Validation Tests, NIOSH Contract No. CDC-99-74-45.
3. Final Report, NIOSH Contract HSM-99-71-31. Personal Sampler Pump for Charcoal Tubes, September 15, 1972.

n-BUTYL ACETATE

ACGIH TLV: 150 ppm (710 mg/m^3)
OSHA Standard: 150 ppm

$CH_3COOC_4H_9$

Method	Sampling Duration	Sampling Location	Useful Range (ppm)	System Cost ($)	Test Cost ($)	Manufacturer
DT	Grab	Portable	100-8000	180	2.25	Bendix
IR	Cont	Portable	1-300	4374	0	Foxboro
PI	Cont	Fixed	2-200	4950	0	HNU
GC	1 min	Portable	1-10^6	10,000	0	Microsensor

NIOSH METHOD NO. P&CAM 127

(Use as described for acetone, with 90° C. column temperature).

n-BUTYL ALCOHOL

Synonym: n-butanol
ACGIH TLV: 50 ppm (150 mg/m^3) ceiling - Skin
OSHA Standard: 100 ppm

C_4H_9OH

Method	Sampling Duration	Sampling Location	Useful Range (ppm)	System Cost ($)	Test Cost ($)	Manufacturer
DT	Grab	Portable	100-1500	180	2.25	Bendix
IR	Cont	Portable	1-200	4374	0	Foxboro
PI	Cont	Fixed	2-200	4950	0	HNU
GC	1 min	Portable	1-10^6	10,000	0	Microsensor

NIOSH METHOD NO. S66

Principle

A known volume of air is drawn through a charcoal tube to trap the organic vapors present. The charcoal in the tube is transferred to a small, stoppered sample container and the analyte is desorbed with carbon disulfide containing 1% 2-propanol. An aliquot of the desorbed sample is injected into a gas chromatograph. The area of the resulting peak is determined and compared with areas obtained from the injection of standards.

Range and Sensitivity

This method was validated over the range of 170-610 mg/m^3 at an atmospheric temperature and pressure of 25°C. and 751 mm Hg, using a 10 L sample. Under the conditions of sample size (10 L) the probable range of this method is 30-900 mg/m^3 at a detector sensitivity that gives nearly full deflection on the strip chart recorder for a 9 mg sample. The method is capable of measuring much smaller amounts if the desorption efficiency is adequate. Desorption efficiency must be determined over the range used.

The upper limit of the range of the method is dependent on the adsorptive capacity of the charcoal tube. This capacity varies with the concentrations of the analyte and other substances in the air. The first section of the charcoal tube was found to hold 21 mg of the analyte when a test atmosphere of 600 mg/m^3 of the analyte in dry air was sampled at 0.2 L/min for 2 hours, 55 min; breakthrough was observed at this time, i.e., the concentration of the analyte in the effluent was 5% of that in the influent. (The charcoal tube consists of two sections of activated charcoal separated by a section of urethane foam.) If a particular atmosphere is suspected of containing a large amount of contaminant, a smaller sampling volume should be taken.

Interferences

When the amount of water in the air is so great that condensation actually occurs in the tube, organic vapors will not be trapped efficiently. Preliminary experiments with toluene indicate that high humidity severely decreases the breakthrough volume. When two or more compounds are known or suspected to be present in the air, such information, including their suspected identities, should be transmitted with the sample.

It must be emphasized that any compound which has the same retention time as the specific compound under study at the operating conditions described in this method is an interference. Retention time data on a single column cannot be considered as proof of chemical identity. If the possibility of interference exists, separation conditions (column packing, temperatures, etc.) must be changed to circumvent the problem.

Precision and Accuracy

The coefficient of variation for the total analytical and sampling method in the range of 170 to 610 mg/m^3 was 0.065. This value corresponds to a standard deviation of 20 mg/m^3 at the OSHA standard level. Statistical information and details of the validation and experimental test procedures can be found in reference 2. The average values obtained using the overall sampling and analytical method were 6.2% lower than the "true" value at the OSHA standard level. The above data are based on validation experiments using the internal standard method (reference 2).

Advantages and Disadvantages

The sampling device is small, portable, and involves no liquids. Interferences are minimal, and most of those which do occur can be eliminated by altering chromatographic conditions. The tubes are analyzed by means of a quick, instrumental method. The method can also be used for the simultaneous analysis of two or more compounds suspected to be present in the same sample by simply changing gas chromatographic conditions from isothermal to a temperature-programmed mode of operation.

One disadvantage of the method is that the amount of sample which can be taken is limited by the number of mg that the tube will hold before overloading. When the sample value obtained for the backup section of the charcoal tube exceeds 25% of that found on the front section, the possibility of sample loss exists. Furthermore, the precision of the method is limited by the reproducibility of the pressure drop across the tubes. This drop will affect the flow rate and cause the volume to be imprecise, because the pump is usually calibrated for one tube only.

Apparatus

Calibrated personal sampling pump. whose flow can be determined within 5% at the recommended flow rate.

Charcoal tubes. Glass tube with both ends flame sealed, 7 cm long with 6 mm O.D. and 4 mm I.D., containing 2 sections of 20/40 mesh activated charcoal separated by a 2 mm portion of urethane foam. The activated charcoal is prepared from coconut shells and is fired at 600° C. prior to packing. The adsorbing section contains 100 mg of charcoal, the backup section 50 mg. A 3 mm portion of urethane foam is placed between the outlet end of the tube and the backup section. A plug of silylated glass wool is placed in front of the adsorbing section. The pressure drop across the tube must be less than one inch of mercury at a flow rate of 1 L/min.

Gas chromatograph. Equipped with a flame ionization detector.

Column. 10 ft × 1/8 in stainless steel packed with 10% FFAP on 80/100 Chromosorb W-AW.

Electronic integrator. Or some other suitable method for determining peak size areas.

Glass sample containers. 2 mL, with glass stoppers or Teflon-lined caps. If an automatic sample injector is used, the sample injector vials can be used.

Syringes. 10 µL, and other convenient sizes for making standards.

Pipets. 1.0 mL delivery type.

Volumetric flasks. 10 mL or convenient sizes for making standard solutions.

Reagents

Eluent. Carbon disulfide (chromatographic grade) containing 1% 2-propanol (reagent grade).

1-Butanol (reagent grade).

Internal standard. n-Dodecane (99+ %) or other suitable standard.

n-Heptane (reagent grade).

Purified nitrogen.

Prepurified hydrogen.

Filtered compressed air.

Procedure

Cleaning of equipment. All glassware used for the laboratory analysis should be detergent washed and thoroughly rinsed with tap water and distilled water.

Calibration of personal pumps. Each personal pump must be calibrated with a representative charcoal tube in the line. This will minimize errors associated with uncertainties in the sample volume collected.

Collection and shipping of samples. Immediately before sampling, break the ends of the tube to provide an opening at least one-half the internal diameter of the tube (2 mm). The smaller section of charcoal is used as a backup and should be positioned nearest the sampling pump. The charcoal tube should be placed in a vertical direction during sampling to minimize channeling through the charcoal. Air being sampled should not be passed through any hose or tubing before entering the charcoal tube.

A maximum sample size of 10 L is recommended. Sample at a flow of 0.20 L/min or less. The flow rate should be known with an accuracy or at least 5%. The temperature and pressure of the atmosphere being sampled should be recorded. If pressure reading is not available, record the elevation. The charcoal tubes should be capped with the supplied plastic caps immediately after sampling. Under no circumstances should rubber caps be used. One tube should be handled in the same manner as the sample tubes (break, seal, and transport), except that no air is sampled through this tube. This tube should be labeled as a blank. Capped tubes should be packed tightly and padded before they are shipped to minimize tube breakage during shipping.

Analysis of samples. In preparation for analysis, each charcoal tube is scored with a file in front of the first section of charcoal and broken open. The glass wool is removed and discarded. The charcoal in the first (larger) section is transferred to a 2 mL stoppered sample container or automatic sample injector vial. The separating section of foam is removed and discarded; the second section is transferred to another sample container or vial. These two sections are analyzed separately.

Desorption of samples. Prior to analysis, 1.0 mL of the eluent is pipetted into each sample container. For the internal standard method of 0.2% solution of internal standard in the eluent used. (All work with carbon disulfide should be performed in a hood because of its high toxicity.) Desorption should be done for 30 minutes. Tests indicate that this is adequate if the sample is agitated occasionally during this period. The sample vials should be capped as soon as the solvent is added to minimize volatilization.

GC conditions. The typical operating conditions for the gas chromatograph are:

30 mL/min (80 psig) nitrogen carrier gas flow
30 mL/min (50 psig) hydrogen gas flow to detector
300 mL/min (50 psig) air flow to detector
200° C. injector temperature
300° C. detector temperature
80° C. column temperature

Injection. The first step in the analysis is the injection of the sample into the gas chromatograph. To eliminate difficulties arising from blow back or distillation within the syringe needle, one should employ the solvent flush injection technique. The 10 µL syringe is first flushed with solvent several times to wet the barrel and plunger. Three µL of solvent are drawn into the syringe to increase the accuracy and reproducibility of the injected sample volume. The needle is removed from the solvent, and the plunger is pulled back about 0.2 µL to separate the solvent flush from the sample with a pocket of air to be used as a marker. The needle is then immersed in the sample, and a 5 µL aliquot is withdrawn, taking into consideration the volume of the needle, since the sample in the needle will be completely injected. After the needle is removed from the sample and prior to injection, the plunger is pulled back 1.2 µL to minimize evaporation of the sample from the tip of the needle. Observe that the sample occupies 4.9-5.0 µL in the barrel of the syringe. Duplicate injections of each sample and standard should be made. No more than a 3% difference in area is to be expected. An automatic sample injector can be used if it is shown to give reproducibility at least as good as the solvent flush technique. In this case 2 µL injections are satisfactory.

Measurement of area. The area of the sample peak is measured by an electronic integrator or some other suitable form of area measurement, and preliminary results are read from a standard curve prepared as discussed below.

Determination of desorption efficiency. The desorption efficiency of a particular compound can vary from one laboratory to another and also from one batch of charcoal to another. Thus, it is necessary to determine at least once the percentage of the specific compound that is removed in the desorption process, provided the same batch of charcoal is used.

Activated charcoal equivalent to the amount in the first section of the sampling tube (100 mg) is measured into a 2.0 mL sample container. This charcoal must be from the same batch as that used in obtaining the samples and can be obtained from unused charcoal tubes. A 750 mg/mL stock solution of the analyte in n-heptane is prepared. A known amount of this solution is injected directly into the activated charcoal with a 10 µL syringe, and the container is capped. The amount injected is equivalent to that present in a 10 L sample at the selected level. It is not practical to inject the neat liquid directly because the amounts to be added would be too small to measure accurately.

At least six tubes at each of three levels (0.5X, 1X, and 2X the OSHA standard) are prepared in this manner and allowed to stand for at least overnight to assure complete adsorption of the analyte onto the charcoal. These six tubes are referred to as the samples. A parallel blank tube should be treated in the same manner except that no sample is added to it. The sample and blank tubes are desorbed and analyzed in exactly the same manner as the sampling tube described above. The weight of analyte found in each tube is determined from the standard curve. Desorption efficiency is determined by the following equation:

D.E. = average wt (mg) recovered ÷ wt (mg) added

The desorption efficiency is dependent on the amount of analyte collected on the charcoal. Plot the desorption efficiency versus the weight of analyte found. The curve is used to correct for adsorption losses.

Calibration and Standards

It is convenient to express concentration of standards in terms of mg/mL of eluent. To minimize error due to the volatility of the eluent, one can add 10 times the weight to 10 mL of the eluent. (For the internal standard method use eluent containing 0.2% of the internal standard.) A series of standards, varying in concentration over the range of interest, is prepared and analyzed under the same GC conditions and during the same time period as the unknown samples. Curves are established by plotting concentrations in mg/mL versus peak area. In the case of the internal standard method, plot the concentration versus the ratio of peak area of analyte to peak area of internal standard. Note: Whether the absolute area or internal standard method is used standard solutions should be analyzed at the same time that the sample analysis is done. This will minimize the effect of variations of FID response.

Calculations

Read the weights, in mg, corresponding to each peak area (area ratio in case of the internal standard method) from the standard curve. No volume corrections are needed, because the standard curve is based on mg/mL eluent and the volume of sample injected is identical to the volume of the standards injected. Corrections for the blank must be made for each sample:

mg = mg sample - mg blank

where:

mg sample = mg found in front section of sample tube
mg blank = mg found in front section of blank tube

A similar procedure is followed for the backup sections. Add the weights present in the front and backup sections of the same sample tube to determine the total weight in the sample. Read the desorption efficiency from the curve for the amount of analyte found in the front section. Divide the total weight by this desorption efficiency to obtain the corrected mg/sample:

Corrected mg/sample = total weight ÷ D.E.

The concentration of the analyte in the air sampled can be expressed in mg/m^3, which is numerically equal to μg/L of air:

mg/m³ = (corrected mg)(1000 L/m³) ÷ (air volume sampled in L)

Another method of expressing concentration is ppm:

ppm = (mg/m³)(24.45)(760)(T + 273) ÷ (M.W.)(P)(298)

where:

P = pressure (mm Hg) of air sampled
T = temperature (°C.) of air sampled
24.45 = molar volume (L/mol) at 25° C. and 760 mm Hg
M.W. = molecular weight (g/mol) of analyte
760 = standard pressure (mm Hg)
298 = standard temperature (°K)

References

1. White, L.D. et al. A convenient optimized method for the analysis of selected solvent vapors in the industrial atmosphere. Amer. Ind. Hyg. Assoc. J. 31: 225, 1970.
2. Documentation of NIOSH Validation Tests, NIOSH Contract No. CDC-99-74-45.
3. Final Report, NIOSH Contract No. HSM-99-71-31, Personal Sampler Pump for Charcoal Tubes, September 15, 1972.

sec-BUTYL ALCOHOL

Synonym: sec-butanol $\quad$ CH₃CH₂CHOHCH₃
ACGIH TLV: 100 ppm (305 mg/m³)
OSHA Standard: 150 ppm

Method	Sampling Duration	Sampling Location	Useful Range (ppm)	System Cost ($)	Test Cost ($)	Manufacturer
DT	Grab	Portable	100-3500	180	2.25	Bendix
IR	Cont	Portable	3-300	4374	0	Foxboro
PI	Cont	Fixed	2-200	4950	0	HNU

NIOSH METHOD NO. S66

(Use as described for n-butyl alcohol, with 70° C. column temperature)

tert-BUTYL ALCOHOL

Synonym: tert-butanol $(CH_3)_3COH$
ACGIH TLV: 100 ppm (305 mg/m^3)
OSHA Standard: 100 ppm

Method	Sampling Duration	Sampling Location	Useful Range (ppm)	System Cost ($)	Test Cost ($)	Manufacturer
DT	Grab	Portable	1000-8900	180	2.25	Bendix
IR	Cont	Portable	2-200	4374	0	Foxboro
PI	Cont	Fixed	2-200	4950	0	HNU

NIOSH METHOD NO. S66

(Use as described for n-butyl alcohol, with 70° C. column temperature)

n-BUTYLAMINE

ACGIH TLV: 5 ppm (15 mg/m^3) ceiling - Skin $C_4H_9NH_2$
OSHA Standard: 5 ppm ceiling

Method	Sampling Duration	Sampling Location	Useful Range (ppm)	System Cost ($)	Test Cost ($)	Manufacturer
DT	Grab	Portable	1-60	180	2.25	Bendix
CS	Grab	Portable	0.5-10	2465	0.80	MDA
IR	Cont	Portable	1-10	4374	0	Foxboro
PI	Cont	Fixed	0.2-20	4950	0	HNU

NIOSH METHOD NO. P&CAM 221

Principle

A known volume of air is drawn through a tube containing silica gel to trap the aliphatic amines. The silica gel is transferred to glass-stoppered tubes and treated with sulfuric acid. A portion of the resulting acid solution is made alkaline with an excess of sodium hydroxide and an aliquot of the alkaline solution is analyzed by gas chromatography with a flame ionization detector.

The method may be used to determine a single aliphatic amine or to determine two or more amines in a single sample. The method has been applied to the following individual compounds:

methylamine, ethylamine, dimethylamine, isopropylamine, butylamine, diethylamine, diisopropylamine, triethylamine, and cyclohexylamine.

Range and Sensitivity

The limit of detection of the method is 0.01 mg of amine (0.02 of methylamine) per sample when the analyte is desorbed from 150 mg of silica gel with 2 mL of sulfuric acid and a 3 μL aliquot of the alkaline mixture is analyzed. This limit corresponds approximately to 1 mg/m^3 (2 mg/m^3 for methylamine) in a 10 L sample of air.

The upper limit is at least 24 mg of amine per sample. This amount of methylamine is retained by the 500 mg section of silica gel in the sorbent tube before breakthrough occurs. Larger amounts of other amines of higher molecular weight are retained. This limit for methylamine corresponds approximately to 2400 mg/m^3 in a 10 L sample of air.

Interferences

The most common possible sampling interference is water vapor. With the sorbent tube described, 96 L of air of 100% relative humidity can be sampled over an 8 hour period at 200 mL/min without displacement of the collected aliphatic amines by water vapor.

Any compound which has nearly the same retention time on the GC column as one of the aliphatic amines is an interferent. This type of interference can often be overcome by changing the GC operating conditions or by selecting another column. Retention time data on a single column, or even on a number of columns, cannot be considered as conclusive proof of chemical identity. It is important, therefore, that a sample of the bulk amine or mixture of amines be submitted at the same time as the sample tubes so that chemical identification of possible interferents can be made. (The bulk sample must NOT be shipped in the same shipping container as the sample tubes.)

Accuracy and Precision

The volume of air sample can be measured within 1% if a pump with a calibrated volume indicator is used. Volumes calculated from initially set flow rates may be less accurate (5%) because of changes in flow rate during sampling. The collection efficiency for aliphatic amines is essentially 100% under most conditions, as demonstrated by the negligible amount of amine collected in the backup sections of the sorbent tube.

The higher molecular weight amines and small amounts of all amines are incompletely desorbed. When a correction is made for desorption efficiency, the accuracy approaches 100%. Storage losses of 16-25% have been observed for primary amines on silica gel in sealed tubes stored for 1 month. Storage losses of 6-7% have been observed for secondary amines over 1 month. Triethylamine was stable for this period. Refrigeration of samples decreases such losses and is recommended if analysis is not to be performed within a few days.

The precision of the analysis is dependent upon the precision and sensitivity of the technique used to quantitate the gas chromatographic peaks of samples and standards. Electronic digital integrators with baseline correction capabilities will maximize analytical precision, particularly at lower concentrations. For the lower concentrations (0.01-0.2 mg/sample) the analytical precision may be 20%; at higher concentrations it approaches 3%. A coefficient of variation of 0.03 has been determined for the analysis of 10 samples of 1.2 mg of methylamine collected from 4 L of air (300 mg/m^3) with a personal sampling pump. The recovery was 99%.

Advantages and Disadvantages

The sampling method uses a small, portable device involving no liquids. The sorbent tube can be used for at least 8 hr to measure an average workday concentration, or for only 15 min to measure an excursion concentration. Desorption of the sample and preparation for analysis are simple. The analysis is accomplished by a rapid instrumental method. Most analytical interferences can be eliminated by altering the GC conditions.

Several aliphatic amines can be collected and determined simultaneously. This is especially useful when the identity and composition of the amine vapors are not known. The major disadvantage of the method is the necessity for measuring desorption efficiencies. When the desorption is less than complete, as for higher molecular weight amines, the effect of the amount of compound absorbed must also be determined.

Apparatus

Sorbent tubes. The sorbent tubes consist of pyrex glass tubes 125 mm long and 8 mm I.D., packed with three separate sections of 45/60 mesh activated silica gel. The weights of the three sections of silica gel are, in order, 600, 150, and 150 mg; these tubes are designed for sample flow in either direction. Plugs of 100 mesh stainless steel screen are used to contain the silica gel sections. These plugs of negligible pressure drop are prepared from 7 mm discs of screen held in place by Teflon rings. Pieces of pyrex tubing 12 mm long and 7 mm O.D. are located between the sorbent sections to inhibit migration of the amines. The ends of the tubes are flame sealed after packing to prevent contamination before use for sampling. The pressure drop across the tubes does not exceed 6 cm of water at a flow rate of 200 mL/min. The silica gel should be the equivalent of silica gel D-08, chromatographic grade, activated and fines free, 45/60 mesh, as supplied by Applied Science Laboratories, Inc., State College, PA. Polyethylene caps should be provided to seal the tubes after sampling has been completed.

Personal sampling pump. The personal sampling pump should be capable of operation at a constant flow rate or 200 mL/min for up to 8 hr. A sample volume indicator is desirable. The pump should be calibrated with a representative sorbent tube in the line. A wet or dry test meter or a glass rotameter capable of measuring a flow rate of 200 mL/min to within 5% may be used in the calibration.

Gas chromatograph. With a flame ionization detector and linear temperature programming.

Column. 1.8 m by 4 mm silanized glass, packed with 60/80 mesh Chromosorb 103 (Johns-Manville) or equivalent.

Precolumn. 15 cm by 4 mm silanized glass, packed with 10 cm of 8/20 mesh Ascarite (Arthur H. Thomas Company) or equivalent, and fitted into the injection port. The precolumn should be replaced when the precision of the analysis becomes poor.

Recorder with an electronic digital integrator.

Glass-stoppered test tubes or flasks.

Glass syringes, 10 µL.

Pipettes and volumetric flasks. For preparation of standard solutions.

Reagents

Sulfuric acid, 1.0 N.

Sodium hydroxide, 1.1 N.

Aliphatic amines. Highest purity available (for use as standards).

Helium, Bureau of Mines grade A.

Hydrogen, prepurified.

Air, compressed and filtered.

Procedure

Cleaning of equipment. Wash all glassware with detergent solution and rinse with tap and distilled water. Dry in an oven.

Collection and shipping of samples. Immediately before beginning the collection of a sample, break each end of the sorbent tube so as to provide openings at least half the inside diameter of the tube. Choose the direction desired for sample flow, mark the inlet end of the tube with a permanent marker, and attach the other end to the sampling pump. For low concentrations of amines, low humidity conditions, or short sampling periods, pump the sample air through the 150 mg sorbent section first, for high concentrations, high humidity, or long sampling periods, pump the sample air through the 600 mg sorbent section first. Sample air must not pass through any hose or tubing before entering the sorbent tube.

With the sorbent tube in a vertical position, sample the air at 200 mL/min for the desired period of time (0.25 to 8 hr). The flow rate and sampling time, or the volume, must be measured as accurately as possible. The temperature and pressure of the air being sampled should be measured and recorded. Immediately after sampling is completed, cap the sorbent tubes with the polyethylene caps provided. Rubber caps must not be used. One tube should be handled in the same manner as the sample tubes (break, seal, and ship) except that no air is pumped through it. Label this tube as a blank.

Pack the tubes tightly before shipping to minimize breakage in transit. After the sample is collected the tubes should not be subjected to extremes of high temperature or low pressure. Refrigeration will reduce sample loss.

Analysis of samples. Remove and discard the stainless steel plugs and glass spacers and transfer each section of silica gel to a separate glass stoppered test tube or flask. Label the samples and analyze each section separately.

Desorption. Desorb the amines from the silica gel by adding 2 mL of 1.0 N sulfuric acid to the 150 mg sections and 8 mL to the 600 mg section. Shake the sample mixtures occasionally over a period of 1 hr. Tests have indicated that desorption reaches a maximum in an hour.

Neutralization. Transfer a 0.5 mL aliquot to another container and add 0.5 mL of 1.1 N sodium hydroxide to make the solution alkaline and regenerate the free amines. An aliquot of this solution is then analyzed by gas chromatography.

Gas chromatographic conditions. Typical operating conditions for the gas chromatograph are as follows:

Helium carrier gas flow rate, 120 mL/min
Hydrogen gas flow rate to detector, 40 mL/min
Air flow rate to detector, 540 mL/min
Injection port temperature, 160° C.
Detector temperature, 190° C.
Column temperature, 115 to 180° C. at 10° C./min; hold for 10 min

For a sample containing only one amine or a sample containing two closely eluting amines, the column temperature may be kept isothermal at an appropriate temperature.

Injection of sample. To eliminate difficulties arising from blowback or distillation within the syringe, use the solvent flush injection technique. Flush the 10 μL syringe several times with distilled water to set the barrel and plunger. Draw in 3 μL of water. With the needle removed from the water, pull the plunger back about 0.2 μL to draw in a pocket of air. Then immerse the needle in the sample solution and withdraw a 3 μL aliquot, taking into consideration the volume in the needle. After the needle is removed from the sample solution and prior to injection, pull the plunger back a short distance to minimize evaporation at the tip of the needle. Inject duplicate 3 μL aliquots of each sample and standard.

Measurement of peak area. Measure the area of each amine peak with an electronic digital integrator or other suitable means of area measurement. Estimate the amount of each individual amine from calibration curves prepared as described below.

Determination of desorption efficiency. Desorption efficiency for a particular compound can vary from one lot of silica gel to another, and also from one laboratory to another. Therefore, it is necessary to determine at least once the desorption efficiency for each amine with each lot of silica gel used. Among the aliphatic amines to which this method is applicable, desorption efficiency decreases with increasing molecular weight. Desorption efficiency increases with increasing ratio of sulfuric acid to silica gel; however, an increase in this ratio results in a reduction in analytical sensitivity. A compromise ratio of 75 mg of silica gel/mL of sulfuric acid is recommended. Desorption efficiency may also vary with the amount of amine present on the silica gel. Measurements of efficiency should, therefore, be made for at least two amounts within the normal range of sample size.

Place 150 mg of silica gel in a 2 mL glass-stoppered tube. The silica gel must be from the same lot as that used in collecting the sample; it can be obtained from unused sorbent tubes. With a μL syringe, inject a known amount of the amine, either pure or in water solution, directly onto the silica gel. Close the tube with the glass stopper and allow it to stand at least overnight to insure complete adsorption of the amine. Prepare at least three tubes for each of at least two different amounts of amine. These tubes are referred to as samples. Prepare a blank in the same manner, omitting the amine. Analyze the samples and blank as described earlier. Also analyze three standards prepared by adding identical amounts of the amine to 2.0 mL of 0.1 N sulfuric acid. Determine the concentrations of the amine in the blank, samples, and standards using calibration curves prepared as described below. The desorption efficiency is calculated by dividing the concentration of amine found in the sample by the concentration obtained for the corresponding standard.

Standards and Calibration

For accuracy in the preparation of standards, it is recommended that an initial standard be prepared in a relatively large volume and at a high concentration. Aliquots of this standard can then be diluted to prepare other standards. Prepare the initial standard by pipetting an appropriate volume of amine into a 100 mL volumetric flask and adding distilled water to the mark. Check the concentration of this solution by titrating with standard sulfuric acid. Make the first dilution by adding an equal volume of 2.0 N sulfuric acid to an aliquot of the initial standard. Make additional dilutions with 1.0 N sulfuric acid so that the final concentration of sulfuric acid will be the same in standards and samples.

For example, to prepare butylamine standards add 1.0 mL of butylamine to a 100 mL volumetric flask and dilute to the mark with water. The concentration of this initial standard is 7.39 mg/mL or 0.101 M. This concentration can be easily checked by titration with 0.100 N sulfuric acid. Dilute a 25 mL aliquot of the initial standard with 25 mL of 2.0 N sulfuric acid to give a second standard of 3.695 mg/mL in 1.0 N sulfuric acid. Dilute 1.0 mL of this second standard to 25 mL with 1.0 N sulfuric acid to give a concentration of 0.148 mg/mL. This concentration in 1.0 N sulfuric acid is comparable to that obtained by sampling butylamine in air at its TLV concentration (15 mg/m^3) on 600 mg of silica gel for 6 hr at 200 mL/min and desorbing into 8 mL of 1.0 N sulfuric acid; if the desorption efficiency is 100%, the resulting concentration of butylamine is 0.135 mg/mL. Various volumes of the second standard (3.695 mg/mL should be diluted to 25 mL to give a series of working standards covering the range of interest.

The standards prepared as described above should be analyzed under the same gas chromatographic conditions and during the same time period as the unknown samples. This will minimize the effect of day-to-day variations in the response of the flame ionization detector. A calibration curve is prepared for each amine by plotting average peak area against concentration of the standard solution in mg/mL (mg of amine/mL of 1.0 N sulfuric acid).

Calculations

From the calibration curve, read the concentration of the amine corresponding to the average peak height measured. Multiply this concentration value by the volume of 1.0 N sulfuric acid used to desorb the amine from the silica gel (8 mL for 600 mg silica gel, or 2 mL for 150 mg). The result is the weight of amine (in mg) in the silica gel section. Correct each value for the amount of amine found in the corresponding blank. Also, correct each value for the desorption efficiency as determined earlier. Add the corrected amounts of amine found in the front and backup sections of the same sample tube to obtain the total weight of amine in the air sample. The concentration of the amine in air may be expressed in mg/m^3:

mg/m^3 = (corrected wt in mg)(1000 L/m^3) ÷ (volume of air sampled in L)

The concentration may also be expressed in terms of parts per million (ppm) by volume:

ppm = (mg/m^3)(24.45)(760)(T+273) ÷ (M.W.)(P)(298)

where:

24.45 = molar volume (L/mol) at 25° C. and 760 mm Hg
M.W. = molecular weight
P = pressure (mm Hg) of air sampled
T = temperature (°C) of air sampled

References

1. Campbell, E.E., G.O. Wood, and R.G. Anderson. Development of Air Sampling Techniques, Los Alamos Scientific Laboratory, Progress Report LA-5634-PR, June 1974, LA-5973-PR, July 1975, and LA-6057-PR, September 1975.
2. Andre, C.E., and A.R. Mosier. Precolumn inlet system for the gas chromatographic analysis of trace quantities of short-chain aliphatic amines. Anal. Chem. 45: 1971, 1973.
3. Wood, G.O., and R.G. Anderson. Development of Air Monitoring Techniques Using Solid Sorbents, Los Alamos Scientific Laboratory Progress Reports LA-6216-PR, February 1976, and LA-6513-PR, September 1976.

n-BUTYL GLYCIDYL ETHER

Synonyms: BGE; 1-n-butoxy-2,3-epoxypropane $\qquad$ C$_4$H$_9$OCH$_2$CH(O)CH$_2$
ACGIH TLV: 25 ppm (135 mg/m^3)
OSHA Standard: 50 ppm (270 mg/m^3)
NIOSH Recommendation: 30 mg/m^3 ceiling (15 min)

Method	Sampling Duration	Sampling Location	Useful Range (ppm)	System Cost ($)	Test Cost ($)	Manufacturer
IR	Cont	Portable	1-100	4374	0	Foxboro

NIOSH METHOD NO. P&CAM 127

(Use as described for acetone, with 130° C. column temperature)

n-BUTYL MERCAPTAN

Synonym: 1-butanethiol $\quad$ C_4H_9SH
ACGIH TLV: 0.5 ppm (1.5 mg/m³)
OSHA Standard: 10 ppm

Method	Sampling Duration	Sampling Location	Useful Range (ppm)	System Cost ($)	Test Cost ($)	Manufacturer
IR	Cont	Portable	0.2-20	4374	0	Foxboro
GC	1 min	Portable	1-10⁶	10,000	0	Microsensor

NIOSH METHOD NO. S350

Principle

A known volume of air is drawn through a glass tube containing Chromosorb 104 to trap n-butyl mercaptan vapors. n-Butyl mercaptan is desorbed from the Chromosorb 104 with acetone, and the sample is analyzed by gas chromatography.

Range and Sensitivity

This method was validated over the range of 16.8-74.2 mg/m³ at an atmospheric temperature of 22° C. and atmospheric pressure of 759 mm Hg, using a 1.5 L sample. This sample size is based on the capacity of the Chromosorb 104 to collect vapors of n-butyl mercaptan in air at high relative humidity. The method may be capable of measuring smaller amounts if the desorption efficiency is adequate. Desorption efficiency must be determined over the range used.

The upper limit of the range of the method depends on the adsorptive capacity of the Chromosorb 104. This capacity may vary with the concentrations of n-butyl mercaptan and other substances in the air. Breakthrough is defined as the time that the effluent concentration from the collection tube (containing 150 mg of Chromosorb 104) reaches 5% of the concentration in the test gas mixture. Breakthrough occurred after sampling for 2.9 hr at an average sampling rate of 0.023 L/min and relative humidity of 94% and temperature of 25° C. The breakthrough test was conducted at a concentration of 74.2 mg/m³.

Interferences

When other compounds are known or suspected to be present in the air, such information, including their suspected identities, should be transmitted with the sample. Any compound that has the same retention time as n-butyl mercaptan at the operating conditions described in this method is an interference. Retention time data on a single column cannot be considered proof of chemical identity.

Precision and Accuracy

The coefficient of variation for the total analytical sampling method in the range of 16.8-74.2 mg/m³ was 0.062. This value corresponds to a 2.2 mg/m³ standard deviation at the OSHA standard level. Statistical information can be found in reference 1. Details of the test procedures are found in reference 2. On the average the concentrations obtained in the laboratory validation study at 0.5X, 1X, and 2X the OSHA standard level were 2% lower than the "true" concentration for 18 samples. Any difference between the "found" and "true" concentrations may not represent a bias in the sampling and analytical method, but rather a random variation from the experimentally determined "true" concentration. Therefore, the method has no bias. The coefficient of variation is a good measure of the accuracy of the method since the recoveries and storage stability were good. Storage stability studies on samples collected from a test atmosphere at a concentration of 35.9 mg/m^3 indicate that collected samples are stable for at least 7 days.

Advantages and Disadvantages

The sampling device is small, portable, and involves no liquids. Interferences are minimal, and most of those that occur can be eliminated by altering chromatographic conditions. The tubes are analyzed by means of a quick, instrumental method. One disadvantage of the method is that the amount of sample that can be taken is limited by the number of mg that the tube will hold before overloading. When the amount of n-butyl mercaptan found on the backup section of the sorbent tube exceeds 25% of that found on the front section, the probability of sample loss exists. The precision of the method is limited by the reproducibility of the pressure drop across the tubes. This drop will affect the flow rate and cause the volume to be imprecise, because the pump is usually calibrated for one tube only.

Apparatus

Personal sampling pump. A calibrated personal sampling pump whose flow can be determined within 5% at the recommended flow rate.

Chromosorb 104 tubes. Glass tubes with both ends unsealed, 8.5 cm long with 6 mm O.D. and 4 mm I.D., containing 2 sections of 60/80 mesh Chromosorb 104 separated by a 2 mm portion of urethane foam. The adsorbing section of the tube contains 150 mg of Chromosorb 104, and the backup section contains 75 mg. A plug of silylated glass wool is placed at the ends of the tube. The pressure drop across the tube must be less than 10 mm Hg at a flow rate of 0.025 L/min. Immediately prior to packing, the tubes should be acetone rinsed and dried at room temperature to eliminate the problem of Chromosorb 104 adhering to the walls of the glass tubes. The Chromosorb 104 tubes are capped with plastic caps at each end.

Sorbent washing procedure. Prior to usage, Chromosorb 104 is washed and dried to reduce or eliminate the effects of unreacted monomers, solvents, and manufacturer's batch to batch differences in production. A quantity of Chromosorb 104 is placed in a sintered glass filter fitted to a large vacuum flask. Reagent grade acetone equal to twice the volume of sorbent is added to the sorbent and mixed, and a vacuum is applied. Repeat the operation of wash-mix-vacuum six times. The sorbent is then transferred to an evaporating dish and dried in a vacuum oven at 120° C. under 25 in Hg vacuum for 4 hr.

Gas chromatograph. Equipped with a flame photometric detector with a sulfur filter. The gas chromatograph should be equipped with a venting valve to vent the solvent to the air before it reaches the detector.

Column. 4 ft × 1/4 in O.D. glass, packed with 60/80 mesh Chromosorb 104.

Electronic integrator. Or some other suitable method for determining peak areas.

Sample containers. 2 mL glass sample containers with glass stoppers or Teflon-lined caps.

Syringes. 10 µL, and other convenient sizes for preparing standards.

Pipets. Delivery type, 1.0 mL and other convenient sizes.

Volumetric flasks. 10 mL and other convenient sizes for preparing standard solutions.

Stopwatch.

Manometer.

Reagents

Acetone, reagent grade.
1-Butanethiol, reagent grade.
n-Hexane.
Nitrogen, purified.
Hydrogen, prepurified.
Oxygen, purified.
Air, filtered, compressed.

Procedure

Cleaning of equipment. All glassware used for the laboratory analysis should be detergent washed and thoroughly rinsed with tap water and distilled water, and dried.

Calibration of sampling pumps. Each personal sampling pump must be calibrated with a representative Chromosorb 104 tube in the line to minimize errors associated with uncertainties in the volume sampled.

Collection and shipping of samples. Immediately before sampling, remove the caps from the ends of the Chromosorb 104 tube. All tubes must be packed with Chromosorb 104 from the same manufacturer's lot and that has been prewashed in acetone. The smaller section of Chromosorb 104 is used as a backup and should be positioned nearest the sampling pump. The tube should be placed in a vertical direction during sampling to minimize channeling through the Chromosorb 104 tube. Air being sampled should not be passed through any hose or tubing before entering the Chromosorb 104 tube.

A sample size of 1.5 L is recommended. Sample at a flow rate between 0.010 and 0.025 L/min. Do not sample at a flow rate less than 0.010 L/min. Record sampling time, flow rate, and type of sampling pump used. The temperature, pressure and relative humidity of the atmosphere being sampled should be recorded. If pressure reading is not available, record the elevation. The Chromosorb 104 tube should be capped with plastic caps immediately after sampling. Under no circumstances should rubber caps be used. With each batch of ten samples, submit one tube from the same lot of tubes used for sample collection. This tube must be subjected to exactly the same handling as the samples except that no air is drawn through it. This tube should be labeled as a blank.

Capped tubes should be packed tightly and padded before they are shipped to minimize tube breakage during shipping. A sample of the bulk material should be submitted to the laboratory in a glass container with a Teflon-lined cap. This sample should not be transported in the same container as the Chromosorb 104 tubes. Because of postal and DOT regulations, the bulk sample should be shipped by surface mail. A minimum of 18 extra Chromosorb 104 tubes should be provided to the analyst for desorption efficiency determinations.

Preparation of samples. Remove the plastic cap from the inlet end of the Chromosorb 104 tube. Remove the glass wool plug and transfer the first (larger) section of Chromosorb 104 to a 2 mL stoppered sample container. Remove the separating section of urethane foam and transfer the backup section of Chromosorb 104 to another stoppered container. Analyze these two sections separately. Firm tapping of the tube may be necessary to effect complete transfer of the Chromosorb 104.

Desorption of samples. Prior to analysis, pipet 1.0 mL of the acetone into each sample container. Cap and shake the sample vigorously. Desorption is complete in 15 minutes. Analyses should be completed within one day after the n-butyl mercaptan is desorbed.

GC conditions. The typical operating conditions for the gas chromatograph are:

50 mL/min (60 psig) nitrogen carrier gas flow
150 mL/min (30 psig) hydrogen gas flow to detector
20 mL/min (20 psig) oxygen gas flow to detector
30 mL/min (20 psig) air flow to detector
150° C. injector manifold temperature
200° C. detector manifold temperature
140° C. column temperature

A retention time of approximately 5 minutes is to be expected for n-butyl mercaptan under these conditions and using the column recommended. The acetone will elute from the column before the n-butyl mercaptan.

Injection. The first step in the analysis is the injection of the sample into the gas chromatograph. To eliminate difficulties arising from blow back or evaporation of solvent within the syringe needle, one should employ the solvent flush injection technique. The 10 μL syringe is first flushed with solvent several times to wet the barrel and plunger. Three μL of solvent are drawn into the syringe to increase the accuracy and reproducibility of the injected sample volume. The needle is removed from the solvent, and the plunger is pulled back about 0.2 μL to separate the solvent flush from the sample with a pocket of air to be used as a marker. The needle is then immersed in the sample, and a 5 μL aliquot is withdrawn, taking into consideration the volume of the needle, since the sample in the needle will be completely injected. After the needle is removed from the sample and prior to injection, the plunger is pulled back 1.2 μL to minimize evaporation of the sample from the tip of the needle. Observe that the sample occupies 4.9-5.0 μL in the barrel of the syringe. Duplicate injections of each sample and standard should be made. No more than a 3% difference in area is to be expected. At the GC conditions described above, venting the acetone solvent for 60 seconds after injection is required. If the solvent is not vented, the flame may be extinguished and the detector may temporarily malfunction. It is not advisable to use an automatic sample injector because of possible plugging of the syringe needle with Chromosorb 104.

Measurement of area. The area of the sample peak is measured by an electronic integrator or some other suitable form of area measurement, and results are read from a standard curve prepared as discussed below.

Determination of desorption efficiency. The desorption efficiency of a particular compound can vary from one laboratory to another and also from one batch of Chromosorb 104 to another. Thus, it is necessary to determine the fraction of the specific compound that is removed in the desorption process for a particular batch of Chromosorb 104.

Chromosorb 104 equivalent to the amount in the first section of the sampling tube (150 mg) is measured into a 64 mm, 4 mm I.D. glass tube flame sealed at one end. This Chromosorb 104 must be from the same batch as that used in obtaining the samples. The Chromosorb 104 must also be prewashed with acetone as described earlier. The open end is capped with Parafilm. A known amount of hexane solution of analyte containing 13.73 mg/mL is injected directly into the Chromosorb 104 with a μL syringe, and the tube is capped with more Parafilm. The amount injected is equivalent to that present in a 1.5 L air sample at the selected level. It is not practical to inject the neat liquid directly onto the Chromosorb 104, because the amounts to be added would be too small to measure accurately.

Six tubes at each of three levels (0.5X, 1X, and 2X the OSHA standard) are prepared in this manner and allowed to stand for at least overnight to assure complete adsorption of the n-butyl mercaptan onto the Chromosorb 104. These tubes are referred to as the samples. A parallel blank tube should be treated in the same manner except that no sample is added to it. The sample and blank tubes are desorbed and analyzed in exactly the same manner as the sampling tube described above.

Since the response of the flame photometric detector is non-linear, a series of standards are prepared to cover variations over each of the three levels (0.5X, 1X, and 2X the OSHA standard). These are analyzed with the samples. The standards are used to confirm the calibration of the gas chromatograph. The desorption efficiency (D.E.) equals the average weight in mg recovered from the tube divided by the weight in mg added to the tube, or:

D.E. = average wt recovered in mg ÷ wt added in mg

The desorption efficiency is dependent on the amount of n-butyl mercaptan collected on the Chromosorb 104. Plot the desorption efficiency versus the weight of n-butyl mercaptan found. The curve is used to correct for adsorption losses.

Calibration and Standards

A series of standards, varying in concentration over the range corresponding to approximately 0.1 to 3 times the OSHA standard for the sample under study, is prepared and analyzed under the same GC conditions and during the same time period as the unknown samples. Curves are established by plotting concentration in mg/1.0 mL versus peak area. Note: since no internal standard is used in this method, standard solutions must be analyzed at the same time that the sample analysis is done. This will minimize the effect of known day-to-day variations and variations during the same day of the FPD response.

Prepare a stock standard solution containing 13.73 mg/mL n-butyl mercaptan in acetone. From this stock solution, appropriate aliquots are withdrawn and dilutions are made in acetone. Prepare at least 5 working standards to cover the range of 0.0055-0.165 mg/1.0 mL. This range is based on a 1.5 L sample. Prepare a standard calibration curve by plotting concentration of n-butyl mercaptan in mg/1.0 mL versus peak area.

Calculations

Read the weight, in mg, corresponding to each peak area from the standard curve. No volume corrections are needed, because the standard curve is based on mg/1.0 mL acetone and the volume of sample injected is identical to the volume of the standards injected. Corrections for the blank must be made for each sample:

mg = mg sample - mg blank

where:

mg sample = mg found in front section of sample tube
mg blank = mg found in front section of blank tube

A similar procedure is followed for the backup sections. Add the weights found in the front and backup sections to determine the total weight of the sample. Read the desorption efficiency from the curve for the amount found in the front section. Divide the total weight by this desorption efficiency to obtain the corrected mg/sample:

Corrected mg/sample = total weight ÷ D.E.

For personal sampling pumps with rotameters only, the following correction should be made:

Corrected volume = $ft(P_1T_2/P_2T_1)^{1/2}$

where:

f = flow rate sampled
t = sampling time
P_1 = pressure during calibration of sampling pump (mm Hg)
P_2 = pressure of air sampled (mm Hg)
T_1 = temperature during calibration of sampling pump (°K)
T_2 = temperature of air sampled (°K)

The concentration of n-butyl mercaptan in the air sampled can be expressed in mg/m³:

$$mg/m^3 = (corrected\ mg)(1000\ L/m^3) \div (air\ volume\ sampled\ in\ L)$$

Another method of expressing concentration is ppm:

$$ppm = (mg/m^3)(24.45)(760)(T + 273) \div (M.W.)(P)(298)$$

where:

P = pressure (mm Hg) of air sampled
T = temperature (°C.) of air sampled
24.45 = molar volume (L/mol) at 25° C. and 760 mm Hg
M.W. = molecular weight (g/mol) of n-butyl mercaptan
760 = standard pressure (mm Hg)
298 = standard temperature (°K)

References

1. Documentation of NIOSH Validation Tests, NIOSH Contract CDC-99-74-45.
2. Backup Data Report for n-Butyl Mercaptan, prepared under NIOSH Contract No. 210-76-0123.
3. Final Report, NIOSH Contract HSM-99-71-31. Personal Sampler Pump for Charcoal Tubes, September 15, 1972.

p-tert-BUTYLTOLUENE

ACGIH TLV: 10 ppm (60 mg/m³)
OSHA Standard: 10 ppm

$CH_3C_6H_5C(CH_3)_3$

Method	Sampling Duration	Sampling Location	Useful Range (ppm)	System Cost ($)	Test Cost ($)	Manufacturer
IR	Cont	Portable	2-20	4374	0	Foxboro

NIOSH METHOD NO. P&CAM 127

(Use as described for acetone, with 115° C. column temperature)

CARBON DIOXIDE

ACGIH TLV: 5000 ppm (9000 mg/m³) CO_2
OSHA Standard: 5000 ppm
NIOSH Recommendation: 10,000 ppm

Method	Sampling Duration	Sampling Location	Useful Range (ppm)	System Cost ($)	Test Cost ($)	Manufacturer
DT	Grab	Portable	100-600,000	150	2.50	Nat'l Draeger
DT	Grab	Portable	500-200,000	165	1.70	Matheson
DT	Grab	Portable	300-200,000	180	2.25	Bendix
DT	4-hr	Personal	500-6000	850	3.10	Nat'l Draeger
IR	Cont	Portable	100-10,000	4374	0	Foxboro
PI	Cont	Fixed	200-20,000	4950	0	HNU
GC	1 min	Portable	$1-10^6$	10,000	0	Microsensor

NIOSH METHOD NO. S249

Principle
A known volume of air is collected in a five-layer gas sampling bag by means of a low flow rate personal sampling pump capable of filling a bag. The carbon dioxide content of the samples is determined by gas chromatography using a thermal conductivity detector.

Range and Sensitivity
This method was validated over the range of 2270-9990 ppm at an atmospheric temperature of 20.5° C. and atmospheric pressure of 757 mm Hg using a 3.5 L sample volume. The working range of the method is estimated to be 500-15,000 ppm, under the experimental conditions cited. The upper limit of the range of the method and the absolute sensitivity have not been established.

Interferences
When two or more compounds are known or suspected to be present in the air, such information, including their suspected identities, should be transmitted with the sample. It must be emphasized that any compound which has the same retention time as the analyte at the operating conditions described in this method is an interference. Retention time data on a single column cannot be considered as proof of chemical identity. If the possibility of interference exists, separation conditions (column packing, temperature, etc.) must be changed to circumvent the problem.

Precision and Accuracy
The coefficient of variation for the total analytical and sampling method in the range of 2270-9990 ppm was 0.014. This value corresponds to a 69 ppm standard deviation at the OSHA standard level. Statistical information and details of the validation and experimental test procedures can be found in references 1 and 2.

On the average, the concentrations obtained at the OSHA standard level using the overall sampling and analytical method were 2.5% lower than the "true" concentrations for a limited number of laboratory experiments. Any difference between the "found" and "true" concentrations may not represent a bias in the sampling and analytical method, but rather a random variation from the experimentally determined "true" concentration. Therefore, no recovery correction should be applied to the final result.

Advantages and Disadvantages

The sampling device is small, portable, and involves no liquids. Interferences are minimal, and most of those which do occur can be eliminated by altering chromatographic conditions. The samples in bags are analyzed by means of a quick instrumental method. One disadvantage of the method is that the gas sampling bag is rather bulky and may be punctured during sampling and shipping.

Apparatus

Personal sampling pump. A personal sampling pump capable of filling a bag at approximately 0.05 L/min is required. This pump should be calibrated to within 5%.

Gas sampling bag, 5 L capacity. Only the five-layer sampling bags manufactured by Calibrated Instruments, Inc. (731 Saw Mill Road, Ardsley, NY 10502) were found to be satisfactory for sample collection and storage for at least 7 days. The bag is fitted with a metal valve and hose bib. For the preparation of calibration standards in the laboratory, 5 L Saran or Tedlar bags could be used.

Gas chromatograph. The unit must be equipped with a thermal conductivity detector and a 5 mL gas sampling loop or equivalent. A portable unit with no column temperature control is adequate.

Column. 5 ft × 1/4 in stainless steel, packed with 80/100 mesh Porapak QS.

Area integrator. An electronic integrator or some other suitable method for measuring peak areas.

Gas-tight syringes. 10 mL and other convenient sizes for making standards.

Calibrated rotameters. Convenient sizes for making standards.

Reagents

Carbon dioxide, 99% or higher purity.
Nitrogen, purified.
Helium, purified.
Air, filtered compressed.

Procedure

Cleaning of sampling bags and checking for leaks. The bags are cleaned by opening the closure mechanism and bleeding out the air sample. The use of a vacuum pump is recommended although this procedure can be carried out by manually flattening the bags. The bags are then flushed with carbon dioxide-free air and evacuated. This procedure is repeated at least twice. Bags may be checked for leaks by filling the bag with air until taut, sealing and applying gentle pressure to the bag. Observe for any discernable leaks and any volume changes or slackening of the bag, preferably over at least a one hour period.

Calibration of personal pumps. Each personal pump should be calibrated to minimize errors associated with uncertainties in the sample volume collected. Although sample volume is not actually used in this determination, the pump should be calibrated to avoid over filling the bags; i.e., a maximum sampling time can be determined based on flow rate and sample volume, which is approximately equal to 80% volume of bag.

Collection and shipping of samples. Immediately before sampling, attach a small piece of Tygon or plastic tubing to the hose bib of the five-layer gas sampling bag. Unscrew the valve fitting and attach the tubing to the outlet of the sampling pump. Air being sampled will pass through the pump and tubing before entering the sampling bag, since a "push" type pump is required. A sample size of 3-4 L is recommended. Sample at a flow rate of 0.05 L/min or less, but not less than 0.01 L/min. The flow rate should be known with an accuracy of at least 5%.

The temperature and pressure of the atmosphere being sampled should be recorded. If pressure reading is not available, record the elevation. The gas sampling bag should be labeled appropriately and sealed tightly. Gas sampling bags should be packed loosely and padded before they are shipped to minimize the danger of getting punctured during shipping.

GC conditions. The typical operating conditions for the gas chromatograph are:

100 mL/min (25 psig) helium carrier gas flow
Ambient injector temperature
70° C. detector temperature
Ambient column temperature

A retention time of approximately 2 min is to be expected for the analyte under these conditions and at ambient temperatures of 20-25° C. using the column recommended. The carbon dioxide elutes after oxygen and nitrogen.

GC analysis. The gas sampling bag is attached to the sample loop of the GC unit via a short piece of tubing. Open the closure valve of the gas sampling bag and fill the 5 mL sample loop by gently squeezing the sample bag. To allow the sample in the loop to attain atmospheric pressure, release the pressure applied to the sample bag just prior to turning the sample loop valve to inject the sample onto the column.

Measurement of area. The area of the sample peak is measured by an electronic integrator or some other suitable form of area measurement, and the results are read from a standard curve as discussed below.

Calibration and Standards

Completely evacuate a 5 L gas sampling bag, preferably with the aid of a vacuum pump. Introduce 1.0 L of filtered air via a septum into the bag; this can be done using a calibrated rotameter. Then add a known volume of carbon dioxide gas through a septum and add more air to a total accurately known volume of between 3-4 L. It is necessary to known accurately the volume of carbon dioxide added and the total volume of air to determine concentration in ppm. The concentration in ppm is equal to the volume of carbon dioxide divided by the sum of the volume of carbon dioxide and the volume of air.

A series of standards, varying in concentration over the range of interest, is prepared as described above and analyzed under the same GC conditions and during the same time period as the unknown samples. Curves are established by plotting concentration in ppm versus peak area. Corrections for the unknown carbon dioxide concentration in the filtered air must be made if necessary. The carbon dioxide correction factor can be determined by filling an evacuated bag with 3-4 L of the filtered air used for preparing the calibration standards. This "blank" air is analyzed under the same conditions as the calibration standards and the samples. The "blank" area thus determined is subtracted from the peak area of each calibration standard. A calibration curve is established by plotting concentration in ppm versus corrected peak area. Note: calibration standards should be analyzed at the same time the sample analysis is done. This will minimize the effect of variations in detector response.

Calculations

Read the concentration in ppm, corresponding to each peak area from the standard curve. Another method of expressing concentration is mg/m³ (corrected to standard conditions of 25° C. and 760 mm Hg):

$$mg/m^3 = (ppm)(M.W.)(760)(T+273) \div (24.45)(P)(298)$$

where:

P = pressure (mg Hg) of air sampled
T = temperature (°C.) of air sampled
24.45 = molar volume (L/mol) at 25° C. and 760 mm Hg
M.W. = molecular weight
760 = standard pressure (mm Hg)
298 = standard temperature (°K)

References

1. Memoranda from Kenneth A. Busch, Chief, Statistical Services, KLCD, to Deputy Director, DLCD, January 16, 1975 and November 8, 1974. Statistical Protocol for Analysis of Data from Contract CDC-99-74-45.
2. Backup Data Report for Carbon Dioxide, prepared under NIOSH Contract No. 210-76-0123.

CARBON DISULFIDE

ACGIH TLV: 10 ppm (30 mg/m³) - Skin CS_2
OSHA Standard: 20 ppm
NIOSH Recommendation: 1 ppm

Method	Sampling Duration	Sampling Location	Useful Range (ppm)	System Cost ($)	Test Cost ($)	Manufacturer
DT	Grab	Portable	2.5-3200	150	2.60	Nat'l Draeger
DT	Grab	Portable	5-60	165	4.60	Matheson
DT	Grab	Portable	2-4000	180	2.25	Bendix
DT	8-hr	Personal	1.25-50	850	3.10	Nat'l Draeger
IR	Cont	Portable	1-40	4374	0	Foxboro
PI	Cont	Fixed	0.2-20	4950	0	HNU
GC	1 min	Portable	$1-10^6$	10,000	0	Microsensor

NIOSH METHOD NO. S248

Principle

A known volume of air is drawn through a charcoal tube to trap the organic vapors present. The charcoal in the tube is transferred to a small, stoppered sample container, and the analyte is desorbed with benzene. An aliquot of the desorbed sample is injected into a gas chromatograph. The area of resulting peak is determined and compared with areas obtained for standards.

Range and Sensitivity

This method was validated over the range of 14.7-58.8 ppm at an atmospheric temperature and pressure of 22° C. and 766 mm Hg, using a 6 L sample. Under the conditions of sample size (6 L) the probable useful range of this method is 5-90 ppm. The method is capable of measuring much smaller amounts if the desorption efficiency is adequate. Desorption efficiency must be determined over the range used.

The upper limit of the range of the method is dependent on the adsorptive capacity of the charcoal tube. This capacity varies with the concentrations of carbon disulfide and other substances in the air. The first section of the charcoal tube was found to hold 6.0 mg of carbon disulfide when a test atmosphere containing 188 mg/m^3 of carbon disulfide in air was sampled at 0.196 L/min for 162 min; breakthrough was observed at this time, i.e., 0.056 mg of carbon disulfide had broken through the front section of the charcoal tube during this time period. (The charcoal tube consists of two sections of activated charcoal separated by a section of urethane foam.) If a particular atmosphere is suspected of containing a large amount of contaminant, a smaller sampling volume should be taken.

Interferences

When the amount of water in the air is so great that condensation actually occurs in the tube, carbon disulfide vapors may not be trapped efficiently. Experiments showed that there was increasing loss of carbon disulfide with an increase in relative humidity. In order to correct this problem, it is necessary to use a dessicant to remove the moisture.

When interfering compounds are known or suspected to be present in the air, such information, including their suspected identities, should be transmitted with the sample. It must be emphasized that any compound which has the same retention method is an interference. Retention time data on a single column cannot be considered proof of chemical identity. If the possibility of interference exists, separation conditions (column packing, temperature, etc.) must be changed to circumvent the problem.

Precision and Accuracy

The coefficient of variation for the total analytical and sampling method in the range of 14.7-58.8 ppm was 0.059. This value corresponds to a 1.8 ppm standard deviation at the ceiling OSHA standard level. Statistical information and details of the validation and experimental test procedures can be found in reference 2.

On the average the concentrations obtained at the ceiling OSHA standard level using the overall sampling and analytical method were 0.7% higher than the "true" concentrations for a limited number of laboratory experiments. Any difference between the "found" and "true" concentrations may not represent a bias in the sampling and analytical method, but rather a random variation from the experimentally determined "true" concentration. Therefore, no recovery correction should be applied to the final result.

Advantages and Disadvantages

The sampling device is small, portable, and involves no liquids. Interferences are minimal, and most of those which do occur can be eliminated by altering chromatographic conditions. The tubes are analyzed by means of a quick, instrumental method. The method may also be used for the simultaneous analysis of two or more substances suspected to be present in the same sample.

One disadvantage of the method is that the amount of sample which can be taken is limited by the number of mg that the tube will hold before overloading. When the sample value obtained for the backup section of the charcoal tube exceeds 25% of that found on the front section, the possibility of sample loss exists. Furthermore, the precision of the method is limited by the reproducibility of the pressure drop across the tubes. This drop will affect the flow rate and cause the volume to be imprecise, because the pump is usually calibrated for one tube only.

Apparatus

Calibrated personal sampling pump. Whose flow can be determined within 5% at the recommended flow rate.

Charcoal tubes. Glass tube with both ends flame sealed, 7 cm long with 6 mm O.D. and 4 mm I.D., containing 2 sections of 20/40 mesh activated charcoal separated by a 2 mm portion of urethane foam. The activated charcoal is prepared from coconut shells and is fired at 600° C. prior to packing. The adsorbing section contains 100 mg of charcoal, the backup section 50 mg. A 3 mm portion of urethane foam is placed between the outlet end of the tube and the backup section. A plug of silylated glass wool is placed in front of the adsorbing section. The pressure drop across the tube must be less than one inch of mercury at a flow rate of 1 L/min.

Drying tubes. Glass tube with both ends open, 7 cm long with 6 mm O.D. and 4 mm I.D. To add the dessicant into the tube, a plug of silylated glass wool is placed into one end of the tube, and the tube is filled with 270 mg of granular anhydrous sodium sulfate. Another plug of silylated glass wool is placed over the sodium sulfate, and the tube is capped at both ends.

Gas chromatograph. Equipped with a flame photometric detector with a sulfur filter. The GC should be equipped with a vent as described later.

Column. 6 ft × 1/4 in O.D. glass, packed with 5% OV-17 on 80/100 mesh Gas Chrom Q.

Electronic integrator. Or some other suitable method for measuring peak area.

Syringes. 10 µL, and other convenient sizes for preparing standards.

Pipets. 10 mL delivery pipets.

Volumetric flasks. 25 mL or convenient sizes for preparing standard solutions.

Reagents

Chromatographic quality carbon disulfide.
Benzene, reagent grade.
Purified oxygen.
Purified nitrogen.
Prepurified hydrogen.
Filtered compressed air.

Procedure

Cleaning of equipment. All glassware used for the laboratory analysis should be detergent washed and thoroughly rinsed with tap water and distilled water, and dried.

Calibration of personal pumps. Each personal pump must be calibrated with a representative charcoal tube and a drying tube in the line. This will minimize errors associated with uncertainties in the sample volume collected.

Collection and shipping of samples. Immediately before sampling, break the ends of the tube to provide an opening at least one-half the internal diameter of the tube (2 mm). Uncap both ends of the drying tube and connect this tube in series to the front section of the charcoal tube with 20 mm of Teflon tubing. The smaller section of charcoal is used as a backup and should be positioned nearer the sampling pump. The charcoal tube should be placed in a vertical direction during sampling to minimize channeling through the charcoal.

At the ceiling OSHA standard concentration, a sample size of 6 L is recommended. Sample for

30 min at a flow rate of 0.20 L/min. At the 8 hour time weighted average OSHA standard, a sample size of 12 L is recommended. Sample at a flow rate of 0.20 L/min. The temperature and pressure of the atmosphere being sampled should be recorded. If pressure reading is not available, record the elevation.

The drying tube should remain connected to the charcoal tube during shipping. Cap the open end of the drying tube with a plastic cap. The charcoal tubes should be capped with the supplied plastic caps immediately after sampling. Under no circumstances should rubber caps be used. With each batch of ten samples submit one charcoal tube and one drying tube from the same lot of tubes which were used for sample collection and which are subjected to exactly the same handling as the samples except that no air is drawn through it. Label this as a blank. Capped tubes should be packed tightly and padded before they are shipped to minimize tube breakage during shipping.

A sample of the bulk material should be submitted to the laboratory in a glass container with a Teflon-lined cap. This sample should not be transported in the same container as the charcoal tubes. It has been found that carbon disulfide tends to migrate within the charcoal tube from the front section to the backup section when held at ambient temperatures for prolonged periods of time. This migration can effectively be retarded by storing the samples at refrigerator temperatures. The tubes appear to be unaffected by short storage at elevated temperatures or by shipping under reduced pressures. It is recommended that the samples be refrigerated if sample analysis cannot be performed within one week.

Preparation of samples. In preparation for analysis, each charcoal tube is scored with a file in front of the first section of charcoal and broken open. The glass wool is removed and discarded. The charcoal in the first (larger) section is transferred to a 25 mL stoppered sample container. The separating section of foam is removed and discarded; the second section is transferred to another stoppered container. These two sections are analyzed separately.

Desorption of samples. Prior to analysis, 10 mL of benzene is pipetted into each sample container. (All work with benzene should be performed in a hood because of its high toxicity.) Desorption should be done for 30 min. Tests indicate that this is adequate if the sample is agitated occasionally during this period.

GC conditions. The typical operating conditions for the gas chromatograph are:

20 mL/min nitrogen carrier gas flow
150 mL/min hydrogen gas flow to detector
35 mL/min air flow to detector
20 mL/min oxygen gas flow to detector
150° C. injector temperature
145° C. detector temperature
30° C. column temperature

Injection. The first step in the analysis is the injection of the sample into the gas chromatograph. To eliminate difficulties arising from blow back or distillation within the syringe needle, one should employ the solvent flush injection technique. The 10 μL syringe is first flushed with solvent several times to wet the barrel and plunger. Three μL of solvent are drawn into the syringe to increase the accuracy and reproducibility of the injected sample volume. The needle is removed from the solvent, and the plunger is pulled back about 0.2 μL to separate the solvent flush from the sample with a pocket of air to be used as a marker. The needle is then immersed in the sample, and a 5 μL aliquot is withdrawn, taking into consideration the volume of the needle, since the sample in the needle will be completely injected. After the needle is removed from the sample and prior to injection, the plunger is pulled back 1.2 μL to minimize evaporation of the sample from the tip of the needle. Observe that the sample occupies 4.9-5.0 µL in the barrel of the syringe. Duplicate injections of each sample and standard should be made. No more than a 3% difference in area is to be expected. The gas chromatograph is equipped with a valve to vent the solvent peak after it passes through a GC column. To avoid exposing the detector to benzene the venting valve should be opened at 2-3 minutes after sample injection to elute the benzene and closed after the benzene is eluted.

Measurement of area. The area of the sample peak is measured by an electronic integrator or some other suitable form of area measurement, and preliminary results are read from a standard curve prepared as discussed below.

Determination of desorption efficiency. The desorption efficiency of a particular compound can vary from one laboratory to another and also from one batch of charcoal to another. Thus, it is necessary to determine at least once the percentage of the specific compound that is removed in the desorption process, provided the same batch of charcoal is used.

Activated charcoal equivalent to the amount in the first section of the sampling tube (100 mg) is measured into a 64 mm by 4 mm I.D. glass tube, flame sealed at one end. This charcoal must be from the same batch as that used in obtaining the samples and can be obtained from unused charcoal tubes. The open end is capped with Parafilm. A known amount of benzene solution of carbon disulfide containing 0.14 mg/µL is injected directly into the activated charcoal with a microliter syringe, and the tube is capped with more Parafilm.

The amount injected is equivalent to that present in a 6 L air sample at the selected level. Six tubes at each of three levels (0.5X, 1X, and 2X of the ceiling OSHA standard) are prepared in this manner and allowed to stand for at least overnight to assure complete adsorption of the analyte onto the charcoal. These tubes are referred to as the samples. A parallel blank tube should be treated in the same manner except that no sample is added to it. The sample and blank tubes are desorbed and analyzed in exactly the same manner as the sampling tube described earlier.

Two or three standards are prepared by injecting the same volume of compound into 10 mL of benzene with the same syringe used in the preparation of the samples. These are analyzed with the samples. The desorption efficiency (D.E.) equals the average weight in mg recovered from the tube divided by the weight in mg added to the tube, or:

D.E. = average weight recovered in mg ÷ weight added in mg

The desorption efficiency is dependent on the amount of analyte collected on the charcoal. Plot the desorption efficiency versus weight of the analyte found. This curve is used to correct for adsorption losses.

Calibration and Standards

It is convenient to express concentration of standards in terms of mg/10 mL benzene, because samples are desorbed in this amount of benzene. the density of the analyte is used to convert mg into µL for easy measurement with a µL syringe. A series of standards, varying in concentration over the range of interest, is prepared and analyzed under the same GC conditions and during the same time period as the unknown samples. Curves are established by plotting concentration in mg/10 mL versus peak area. Note: Since no internal standard is used in this method, standard solutions must be analyzed at the same time that the sample analysis is done. This will minimize the effect of known day-to-day variations and variations during the same day of the flame photometric detector response.

Calculations

Read the weight, in mg, corresponding to each peak area from the standard curve. No volume corrections are needed, because the standard curve is based on mg/10 mL benzene, and the volume of sample injected is identical to the volume of the standards injected. Corrections for the blank must be made for each sample:

mg = mg sample - mg blank

where:

mg sample = mg found in front section of sample tube
mg blank = mg found in front section of blank tube

A similar procedure is followed for the backup sections. Add the weights found in the front and backup sections to get the total weight in the sample. Read the desorption efficiency from the curve for the amount found in the front section. Divide the total weight by this desorption efficiency to obtain the corrected mg/sample:

Corrected mg/sample = total weight ÷ D.E.

The concentration of the analyte in the air sampled can be expressed in mg/m^3:

mg/m^3 = (corrected mg)(1000 L/m^3) ÷ (air volume sampled in L)

Another method of expressing concentration is ppm:

ppm = (mg/m^3)(24.45)(760)(T + 273) ÷ (M.W.)(P)(298)

where:

P = pressure (mm Hg) of air sampled
T = temperature (°C.) of air sampled
24.45 = molar volume (L/mol) at 25° C. and 760 mm Hg
M.W. = molecular weight (g/mol) of analyte
760 = standard pressure (mm Hg)
298 = standard temperature (°K)

References

1. McCammon, C., Quinn, P., and Kupel, R. A charcoal sampling method and a gas chromatographic analytical procedure for carbon disulfide. Amer. Ind. Hyg. Assoc. J. 36: 618, 1975.
2. Documentation of NIOSH Validation Tests, NIOSH Contract CDC-99-74-75.
3. Final Report, NIOSH Contract HSM-99-71-31. Personal Sampler Pump for Charcoal Tubes, September 15, 1972.

CARBON MONOXIDE

ACGIH TLV: 50 ppm (55 mg/m³) CO
OSHA Standard: 50 ppm
NIOSH Recommendation: 35 ppm

Method	Sampling Duration	Sampling Location	Useful Range (ppm)	System Cost ($)	Test Cost ($)	Manufacturer
DT	Grab	Portable	5-70,000	150	2.60	Nat'l Draeger
DT	Grab	Portable	5-1000	165	1.70	Matheson
DT	Grab	Portable	5-100,000	180	2.25	Bendix
ES	Cont	Personal	10-100	450	0	Bendix
CS	Cont	Personal	1-500	595	0	Dynamation
CS	Cont	Portable	1-300	650	0	Dynamation
ES	Cont	Personal	1-2000	695	0	Energetics
DT	4 hr	Personal	2.5-500	850	3.10	Nat'l Draeger
CS	Cont	Fixed	1-500	995	0	Dynamation
ES	Cont	Portable	2-500	1035	0	Bacharach
ES	Cont	Portable	5-500	1350	0	Rexnord
ES	1 min	Personal	1-500	1550	0	Interscan
ES	Cont	Portable	1-100(3000)	1675	0	Interscan
ES	Cont	Portable	1-600	1900	0	Energetics
ES	Cont	Fixed	1-50(500)	2000	0	Interscan
ES	Cont	Port/Fixed	1-100(500)	2528	0	Interscan
IR	Cont	Portable	1-100	4374	0	Foxboro
PI	Cont	Fixed	2-200	4950	0	HNU
GC	1 min	Portable	1-10^6	10,000	0	Microsensor

NIOSH METHOD NO. P&CAM 112

Principle

Air samples are obtained in 5 L or larger inert plastic bags and analyzed by infrared spectrophotometry in a 10 meter pathlength gas cell. Compressed air tanks (scuba tanks) may also be analyzed.

Range and Sensitivity

The analytical range extends from 10 to 500 ppm. The upper limit may be extended by using an aliquot of the sample.

Interferences

Any gas that absorbs infrared radiation at the analytical wavelength will interfere. Possible interferences include: acetylene, aldehyde, cyanogen, diazomethane, hydrogen, sulfide, nitrosylchloride, nitrous oxide, olefins and propyne. These gases are not usually present in significant concentration in normal atmospheres, but they can readily be ruled out by reference to the complete spectrum of the gas sample scanned from 2-14 microns. Each of the interfering gases has additional absorption lines in other areas of the spectrum which readily denote its presence. Carbon dioxide in concentrations as high as 3% does not interfere. Water vapor is eliminated by the use of drying agent traps on the inlet of the IR cell.

Precision and Accuracy

The analytical method has a precision of 10%. No collaborative testing has been done with this method.

Advantages and Disadvantages

The major advantage of the method over CO meters and detector tubes is that the analysis is more accurate. Approved detector tubes are accurate to only 25%. The disadvantage is that samples must be returned to the laboratory for analysis whereas immediate results are available with CO meters and detector tubes.

Apparatus

Sampling pump. Peristaltic pump, diaphragm pump, or vacuum pump with filtered outlet to remove oil.

Tedlar bags. 5 L or larger.

Infrared spectrometer. 2 to 15 micron range, wavelength accuracy 0.015 micron, resolving power 0.01 micron, wavelength reproducibility 0.005 micron, transmission accuracy 5%.

10 meter pathlength gas cell.

Manometer.

Vacuum pump. Capable of reducing pressure to 1 mm.

Gas-tight syringes. 100, 500, and 1000 μL.

Gas tank regulators, connections, and needle valves. For introducing dilution gas, CO, and samples.

Reagents

Anhydrous calcium chloride or indicating drierite.

Carbon monoxide.

Tank of dry nitrogen or CO-free air. CO-free air can be obtained by using cylinder air known to be free of CO, or by allowing dilution air to enter the cell via a Hopcalite absorption trap connected to the inlet of the cell.

Hopcalite. Used to remove CO if room air is used for dilution.

Procedure

Cleaning of equipment. Equipment must be properly evacuated or flushed to minimize any background interference.

Collection and shipping of samples. The sampling containers, Tedlar and Mylar bags, are flushed or evacuated prior to use. Samples are obtained by filling the sampling containers with air either by using a sampling pump to push air in or by simply opening the valve on an evacuated container and allowing it to reach atmospheric pressure. After the sample has been taken, all parts are sealed to minimize leakage in or out of the containers.

Care must be taken in transporting the samples to the laboratory, so as not to damage or alter the existing sample. CO samples do not deteriorate on standing. If samples collected for analysis are under reduced pressure, care should be taken to see that dilution does not occur because of leaks in the container. Atmospheric samples can generally be stored at atmospheric pressure to avoid dilution. Compressed air tanks (scuba tanks) may also be analyzed for CO content.

Analysis of samples. The 10 meter pathlength gas cell is connected to a manometer via a T connection. The cell is then evacuated to approximately 1.0 mm Hg. The Tedlar bag sample is introduced into the cell through a calcium chloride or drierite tube. If a rigid sample container is used, the equilibrium pressure (Pe) must be noted and the cell filled to atmospheric pressure with CO-free air or dry nitrogen. A 15 min waiting period before analysis is necessary for equilibrium to be established. If the sample is a compressed air tank, fill the evacuated gas cell to atmospheric pressure with air from the tank. The spectrum is scanned from 4-6 micron (2500-1670 cm^{-1}) and the absorbance at 4.67 micron (2143 cm^{-1}) is measured by the baseline technique.

Calibration and Standards

The volume of the 10 meter cell is determined by standard techniques. The simplest procedure is to evacuate the cell, and then bring to atmospheric pressure by permitting air to enter via a calibrated wet-test meter. The volume of the cell is the volume of air shown on the meter. After the volume has been ascertained, the cell is again evacuated, and known volumes of CO added with the aid of standard gas-tight syringes, through a rubber serum bottle cap attached to the inlet of the cell, or through the rubber tubing which connects the gas cell with the tank of dilution gas. The pressure in the cell is then brought to atmospheric pressure with CO-free dry air or nitrogen. The absorbance at 4.67 micron is measured as described in the procedure.

The CO concentration is calculated from the quantity of CO added, and the volume of the cell. For example, 1.0 mL CO in a 3.85 L cell gives a concentration of 260 ppm CO. A calibration curve relating absorbance and concentration is prepared for a series of known volumes of CO introduced into the IR cell.

Calculations

The concentration of the unknown is read from the calibration curve. To calculate the standard curve concentrations in parts per million (ppm), one may use the following equation:

Sample concentration (ppm) = volume of CO added in μL $\div$ volume of IR cell in L

The observed concentration from the calibration curve is corrected for the volume of sample actually introduced into the cell. For samples introduced into the cell with syringes, the volume is readily known and the correction applied using the following equation:

Sample concentration (ppm) = (C)(volume of IR cell in L) $\div$ (volume of sample in L)

where:

C = observed concentration from the calibration curve (ppm)

For samples introduced from non-rigid Tedlar bags with volumes greater than the volume of the IR cell, the volume of the cell equals the volume of the sample and the sample concentration = C. For samples introduced by pressure measurements, the sample concentration is calculated as follows:

Sample concentration (ppm) = (C)(Pa) $\div$ (Pa - Pe)

where:

C = observed concentration from the calibration curve (ppm)
Pe = equilibrium pressure after connecting the sample container to the IR cell
Pa = atmospheric pressure

CARBON TETRACHLORIDE

Synonym: tetrachloromethane   CCl$_4$
ACGIH TLV: 5 ppm (30 mg/m^3) - Skin, suspected carcinogen
OSHA Standard: 10 ppm
NIOSH Recommendation: 2 ppm ceiling (60 min)

Method	Sampling Duration	Sampling Location	Useful Range (ppm)	System Cost ($)	Test Cost ($)	Manufacturer
DT	Grab	Portable	5-50	150	2.60	Nat'l Draeger
DT	Grab	Portable	1-60	165	2.30	Matheson
DT	Grab	Portable	1-60	180	2.25	Bendix
IR	Cont	Portable	0.2-20	4374	0	Foxboro
PI	Cont	Fixed	0.2-20	4950	0	HNU
GC	1 min	Portable	1-10^6	10,000	0	Microsensor

NIOSH METHOD NO. P&CAM 127

(See acetone)

CHLORINE

ACGIH TLV: 1 ppm (3 mg/m³) Cl_2
OSHA Standard: 1 ppm
NIOSH Recommendation: 0.5 ppm ceiling (15 min)

Method	Sampling Duration	Sampling Location	Useful Range (ppm)	System Cost ($)	Test Cost ($)	Manufacturer
DT	Grab	Portable	0.3-500	150	2.60	Nat'l Draeger
DT	Grab	Portable	0.1-20	165	2.30	Matheson
DT	Grab	Portable	0.33-1000	180	2.25	Bendix
DT	8 hr	Personal	0.1-20	850	3.10	Nat'l Draeger
ES	1 min	Personal	0.02-2	1795	0	Interscan
ES	Cont	Portable	0.02-10	2156	0	Interscan
CS	Grab	Portable	0.25-4	2465	0.80	MDA
ES	Cont	Fixed	0.02-2(10)	1775	0	Interscan
ES	Cont	Port/Fixed	0.02-2(50)	3440	0	Interscan
CS	Cont	Fixed	0.2-4	4950	Variable	MDA
CM	Cont	Port/Fixed	0.002-5	4995	Variable	CEA

NIOSH METHOD NO. P&CAM 209

Principle

Sampling is performed by passing a measured volume of air through a fritted bubbler containing 100 mL of dilute methyl orange. Near a pH of 3.0, the color of methyl orange solution ceases to vary with acidity. The dye is quantitatively bleached by free chlorine, and the extent of bleaching can be determined colorimetrically (references 1 and 6). The optimum concentration range is 0.05-1.0 ppm in ambient air (145-2900 µg/m³ at 25° C. and 760 Torr).

Range and Sensitivity

The procedure given is designed to cover the range of 5-100 µg of free chlorine/100 mL of sampling solution. For a 30 L air sample, this corresponds to approximately 0.05-1.0 ppm in air, which is the optimum range. Increasing the volume of air sampled will extend the range at the lower end, but only within limits, since 50 L of chlorine-free air produce the same effect as about 0.01 ppm of chlorine. By using a sampling solution more dilute in methyl orange, a concentration of 1 µg/100 mL of solution may be measured. But beyond this, problems are encountered because of the absorption of ammonia and other gases from the air and by the presence of minute amounts of chlorine-consuming materials even in distilled water.

Interferences

Free bromine, which gives the same reaction, interferes in a positive direction (reference 4). Manganese (III, IV) in concentrations of 0.1 ppm or above also interferes positively (reference 3). In the gaseous state, interference from SO_2 is minimal, but in solution, negative interference is significant. Nitrates impart an off-color orange to the methyl orange reagent. NO_2 interferes positively, reacting as 20% chlorine. SO_2 interferes negatively, decreasing the chlorine by an amount equal to one-third the SO_2 concentration. Ozone may also interfere positively (reference 2).

Precision and Accuracy

The data available (reference 5) indicate that 26 chlorine concentrations produced by two different methods (flowmeter calibrated by KI absorption, and gas-tight syringe) were measured by this procedure with an average error of less than 5% of the amount present.

Advantages and Disadvantages

The method is relatively simple with direct bleaching of the methyl orange reagent. Interfering substances in the air sample may affect the accuracy of the method. The sampled solutions must be protected from direct sunlight if the color is to be preserved for 24 hr.

Apparatus

Spectrophotometer. Suitable for measurement at 505 nm, preferably accomodating 5 cm cells.

Fritted bubbler. Coarse porosity, of 250-350 mL capacity. A small bubbler of 50-60 mL capacity may be more convenient for industrial hygiene sampling; volumes of reagents are then reduced proportionally.

Reagents

Reagents must be ACS analytical grade quality. Distilled water should conform to ASTM Standard for Referee Reagent Water.

Chlorine-demand-free water. Add sufficient chlorine to distilled water to destroy the ammonia. The amount of chlorine required will be about ten times the amount of ammoniacal nitrogen present. In no case should the initial residual be less than 1.0 mg/L free chlorine. Allow the chlorinated distilled water to stand overnight or longer, then expose to direct sunlight for one day or until all residual chlorine is discharged. A UV lamp may also be used to discharge the chlorine.

Methyl orange stock solution, 0.05%. Dissolve 0.500 g reagent grade methyl orange in distilled water and dilute to 1 L. This solution is stable indefinitely if freshly boiled and cooled distilled water is used.

Methyl orange reagent, 0.005%. Dilute 100 mL of stock solution to 1 L with distilled water. Prepare fresh for use.

Sampling solution. 6 mL of 0.005% methyl orange reagent is diluted to 100 mL with distilled water and 3 drops (0.15-0.20 mL) of 5.0 N HCl added. One drop of butanol may be added to induce foaming and increase collection efficiency.

Acidified water. To 100 mL of distilled water, add 3 drops (0.15-0.20 mL) of 5 N HCl.

Potassium dichromate solution, 0.1000 N. Dissolve 4.904 g anhydrous $K_2Cr_2O_7$, primary standard grade, in distilled water and dilute to 1 L.

Starch indicator solution. Prepare a thin paste of 1 g of soluble starch in a few mL of distilled water. Bring 200 mL of distilled water to a boil, remove from heat, and stir in the starch paste. Prepare fresh before use.

Potassium iodide, reagent grade.

Sodium thiosulfate solution, 0.1 N. Dissolve 25 g of $Na_2S_2O_3 \cdot 5H_2O$ in freshly boiled and cooled distilled water and dilute to 1 L. Add 5 mL chloroform as preservative and allow to age for 2 weeks before standardizing as follows: to 80 mL of distilled water, add, with constant stirring, 1 mL conc. sulfuric acid, 10.00 mL 0.1000 N $K_2Cr_2O_7$, and approximately 1 g of KI. Allow to stand in the dark for 6 min. Titrate with 0.1 N thiosulfate solution. Upon approaching the end-point (brown color changing to yellowish green), add 1 mL of starch indicator solution and continue titrating to the end-point (blue to light green).

Normality $Na_2S_2O_3 = 1.000 \div$ mL of $Na_2S_2O_3$ used

Sodium thiosulfate solution, 0.01 N. Dilute 100 mL of the aged and standardized 0.1 N sodium thiosulfate solution to 1 L with freshly boiled and cooled distilled water. Add 5 mL chloroform as preservative and store in a glass stoppered bottle. Standardize frequently with 0.0100 N potassium dichromate.

Chlorine solution, approximately 10 ppm. Prepare by serial dilution of household bleach (approx 50,000 ppm), or by dilution of strong chlorine water made by bubbling chlorine gas through cold distilled water. The diluted solution should contain approximately 10 ppm of free (available) chlorine. Prepare 1 L.

Procedure

Place 100 mL of sampling solution in the fritted bubbler. A measured volume of air is drawn through a rate of 1-2 L/min for a period of time appropriate to the estimated chlorine concentration. Transfer the solution to a 100 mL volumetric flask and make to volume, if necessary, with acidified water. Measure absorbance at 505 nm in 5 cm cells against distilled water as reference.

The volume of sampling solution, the concentration of methyl orange in the sampling solution, the amount of air sampled, the size of the absorbing vessel and the length of the photometer cell can be varied to suit the needs of the situation as long as proper attention is paid to the corresponding changes necessary in the calibration procedure.

Calibration and Standardization

Prepare a series of six 100 mL volumetric flasks containing 6 mL of 0.005% methyl orange reagent, 75 mL distilled water, and 3 drops (0.15-0.20 mL) of 5.0 N HCl. Carefully and accurately pipet 0, 0.5, 1.0, 5.0 and 9.0 mL of chlorine solution (approximately 10 ppm) into the respective flasks, holding the pipet tip beneath the surface. Quickly mix and make to volume with distilled water.

Immediately standardize the 10 ppm chlorine solution as follows: to a flask containing 1 gm KI and 5 mL glacial acetic acid, add 400 mL of chlorine solution, swirling to mix. Titrate with 0.01 N sodium thiosulfate until the iodine color becomes a faint yellow. Add 1 mL of starch indicator solution and continue the titration to the end-point (blue to colorless). One mL of 0.0100 N sodium thiosulfate = 0.3546 mg of free chlorine. Compute the amounts of free chlorine added to each flask in the previous step. Transfer the standards prepared earlier to absorption cells and measure absorbance vs. μg of chlorine to draw the standard curve.

Calculations

ppm Cl_2 = (mg Cl_2 found)(24,450) $\div$ (L of air sampled)(71)

For different temperatures and atmospheric pressures, proper correction for air volume should be made to standard conditions of 25°C. and 760 Torr.

References

1. Taras, M. Colorimetric determination of free chlorine with methyl orange. Anal. Chem. 19: 3-12, 1947.
2. Boltz, D.F. *Colorimetric Determination of Nonmetals,* Interscience Publishers, New York, 1958, p. 163.
3. *Standard Methods for the Examination of Water and Waste Water,* 12th ed., American Public Health Association, New York, 1965, p. 9.
4. Traylor, P.A. and S.A. Shrader. Determination of Small Amounts of Free Bromine in Air, Dow Chemical Company, Main Laboratory Reference MR4N, Midland, Michigan.

5. Thomas, M.D. and R. Amtower. Unpublished work.
6. Intersociety Committee. Methods of Air Sampling and Analysis, Analysis for Free Chlorine Content of the Atmosphere (42215-01-70T), American Public Health Association, Washington, D.C., 1972, pp. 282-284.

CHLOROBENZENE

Synonym: monochlorobenzene C_6H_5Cl
ACGIH TLV: 75 ppm (350 mg/m^3)
OSHA Standard: 75 ppm

Method	Sampling Duration	Sampling Location	Useful Range (ppm)	System Cost ($)	Test Cost ($)	Manufacturer
DT	Grab	Portable	5-500	150	3.00	Nat'l Draeger
DT	Grab	Portable	5-140	165	1.70	Matheson
DT	Grab	Portable	5-350	180	2.25	Bendix
IR	Cont	Portable	1-150	4374	0	Foxboro
PI	Cont	Fixed	2-200	4950	0	HNU
GC	1 min	Portable	1-10^6	10,000	0	Microsensor

NIOSH METHOD NO. P&CAM 127

(Use as described for acetone, with 105° C. column temperature)

CHLOROBROMOMETHANE

Synonym: bromochloromethane CH_2ClBr
ACGIH TLV: 200 ppm (1050 mg/m^3)
OSHA Standard: 200 ppm

Method	Sampling Duration	Sampling Location	Useful Range (ppm)	System Cost ($)	Test Cost ($)	Manufacturer
DT	Grab	Portable	8-350	180	2.25	Bendix
IR	Cont	Portable	4-400	4374	0	Foxboro
PI	Cont	Fixed	0.2-20	4950	0	HNU

NIOSH METHOD NO. P&CAM 127

(Use as described for acetone, with 80° C. column temperature)

CHLOROFORM

Synonym: trichloromethane $\qquad$ $CHCl_3$
ACGIH TLV: 10 ppm (50 mg/m³) - Suspected carcinogen
OSHA Standard: 50 ppm ceiling
NIOSH Recommendation: 2 ppm ceiling (60 min)

Method	Sampling Duration	Sampling Location	Useful Range (ppm)	System Cost ($)	Test Cost ($)	Manufacturer
DT	Grab	Portable	2-10	150	2.60	Nat'l Draeger
DT	Grab	Portable	23-500	165	3.40	Matheson
DT	Grab	Portable	4-400	180	2.25	Bendix
IR	Cont	Portable	1-100	4374	0	Foxboro
PI	Cont	Fixed	0.2-20	4950	0	HNU
GC	1 min	Portable	$1-10^6$	10,000	0	Microsensor

NIOSH METHOD NO. P&CAM 127

(See acetone)

1-CHLORO-1-NITROPROPANE

ACGIH TLV: 2 ppm (10 mg/m³) $\qquad$ $CH_3CH_2CHClNO_2$
OSHA Standard: 20 ppm

Method	Sampling Duration	Sampling Location	Useful Range (ppm)	System Cost ($)	Test Cost ($)	Manufacturer
IR	Cont	Portable	2-40	4374	0	Foxboro

NIOSH METHOD NO. S211

Principle

A known volume of air is drawn through a tube containing Chromosorb 108 to trap the organic vapors present. The sampling tube consists of a front adsorbing section and a backup section. The Chromosorb 108 in each tube is transferred to respective vials and the 1-chloro-1-nitropropane is desorbed with ethyl acetate. An aliquot of this sample solution is injected into a gas chromatograph equipped with a flame-ionization detector. The area of the resulting peak is determined and compared with areas obtained from the injection of standards.

Range and Sensitivity

This method was validated over the range of 50.8-206.4 mg/m^3 at atmospheric temperatures of 24.5 and 22.1° C., and atmospheric pressures of 765.4 and 767.9 mm Hg using a 12 L sample volume. The method may be capable of measuring smaller amounts if the desorption efficiency is adequate. Desorption efficiency must be determined over the range used.

The upper limit of the range of the method is dependent on the adsorptive capacity of the Chromosorb 108. This capacity varies with the concentrations of 1-chloro-1-nitropropane and other substances in the air. When an atmosphere at 90% relative humidity containing 218.0 mg/m^3 of 1-chloro-1-nitropropane was sampled at 0.2001 L/min, 5% breakthrough was observed after 111 min (capacity = 22.21 L or 4.84 mg). The sample size recommended is less than two-thirds the 5% breakthrough capacity to minimize the probability of overloading the sampling tube. The detection limit was not rigorously determined.

Interferences

When two or more compounds are known or suspected to be present in the air, such information, including their suspected identities, should be transmitted with the sample. It must be emphasized that any compound which has the same retention time as the analyte at the operating conditions described in this method is an interference. Retention time data on a single column cannot be considered as proof of chemical identity.

Precision and Accuracy

The coefficient of variation for the total analytical and sampling method in the range of 50.8-206.4 mg/m^3 was 0.0941. This value corresponds to a 9.41 mg/m^3 standard deviation at the OSHA standard level. Statistical information and details of the validation and experimental test procedures can be found in references 1 and 2.

In validation experiments, this method was found to be capable of coming within 25% of the "true value" on the average 95% of the time over the validation range. The concentrations measured at 0.5, 1, and 2 times the OSHA standard were 2.7% lower than the dynamically generated test concentrations (n=18). The desorption efficiency was determined to be 0.912 for a collector loading of 0.605 mg. In storage stability studies, the mean of samples analyzed after 7 days were within 1.7% of the mean of samples analyzed immediately after collection. Experiments performed in the validation study are described in reference 2.

Advantages and Disadvantages

The sampling device is small, portable, and involves no liquids. Interferences are minimal, and most of those which do occur can be eliminated by altering chromatographic conditions. The collected samples are analyzed by means of a quick, instrumental method.

One disadvantage of the method is that the amount of sample that can be taken is limited by the number of mg that the tube will hold before overloading. When the amount of 1-chloro-1-nitropropane found on the backup Chromosorb 108 section exceeds 25% of that found on the front section, the probability of sample loss exists.

The precision of the method is affected by the reproducibility of the pressure drop across the tubes. This drop will affect the flow rate and may cause the volume to be imprecise because the pump is usually calibrated for one tube only.

Apparatus

Sampling pump. A calibrated personal sampling pump suitable for sampling at 0.2 L/min for 60 min. The pump must be accurate to within 5% at the recommended flow rate.

Sampling tubes. The sampling tube consists of a glass tube, flame-sealed at both ends, 10 cm long with a 10 mm O.D. and 8 mm I.D., packed with two sections of 60/80 mesh cleaned Chromosorb 108. The front adsorbing section contains 400 mg and the backup section contains 200 mg. The two sections are separated by a portion of silylated glass wool. A plug of silylated glass wool is placed at each end of the sorbent tube. The pressure drop across the tube must be less than one inch of mercury at a flow rate of 0.2 L/min.

Gas chromatograph. With a flame-ionization detector.

Column. 20 ft × 1/8 in stainless steel, packed with 10% FFAP stationary phase on 100/200 mesh Supelcoport.

Electronic integrator. Or some other suitable method for measuring peak areas.

Syringes. 10 and 100 μL, and other convenient sizes for making standards and for taking sample aliquots.

Pipettes. 2 mL delivery type.

Volumetric flasks. 10 mL or other convenient sizes for making standard solutions.

Sample vials. 5 mL with Teflon-lined screw caps.

Reagents

1-Chloro-1-nitropropane, practical grade.

Ethyl acetate, reagent grade.

1-Heptanol, 99%. Or other suitable internal standard. The appropriate solution of the internal standard is prepared in ethyl acetate.

Pre-cleaned resin. Chromosorb 108 resin (60/80 mesh) is washed at least three times with methylene chloride in a separatory funnel using approximately 200 mL of solvent per 10 g of resin. The resin is washed next with ethyl acetate in a beaker of appropriate volume. The resin is then air-dried in a hood.

Nitrogen, purified.

Hydrogen, prepurified.

Air, filtered, compressed.

Procedure

Cleaning of equipment. All glassware used for the laboratory analysis should be detergent washed and thoroughly rinsed with tap water and distilled water.

Calibration of personal pumps. Each personal sampling pump must be calibrated with a representative sorbent tube in the line. This will minimize errors associated with uncertainties in the sample volume collected.

Collection and shipping of samples. Immediately before sampling, the ends of the tubes should be broken so as to provide openings approximately one-half the internal diameter of the tubes (4 mm). The section containing 200 mg of Chromosorb 108 is used as a backup and should be positioned nearest the sampling pump. The Chromosorb 108 tube should be maintained in a vertical position during sampling to avoid channeling and subsequent premature breakthrough of the analyte.

Air being sampled should not be passed through any hose or tubing before entering the front section of the Chromosorb 108 tube. A sample size of 12 L is recommended. Sample at a known flow rate between 0.2 and 0.4 L/min. Set the flow rate as accurately as possible using the manufacturer's directions. Record the necessary information to determine flow rate and also record the initial and final sampling time. Record the temperature and pressure of the atmosphere being sampled. If pressure reading is not available, record the elevation.

The Chromosorb 108 tubes should be labeled properly and capped with the supplied plastic caps immediately after sampling. One Chromosorb 108 tube should be handled in the same manner as the sample tubes (break, seal, and transport), except for the taking of an air sample. This tube should be labeled as a blank. Submit one blank for every batch or partial batch of ten samples. A sufficient number of unused Chromosorb 108 tubes should be available for use in desorption efficiency studies in conjunction with these samples, because desorption efficiency may vary from one batch of Chromosorb 108 to another. Record the batch number of the Chromosorb 108 used. Capped Chromosorb 108 tubes should be packed tightly and padded before they are shipped to minimize tube breakage during shipping.

Preparation of samples. In preparation for analysis, each tube is scored with a file and broken open. The glass wool is removed and discarded. The Chromosorb 108 in each tube is transferred to a 5 mL screw-cap sample vial. Each tube is analyzed separately.

Desorption of sample. Prior to analysis, 2.0 mL of ethyl acetate is pipetted into each sample vial. Desorption should be done for 30 min. Tests indicate that this is adequate if the sample is agitated occasionally during this period. The sample vials should be capped as soon as the solvent is added to minimize volatilization. For the internal standard method, desorb using 2.0 mL of ethyl acetate containing a known amount of internal standard.

GC conditions. The typical operating conditions for the gas chromatograph are:

30 mL/min (60 psig) nitrogen carrier gas flow
30 mL/min (25 psig) hydrogen gas flow to detector
300 mL/min (60 psig) air flow to detector
225° C. injector temperature
250° C. detector temperature
130° C. column temperature

A retention time of approximately 11 min is to be expected for the analyte using these conditions and the column recommended. The internal standard elutes in approximately 18 min.

Injection. A 2 μL aliquot of the sample solution is injected into the gas chromatograph. The solvent flush method or other suitable alternative such as an automatic sample injector can be used provided that duplicate injections of a solution agree well. No more than a 3% difference in area is to be expected.

Measurement of area. The signal of the sample peak is measured by an electronic integrator or some other suitable form of measurement such as peak height, and preliminary results are read from a standard curve prepared as discussed later.

Determination of desorption efficiency. The desorption efficiency of a particular compound may vary from one laboratory to another and also from one batch of Chromosorb 108 to another. Thus, it is necessary to determine the percentage of the specific compound that is removed in the desorption process for a particular batch of resin used for sample collection and over the concentration range of interest. The desorption efficiency must be determined over the sample concentration range of interest in order to determine the range which should be tested, the samples are analyzed first and then the analytical samples are prepared based on the amount of 1-chloro-1-nitropropane found in the samples.

The analytical samples are prepared as follows: Chromosorb 108, equivalent to the amount in the front section (400 mg), is measured into a 5 mL screw-cap vial. This resin must be from the same batch used in obtaining the samples. A known amount of the solution of 1-chloro-1-nitropropane in ethyl acetate (spiking solution) is injected directly into the resin by means of a μL syringe. Adjust the concentration of the spiking solution such that no more than a 10 μL aliquot is used to prepare the analytical samples.

Six analytical samples at each of the three concentration levels (0.5, 1, and 2 times the OSHA standard) are prepared by adding an amount of 1-chloro-1-nitropropane equivalent to a 12 L sample at the selected level. A stock solution containing 241.8 mg of 1-chloro-1-nitropropane/mL of ethyl acetate is prepared. Aliquots (2.5, 5.0 and 10.0 μL) of the solution are added to the Chromosorb 108 vials to produce 0.5, 1, and 2 times the OSHA standard level. The analytical samples are allowed to stand overnight to assure complete adsorption of the analyte onto the sorbent. A parallel blank vial is treated in the same manner except that no sample is added to it.

Desorption and analysis. Desorption and analysis experiments are done on the analytical samples as described above. Calibration standards are prepared by adding the appropriate volume of spiking solution to 2.0 mL of ethyl acetate with the same syringe used in the preparation of the samples. Standards should be prepared and analyzed at the same time the sample analysis is done. If the internal standard method is used, prepare calibration standards by using 2.0 mL of ethyl acetate containing a known amount of the internal standard.

The desorption efficiency (D.E.) equals the average weight in mg recovered from the vial divided by the weight in mg added to the vial, or:

D.E. = (average wt in mg recovered - resin blank in mg) ÷ weight in mg added

The desorption efficiency may be dependent on the amount of 1-chloro-1-nitropropane collected on the sorbent. Plot the desorption efficiency versus weight of 1-chloro-1-nitropropane found. This curve is used later to correct for adsorption losses.

Calibration and Standardization

A series of standards varying in concentration over the range corresponding to 12 L collections at 0.1-3 times the OSHA standard is prepared and analyzed under the same GC conditions and during the same time period as the unknown samples. This is done in order to minimize variations in FID response. It is convenient to express concentration of standards in terms of mg/2.0 mL since the samples are desorbed in 2.0 mL of ethyl acetate. A calibration curve is established by plotting peak area versus concentration in mg/2.0 mL.

Prepare a stock standard solution containing about 240 mg/mL of 1-chloro-1-nitropropane in ethyl acetate. From this stock solution, appropriate aliquots are added to 2.0 mL of ethyl acetate. Prepare at least five standards to cover the range of 0.12-3.6 mg/sample. The range is based on a 12 L air sample.

For the internal standard method, use ethyl acetate containing a predetermined amount of the internal standard. The internal standard concentration used should be approximately 70% of the analyte concentration for a standard solution representing a 12 L collection at 2 times the OSHA standard. The area ratio of the analyte to that of the internal standard is plotted against the analyte concentration in mg/2.0 mL.

Calculations

Read the weight, in mg, corresponding to each peak area from the standard curve. No volume corrections are needed, because the standard curve is based on mg/2.0 mL and the volume of the sample injected is identical to the volume of the standards injected. Corrections for the blank must be made for each sample:

mg = mg sample - mg blank

where:

    mg sample = mg found in sample vial
    mg blank = mg found in blank vial

A similar procedure is followed for the backup sections. Add the weights found in the front and backup sections to get the total weight in the sample. Read the desorption efficiency from the curve for the amount found in the front section. Divide the total weight by this desorption efficiency to obtain the corrected mg/sample:

Corrected mg/sample = total weight ÷ D.E.

Determine the volume of air samples at ambient conditions in L based on the appropriate information, such as flow rate in L/min multiplied by sampling time. If a pump using a rotameter for flow rate control was used for sample collection, a pressure and temperature correction must be made for the indicated flow rate. The expression for this correction is:

Corrected volume = $ft(P_1T_2/P_2T_1)^{1/2}$

where:

    f = sampling flow rate
    t = sampling time
    P_1 = pressure during calibration of sampling pump (mm Hg)
    P_2 = pressure of air sampled (mm Hg)
    T_1 = temperature during calibration of sampling pump (°K)
    T_2 = temperature of air sampled (°K)

The concentration of the analyte in the air sampled can be expressed in mg/m^3, which is numerically equal to $\mu g/L$:

mg/m^3 = (corrected mg)(1000 L/m^3) ÷ (air volume sampled in L)

Another method of expressing concentration is ppm (corrected to standard conditions of 25° C. and 760 mm Hg):

ppm = (mg/m^3)(24.45)(760)(T+273) ÷ (123.54)(P)(298)

where:

    P = pressure (mm Hg) of air sampled
    T = temperature (°C.) of air sampled
    24.45 = molar volume (L/mol) at 25° C. and 760 mm Hg
    123.54 = molecular weight of 1-chloro-1-nitropropane
    760 = standard pressure (mm Hg)
    298 = standard temperature (°K)

References

1. *Documentation of NIOSH Validation Tests,* National Institute for Occupational Safety and Health, Cincinnati, Ohio (DHEW-NIOSH Publication No. 77-185), 1977. Available from Superintendent of Documents, U.S. Government Printing Office, Washington, D.C., Order No. 017-033-00231-2.
2. Backup Data Report for 1-Chloro-1-nitropropane, No. S211, prepared under NIOSH Contract No. 210-76-0123.

CHLOROPRENE

Synonym: 2-chloro-1,3-butadiene $CH_2=CClCH=CH_2$
ACGIH TLV: 10 ppm (45 mg/m^3) - Skin
OSHA Standard: 25 ppm
NIOSH Recommendation: 1 ppm ceiling (15 min)

Method	Sampling Duration	Sampling Location	Useful Range (ppm)	System Cost ($)	Test Cost ($)	Manufacturer
DT	Grab	Portable	5-90	150	2.60	Nat'l Draeger
DT	4 hr	Personal	1-100	850	3.10	Nat'l Draeger
IR	Cont	Portable	1-50	4374	0	Foxboro
PI	Cont	Fixed	0.2-20	4950	0	HNU

NIOSH METHOD NO. S112

Principle

A known volume of air is drawn through a charcoal tube containing activated coconut charcoal to trap chloroprene vapors. Chloroprene is desorbed from the charcoal with carbon disulfide, and the sample is analyzed by gas chromatography.

Range and Sensitivity

This method was validated over the range of 44.2-173.9 mg/m^3 at an atmospheric temperature of 21° C. and atmospheric pressure of 760 mm Hg, using a 3 L sample. This maximum sample size is based on the capacity of the charcoal to collect vapors of chloroprene in air. The method may be capable of measuring smaller amounts if the desorption efficiency is adequate. Desorption efficiency must be determined over the range used.

The upper limit of the range of the method depends on the adsorptive capacity of the charcoal. This capacity may vary with the concentrations of chloroprene and other substances in the air. Breakthrough is defined as the time that the effluent concentration from the collection tube (containing 100 mg of charcoal) reaches 5% of the concentration in the test gas mixture. The criterion for acceptance is that the volume of air that has passed through the tube at the time of breakthrough must be greater than 1.5 times the volume of air that would be passed through the tube for a sample at 2 times the OSHA standard level. Breakthrough did not occur after sampling for 4 hr at an average sampling rate of 0.045 L/min and relative humidity of 91% and temperature of 25° C. The breakthrough test was conducted at a concentration of 197 mg/m^3.

Interferences

When other compounds are known or suspected to be present in the air, such information, including their suspected identities, should be transmitted with the sample. Any compound that has the same retention time as chloroprene at the operating conditions described in this method is an interference. Retention time data on a single column cannot be considered proof of chemical identity.

Precision and Accuracy

The coefficient of variation for the total analytical and sampling method in the range of 44.2-173.9 mg/m^3 was 0.071. This value corresponds to a 6.4 mg/m^3 standard deviation at the OSHA standard level. Statistical information can be found in reference 2. Details of the test procedures are found in reference 3.

No recovery correction should be applied to the final result. On the average, the concentrations obtained in the laboratory validation study at 0.5X, 1X, and 2X the OSHA standard level were 1.2% lower than the "true" concentrations for 18 samples. Any difference between the "found" and "true" concentrations may not represent a bias in the sampling and analytical method, but rather a random variation from the experimentally determined "true" concentration. Therefore, the method has no bias. The coefficient of variation is a good measure of the accuracy of the method since the recoveries and storage stability were good. Storage stability studies on samples collected from a test atmosphere at a concentration of 86.0 mg/m^3 indicate that collected samples are stable for at least 7 days.

Advantages and Disadvantages

The sampling device is small, portable, and involves no liquids. Interferences are minimal, and most of those that occur can be eliminated by altering chromatographic conditions. The tubes are analyzed by means of a quick, instrumental method.

One disadvantage of the method is that the amount of sample that can be taken is limited by the number of mg that the tube will hold before overloading. When the amount of chloroprene found on the backup section of the charcoal tube exceeds 25% of that found on the front section, the probability of sample loss exists.

The precision of the method is limited by the reproducibility of the pressure drop across the tubes. This drop will affect the flow rate and cause the volume to be imprecise, because the pump is usually calibrated for one tube only.

Apparatus

Personal sampling pump. A calibrated personal sampling pump whose flow can be determined within 5% at the recommended flow rate.

Charcoal tubes. Glass tube with both ends flame sealed, 7 cm long with 6 mm O.D. and 4 mm I.D., containing 2 sections of 20/40 mesh activated charcoal separated by a 2 mm portion of urethane foam. The activated charcoal is prepared from coconut shells and is fired at 600° C. prior to packing. The adsorbing section contains 100 mg of charcoal, the backup section 50 mg. A 3 mm portion of urethane foam is placed between the outlet end of the tube and the backup section. A plug of silylated glass wool is placed in front of the adsorbing section. The pressure drop across the tube must be less than 25 mm of mercury at a flow rate of 1 L/min.

Gas chromatograph. Equipped with a flame-ionization detector.

Column. 4 ft × 1/4 in O.D. stainless steel, packed with 50/80 mesh Porapak Q.

Electronic integrator. Or some other suitable method for measuring peak area.

Sample containers. 2 mL glass sample containers with glass stoppers or Teflon-lined caps. If an automatic sample injector is used, the sample injector vials can be used.

Syringes. 10 μL, and other convenient sizes for making standards.
Pipets. Delivery type, 1.0 mL and other convenient sizes.
Volumetric flasks. 10 mL and other convenient sizes for making standard solutions.
Micro-distillation apparatus. With provision for vacuum fractional distillation of pure chloroprene for making standards.
Stopwatch.
Manometer.

Reagents

Carbon disulfide, chromatographic quality.
Chloroprene, reagent grade in xylene solution.
n-Pentane, reagent grade.
n-Hexane, reagent grade.
Nitrogen, purified.
Hydrogen, prepurified.
Air, filtered, compressed.

Procedure

Cleaning of equipment. All glassware used for the laboratory analysis should be detergent washed and thoroughly rinsed with tap water and distilled water, and dried.

Calibration of sampling pumps. Each personal sampling pump must be calibrated with a representative charcoal tube in the line. This will minimize errors associated with uncertainties in the sample volume collected.

Collection and shipping of samples. Immediately before sampling, break the ends of the charcoal tube to provide an opening at least one-half the internal diameter of the tube. All tubes must be from the same manufacturer's lot. The smaller section of charcoal is used as a backup and should be positioned nearest the sampling pump. The charcoal tube should be placed in a vertical direction during sampling to minimize channeling through the charcoal.

Air being sampled should not be passed through any hose or tubing before entering the charcoal tube. A maximum sample size of 3 L is recommended. Sample at a flow rate between 0.01 and 0.05 L/min. The collection time and the flow rate with an accuracy of 5% or better should be recorded. Do not sample at a flow rate less than 0.010 L/min. The temperature, pressure and relative humidity of the atmosphere being sampled should be recorded. If pressure reading is not available, record the elevation.

The charcoal tube should be capped with plastic caps immediately after sampling. Under no circumstances should rubber caps be used. With each batch of ten samples, submit one tube from the same lot of tubes used for sample collection. This tube must be subjected to exactly the same handling as the samples except that no air is drawn through it. This tube should be labeled as a blank. Capped tubes should be packed tightly and padded before they are shipped to minimize tube breakage during shipping.

Preparation of samples. In preparation for analysis, each charcoal tube is scored with a file in front of the first section of charcoal and broken open. The glass wool is removed and discarded. The charcoal in the first (larger) section is transferred to a 2 mL stoppered sample container or automatic sample injector vial. The separating section of foam is removed and discarded; the second section is transferred to another sample container or vial. These two sections are analyzed separately.

Desorption of samples. Prior to analysis, 1.0 mL of the eluent is pipetted into each sample container. (All work with carbon disulfide should be performed in a hood because of its high toxicity.) Desorption should be done for 30 min. Tests indicate that this is adequate if the sample is agitated occasionally during this period. The sample vials should be capped as soon as the solvent is added to minimize volatilization.

GC conditions. The typical operating conditions for the gas chromatograph are:

50 mL/min (60 psig) nitrogen carrier gas flow
50 mL/min (24 psig) hydrogen gas flow to detector
500 mL/min (50 psig) air flow to detector
200° C. injector temperature
250° C. detector temperature
125° C. column temperature

A retention time of approximately 10 min is to be expected for chloroprene under these conditions and using the column recommended. The carbon disulfide will elute from the column before the chloroprene.

Injection. The first step in the analysis is the injection of the sample into the gas chromatograph. To eliminate difficulties arising from blow back or distillation within the syringe needle, one should employ the solvent flush injection technique. The 10 μL syringe is first flushed with solvent several times to wet the barrel and plunger. Three μL of solvent are drawn into the syringe to increase the accuracy and reproducibility of the injected sample volume. The needle is removed from the solvent, and the plunger is pulled back about 0.2 μL to separate the solvent flush from the sample with a pocket of air to be used as a marker. The needle is then immersed in the sample, and a 5 μL aliquot is withdrawn, taking into consideration the volume of the needle, since the sample in the needle will be completely injected. After the needle is removed from the sample and prior to injection, the plunger is pulled back 1.2 μL to minimize evaporation of the sample from the tip of the needle. Observe that the sample occupies 4.9-5.0 μL in the barrel of the syringe. Duplicate injections of each sample and standard should be made. No more than a 3% difference in area is to be expected. An automatic sample injector can be used if it is shown to give reproducibility at least as good as the solvent flush technique.

Measurement of area. The area of the sample peak is measured by an electronic integrator or some other suitable form of area measurement, and results are read from a standard curve prepared as discussed below.

Determination of desorption efficiency. The desorption efficiency of a particular compound can vary from one laboratory to another and also from one batch of charcoal to another. Thus, it is necessary to determine the fraction of the specific compound that is removed in the desorption process for a particular batch of charcoal.

Activated charcoal equivalent to the amount in the first section of the sampling tube (100 mg) is measured into a 64 mm × 4 mm I.D. glass tube, flame sealed at one end. This charcoal must be from the same batch as that used in obtaining the samples and can be obtained from unused charcoal tubes. The open end is capped with Parafilm. A known amount of freshly prepared pentane solution of chloroprene containing 67.9 μg/μL is injected directly into the activated charcoal with a μL syringe, and the tube is capped with more Parafilm. When using an automatic sample injector, the sample vials, capped with Teflon-faced septa, may be used in place of glass tubes. The amount injected is equivalent to that present in a 3 L air sample at the selected level. It is not practical to inject the neat liquid directly onto the charcoal because the amounts to be added would be too small to measure accurately.

At least six tubes at each of three levels (0.5X, 1X, and 2X the OSHA standard) are prepared in this manner and allowed to stand for at least overnight to assure complete adsorption of the chloroprene onto the charcoal. These six tubes are referred to as the samples. A parallel blank tube should be treated in the same manner except that no sample is added to it. The sample and blank tubes are desorbed and analyzed in exactly the same manner as the sampling tube described earlier.

Two or three standards are prepared by injecting the same volume of chloroprene into 1.0 mL of carbon disulfide with the same syringe used in the preparation of the samples. These are analyzed with the samples. These standards are used to confirm the calibration of the gas chromatograph. The desorption efficiency (D.E.) equals the average weight in mg recovered from the tube divided by the weight in mg added to the tube or:

D.E. = average mg recovered ÷ mg added

The desorption efficiency is dependent on the amount of chloroprene collected on the charcoal. Plot the desorption efficiency versus the weight of chloroprene found. The curve is used later to correct for adsorption losses.

Calibration and Standardization

A series of standards, varying in concentration over the range is prepared and analyzed under the same GC conditions and during the same time period as the unknown samples. Curves are established by plotting concentration in mg/1.0 mL versus peak area. Note: since no internal standard is used in this method, standard solutions must be analyzed at the same time that the sample analysis is done. This will minimize the effect of known day-to-day variations and variations during the same day of the FID response.

Calculations are based on a molecular weight of 88.54 and a density of 0.958 for pure chloroprene and 25 ppm (90.53 mg/m^3) for the OSHA standard. Since chloroprene polymerizes readily, special precautions must be taken in the preparation and storage of standards.

Stock standard solution. Pure chloroprene is obtained by fractionally distilling commercially available 50% chloroprene in xylene solution under vacuum. (The chloroprene used in the laboratory validation study was distilled at 31° C. at a pressure of 354 mm Hg.) A stock standard solution is prepared from freshly distilled chloroprene. Exactly 1.0 mL (0.958 g at 20° C.) of pure chloroprene is delivered from a delivery type pipet under the surface of pentane in a partially filled 10 mL volumetric flask and then the solution is made up to exactly 10 mL with pentane. This solution may be stable for one day or even longer if stored at -15° C. (Since chloroprene tends to polymerize, even in solution, it may be necessary to monitor its concentration in the standard solutions.)

Working standard solution. Aliquots of the stock standard solution are delivered below the surface of carbon disulfide in partially filled 10 mL volumetric flasks. Each solution is diluted to exactly 10 mL with carbon disulfide and carefully mixed. A calibration curve should be prepared for the concentration range representing 0.1 to 3 times the OSHA standard.

Solutions for desorption efficiency tests. Exactly 0.71 mL of pure chloroprene or an appropriate aliquot of the stock standard solution is delivered from appropriate pipets below the surface of pentane in a partially filled 10 mL volumetric flask and the solution is made up to exactly 10 mL with pentane. Appropriate aliquots are used for desorption efficiency tests. For a 3 L air sample, aliquots of 2, 4, and 8 µL represent 0.5X, 1X, and 2X the OSHA standard.

Standards should be prepared immediately from freshly distilled chloroprene and stored at -15° C. when not in use. A reference standard of hexane in carbon disulfide in a sealed vial with septum cap can be used to monitor the stability of the chloroprene standards. The concentration of the hexane reference standard should be chosen so that its flame-ionization detector response is close to that of the chloroprene standards. When the ratio of the concentration of chloroprene standards to reference standard appears to decrease, prepare new standards.

Chloroprene

Calculations

Read the weight, in mg, corresponding to each peak area from the standard curve. No volume corrections are needed, because the standard curve is based on mg/1.0 mL carbon disulfide and the volume of sample injected is identical to the volume of the standards injected. Corrections for the blank must be made for each sample:

mg = mg sample - mg blank

where:

mg sample = mg found in front section of sample tube
mg blank = mg found in front section of blank tube

A similar procedure is followed for the backup sections. Add the weights found in the front and backup sections to get the total weight in the sample. Read the desorption efficiency from the curve for the amount found in the front section. Divide the total weight by this desorption efficiency to obtain the corrected mg/sample:

Corrected mg/sample = total weight ÷ D.E.

For personal sampling pumps with rotameters only, the following correction should be made:

Corrected volume = $ft(P_1T_2/P_2T_1)^{1/2}$

where:

f = flow rate sampled
t = sampling time
P_1 = pressure during calibration of sampling pump (mm Hg)
P_2 = pressure of air sampled (mm Hg)
T_1 = temperature during calibration of sampling pump (°K)
T_2 = temperature of air sampled (°K)

The concentration of chloroprene in the air sampled can be expressed in mg/m^3:

mg/m^3 = (corrected mg)(1000 L/m^3) ÷ (air volume sampled in L)

Another method of expressing concentration is ppm:

ppm = (mg/m^3)(24.45)(760)(T+273) ÷ (M.W.)(P)(298)

where:

P = pressure (mm Hg) of air sampled
T = temperature (°C.) of air sampled
24.45 = molar volume (L/mol) at 25° C. and 760 mm Hg
M.W. = molecular weight (g/mol) of chloroprene
760 = standard pressure (mm Hg)
298 = standard temperature (°K)

CHROMIC ACID

Synonym: chromate CrO_3
ACGIH TLV: 0.05 mg/m³
OSHA Standard: 0.1 mg/m³ ceiling
NIOSH Recommendation: 0.05 mg/m³

Method	Sampling Duration	Sampling Location	Useful Range (mg/m³)	System Cost ($)	Test Cost ($)	Manufacturer
DT	Grab	Portable	0.1-0.5	150	3.30	Nat'l Draeger

NIOSH METHOD NO. S317

Principle
A known volume of air is drawn through a polyvinyl chloride (PVC) filter to collect the analyte. The analyte is extracted from the filter with 0.25 M sulfuric acid, diphenylcarbazide is added, and additional acid is added to bring the volume to 20 mL. The absorbance of the solution at 540 nm is used as a quantitative measure of the analyte.

Range and Sensitivity
This method was validated over the range of 0.052-0.204 mg/m³ at an atmospheric temperature and pressure of 22° C. and 761 mm Hg, using a 15 min sample. Under the conditions of the sample size (22.5 L), the linear working range of the method is estimated to be 0.025-0.6 mg/m³ CrO_3. When using a 5 cm cell and a final volume of 20 mL for a 22.5 L sample, the sensitivity is estimated to be 0.01 mg/m³.

Interferences
Possible interferences for the diphenylcarbazide method include many of the heavy metals. The ones likely to be encountered at appreciable levels are iron, copper, nickel, and vanadium. Tests show that 10 µg of any of these give a response less than or equal to that obtained from 0.04 µg CrO_3.

Precision and Accuracy
The coefficient of variation for the total analytical and sampling method in the range of 0.052-0.204 mg/m³ was 0.084. This value corresponds to a 0.008 mg/m³ standard deviation at the

OSHA standard level. Statistical information and details of the validation and experimental test procedures can be found in reference 4.

A collection efficiency of 0.945, standard deviation of 0.035, was determined for the collection medium and a correction for collection efficiency should be made. The coefficient of variation is a satisfactory measure of both accuracy and precision of the sampling and analytical method since there are no apparent biases for which corrections have not been made.

Advantages and Disadvantages

The method is simple, specific, and sensitive. The samples collected on PVC filters are stable.

Apparatus

Filter unit. Consists of the filter media and appropriate cassette filter holder, either a two or three piece filter cassette.

Personal sampling pump. A personal sampling pump whose flow can be determined to an accuracy of 5% at a sampling rate of 1.5 L/min. This pump must be properly calibrated so the volume of air sampled can be measured as accurately as possible. The pump must be calibrated with a representative filter unit in the line.

Thermometer.
Manometer.
Stopwatch.
PVC filters. VMI (Gelman) 37 mm diameter and 5.0 micron pore size or equivalent.
Spectrophotometer. This instrument should be capable of measuring the developed color at 540 nm.
Matched glass cells or cuvettes, 5 cm path length.
Polypropylene or Teflon forceps.
Assorted laboratory glassware. Pipets, volumetric flasks, large test tubes, and graduated cylinders of appropriate capacities.
Screw cap bottles. Within 1 hour after the sample has been collected, the filter is transferred to a clean screw cap bottle (a 45 mm tissue sample holder is satisfactory) for shipping.
Buchner funnel.

Reagents

All reagents must be analytical reagent grade or better.

Sulfuric acid, 0.25 M. Add 14 mL of concentrated sulfuric acid to some distilled water in a 1 L volumetric flask and dilute to mark. For uniformity of results it is suggested that the same solution be used for all samples, blanks, and standards. After thorough mixing, it is convenient to transfer part of the solution to a small plastic wash bottle.

Diphenylcarbazide. Dissolve 0.50 g of sym-diphenylcarbazide in a mixture of 100 mL of acetone and 100 mL of double distilled water. Store in a dark bottle in the refrigerator. This solution is good for up to one month.

Chromium (VI) standard. Dissolve 0.2829 g of potassium dichromate in water in a 1 L volumetric flask and dilute to mark. This solution is 100 ppm (by weight) of chromium (VI). The chromium (VI) concentration can also be expressed as 0.1 $\mu g/\mu L$.

Procedure

Cleaning of equipment. Wash all glassware in hot, soapy water, and follow with tap and distilled water rinses. After initial cleaning, soak glassware in concentrated nitric acid (10% nitric acid for plastics) for 30 min. Rinse thoroughly with tap water, distilled water, and double distilled water.

Calibration of personal sampling pumps. Each personal sampling pump must be calibrated with a representative filter cassette in the line. This will minimize errors associated with uncertainties in the sample volume collected.

Collection and shipping of samples. Assemble the filter in the filter cassette holder and close firmly to insure that the center ring seals the edge of the filter. The PVC filter is held in place by a cellulose backup pad. Remove the cassette plugs and attach to the personal sampling pump tubing. Clip the cassette to the worker's lapel.

Air being sampled should not pass through any hose or tubing before entering the filter cassette. A 15 min sample is recommended. Sample at a flow rate of 1.5 L/min. The flow rate should be known with an accuracy of 5%. Turn the pump on and begin sample collection. Since it is possible for a filter to become plugged by heavy particulate loading or by the presence of oil mists or other liquids in the air, the pump rotameter should be observed frequently, and the sampling should be terminated by any evidence of a problem.

Terminate sampling at the predetermined time and note sample flow rate, collection time and ambient temperature and pressure. If pressure reading is not available, record the elevation. The PVC filter should be removed from the cassette filter holder and the cellulose backup pad within 1 hour of sampling and placed in a clean screw cap bottle. Care must be taken to handle the filter only with clean tweezers.

Carefully record the sample identity and all relevant sampling data. With each batch of ten samples, submit one filter from the same lot of filters which was used for sample collection and which is subjected to exactly the same handling as the samples except that no air is drawn through it. Label this as a blank. The screw cap bottles in which the samples are stored should be shipped in a suitable container, designed to prevent damage in transit.

Analysis of samples. Pipet 20 mL of water into each of six large test tubes. Put a piece of tape on the test tube so that the bottom of the tape matches the meniscus. Rinse and dry the test tubes.

To facilitate the analysis of the sample and to remove the possibility of interferences from suspended dusts, a simple apparatus may be constructed. A two-hold stopper sufficiently large to fit the test tubes, if altered by enlarging one hole, can accomodate a small Buchner funnel and vacuum line. Remove the filter from the screw cap bottle and place the filter in the Buchner funnel. Cover the filter with 5 mL 0.25 M sulfuric acid and allow to stand for 5 min. Rinse the screw cap bottle with 2-3 mL of 0.25 M sulfuric acid. Using the vacuum, wash the filter with another 5-7 mL of 0.25 M sulfuric acid and filter the solution into the test tube. After the solution has filtered through, remove the filter and wash the funnel with several mL of 0.25 M sulfuric acid. Standards should be set up along with each set of samples being analyzed. Also, a blank filter through which no air has been drawn, should be analyzed along with the samples.

Add 0.5 mL of the diphenylcarbazide solution to each test tube. Then add additional 0.25 M sulfuric acid until the meniscus matches the bottom edge of the tape. Shake the test tube to mix and transfer to a clean, dry 5 cm cell for analysis. Read the absorbance of the solution on a spectrophotometer at 540 nm against a blank prepared from 0.25 M sulfuric acid in the same fashion as for the samples.

Calibration and Standards

Transfer 6 or 7 mL of 0.25 M sulfuric acid into each of six 5 cm cells. Pipet 5, 10, 20, 50, and 100 μL of the 100 ppm standard into the six cells, respectively. Process one cell as a blank. Add 0.5 mL of the diphenylcarbazide solution and sufficient 0.25 M sulfuric acid to bring to the 20 mL mark. Shake and wipe clean. Adjust the baseline of the spectrophotometer to zero by reading distilled water. Read the blank, then read the five standards at 540 nm. A calibration curve is drawn by plotting the absorbance of the standards against μg of chromium (VI) in the colored solution.

Calculations

Subtract the absorbance of the blank from the absorbance of each sample. Determine from the calibration curve the μg of CrO_3 present in each sample. The concentration of the analyte in the air sampled can be expressed in mg/m^3 ($mg/m^3 = \mu g/L$):

$mg/m^3 = \mu g\ CrO_3 \div$ air volume sampled in L

Divide the concentration of the analyte (mg/m^3) by the collection efficiency. Note: the collection efficiency, when sampled at an atmosphere containing 0.192 mg/m^3 at 1.5 L/min was determined to be 0.945:

Corr mg/m^3 = concentration of analyte ÷ collection efficiency

References
1. Snell and Snell. *Colorimetric Methods of Analysis,* Volume 11A, D. Van Nostrand Company, pp. 212-215.
2. Jacobs, M.B. *The Analytical Toxicology of Industrial Poisons,* Interscience Publishers, pp. 396-402.
3. Abell, M.T. and J.R. Carlberg. A simple reliable method for determination of airborne hexavalent chromium. Amer. Ind. Hyg. Asso. J. 35: 229, 1974.
4. Documentation of NIOSH Validation Tests. Contract no. CDC-99-74-45.

COBALT DUST OR FUME

ACGIH TLV: 0.05 mg/m^3 Co
OSHA Standard: 0.1 mg/m^3

Method	Sampling Duration	Sampling Location	Useful Range (mg/m^3)	System Cost ($)	Test Cost ($)	Manufacturer
LS	Cont	Personal	0.01-100	1695	0	GCA
LS	Cont	Portable	0.001-100	2785	0	MDA
LS	Cont	Fixed	0.001-200	3990	0	GCA
BA	1-4 min	Portable	0.02-150	4890	0	GCA

NIOSH METHOD NO. S203

Principle
Cobalt metal dust and fume, collected on a mixed cellulose ester membrane filter, is dissolved from the sample container filter by treating with aqua regia at room temperature for 30 min followed by wet ashing using aqua regia and nitric acid to destroy the organic matrix. The cobalt is then solubilized in 5% nitric acid. The solutions of samples and standards are aspirated into an oxidizing air-acetylene flame of an atomic absorption (AA) spectrophotometer. A hollow cathode lamp for cobalt is used to provide a characteristic cobalt line at 240.7 nm. The absorbance is proportional to the cobalt concentration.

Range and Sensitivity

This method was independently validated for cobalt fume over the range of 0.031-0.221 mg/m^3, and for cobalt dust over the range of 0.040-0.262 mg/m^3 at an atmospheric temperature and pressure of 18-26° C. and 740-765 mm Hg, using a 270 L and 300 L sample, respectively.

For a sample size of 300 L, the working range of the method is estimated to be 0.01-0.30 mg/m^3. The method may be extended to higher values by further dilution of the sample solution; the absolute sensitivity of the method has not been established.

Interferences

There are no known spectral line interferences for the cobalt AA assay. Other particulate cobalt compounds would interfere with the analysis.

Precision and Accuracy

The coefficient of variation for the total analytical and sampling method in the range of 0.031-0.26 mg/m^3 was 0.070. This value corresponds to a 0.007 mg/m^3 standard deviation at the OSHA standard concentration. Statistical information and details of the validation and experimental test procedures can be found in references 1 and 2.

A collection efficiency of 100% was determined for the collection medium; thus, no bias was introduced in the sample collection step. There was also no bias in the analytical method; the average recovery from the filters was 102%. Thus, the coefficient of variation is a satisfactory measure of both accuracy and precision of the sampling and analytical method. This conclusion is supported by data on a separate set of samples in which the generator concentration was determined by an independent method (reference 1).

Advantages and Disadvantages

The method is simple and specific for cobalt. Care must be taken to initially treat the sample at room temperature with aqua regia before hot ashing so as to avoid passivation of cobalt metal.

Apparatus

Filter unit. Consists of the filter media and 37 mm 3 piece filter holder.

Personal sampling pump. A calibrated personal sampling pump whose flow can be determined to an accuracy of 5% at the recommended flow rate of 1.5 L/min. The pump must be calibrated with a filter holder and filter in the line.

Thermometer.

Barometer.

Stopwatch.

Mixed cellulose ester membrane filter. 37 mm diameter, 0.8 micron pore size.

Atomic absorption spectrophotometer. With a monochromator that has a reciprocal linear dispersion of about 6.5 Angstrom/mm in the ultraviolet region. The instrument must have the burner head for an air-acetylene flame.

Cobalt hollow cathode lamp.

Air.

Fuel. Purified acetylene.

Pressure regulators. Two-stage, for each compressed gas tank used.

Glassware, borosilicate.

50 mL beakers with watchglass covers.

Pipettes. Delivery or graduated, 1, 5, 10 mL and other convenient sizes for making standards.

10 mL volumetric flasks.

Hot plate. Adjustable thermostatically controlled, capable of reaching 400° C.

Reagents

All reagents used must be ACS reagent grade or better.

Water, distilled or deionized.

Nitric acid, concentrated.

Nitric acid, 5%.

Hydrochloric acid, concentrated.

Aqua regia. Mix 3 parts HCl and 1 part nitric acid.

Aqueous standard cobalt solution (1000 µg/mL). Commercially available. This solution can also be prepared by dissolving 1.000 g cobalt metal in a minimum amount of aqua regia and diluting to 1.0 L with 5% HNO_3.

Cobalt working standard solution, 60 µg/mL. Prepare by appropriate dilution of above solution. Prepare fresh each day.

Procedure

Cleaning of equipment. Before use, all glassware should be initially soaked in a mild detergent solution to remove any residual grease or chemicals. After initial cleaning, the glassware should be thoroughly rinsed with warm tap water, concentrated nitric acid, tap water, and distilled water, in that order, and then dried.

Collection and shipping of samples. To collect cobalt metal dust and fume, a personal sampling pump is used to pull air through a cellulose ester membrane filter. The filter holder is held together with tape or a shrinking band. If the middle piece of the filter holder does not fit snugly into the bottom piece of the filter holder, the contaminant will leak around the filter. A piece of flexible tubing is used to connect the filter holder to the pump. Sample at a flow rate of 1.5 L/min with face cap on and small plugs removed. After sampling, replace small plugs.

With each batch of ten samples submit one filter from the same lot of filters which was used for sample collection and which is subjected to exactly the same handling as for the samples except that no air is drawn through it. Label this as a blank. The filter holders should be shipped in a suitable container, designed to prevent damage in transit.

Analysis of samples. Open the filter holder and carefully remove the cellulose membrane filter from the holder and cellulose backup pad with the aid of appropriate tweezers. Transfer filter to a 50 mL beaker.

To solubilize the cobalt metal, treat the sample with 3 mL of aqua regia at room temperature for 30 min. Complete dissolution and destroy the organic filter matrix by heating each beaker with a watchglass on a hot plate (140° C.) in a fume hood until all the filter is dissolved and the volume is reduced nearly to dryness (less than 0.5 mL). Repeat this process two more times using 3 mL of concentrated nitric acid each time. Do not allow the solution to evaporate to less than 1 mL on the last digestion.

Cool solutions and add about 3-4 mL of 5% nitric acid to each one. Quantitatively transfer the clear solution into a 10 mL volumetric flask. Rinse each beaker at least twice with 2 mL portions of 5% nitric acid and quantitatively transfer each rinsing to the solution in the volumetric flask. Dilute all samples to 10 mL with 5% nitric acid.

Aspirate the solutions into an oxidizing air-acetylene flame and record the absorbance at 240.7 nm. The absorbance is proportional to the sample concentration and can be determined from the appropriate calibration curve. When very low metal concentrations are found in the sample, scale expansion can be used to increase instrument response. Note: follow instrument manufacturer's recommendation for specific operating parameters. Appropriate filter blanks must be analyzed by the same procedure used for the samples.

Determination of sample recovery. To eliminate any bias in the analytical method, it is necessary to determine the recovery of the compound. The sample recovery should be determined in duplicate at each test level chosen so as to cover the concentration ranges of interest. If the recovery is less than 95%, the appropriate correction factor should be used to calculate the "true" value.

An aliquot of a standard solution containing a known amount of the analyte, preferably equivalent to the sample concentration expected, is added to a representative cellulose membrane filter and air dried. The analyte is then recovered from the filter and analyzed as described earlier. Duplicate determinations should agree within 5%.

For the validation study conducted to determine the precision and accuracy of this method, an amount of the analyte equivalent to that present in a 270 L sample at the selected level has been used for the recovery studies. Six filters at each of the three levels (0.5X, 1X and 2X the OSHA standard) were spiked accordingly by adding an aliquot of the solution containing known amounts of dissolved cobalt metal in 5% nitric acid. A parallel blank filter was also treated in the same manner except that no sample was added to it. All filters were then extracted and analyzed as described above. The average recovery value obtained was found to be 102%.

The sample recovery equals the average weight in μg recovered from the filter divided by the weight in μg added to the filter, or:

Recovery = average weight recovered ÷ weight added

Calibration and Standards

From the cobalt working standard solution, prepare at least six calibration standards to cover the cobalt concentration range from 10-60 μg/10 mL. Prepare these calibration standards freshly daily by dilution of appropriate aliquots of the working standard solution using 5% nitric acid for all dilutions. Aspirate into the air-acetylene flame and record the absorbance at 240.7 nm. Prepare a calibration curve by plotting on linear graph paper the absorbance versus the concentration of each standard in μg/10 mL. It is advisable to run a set of standards both before and after the analysis of a series of samples to ensure that conditions have not changed.

Calculations

Read the weight, in μg, corresponding to the total absorbance from the standard curve. No volume corrections are needed, because the standard curve is based on μg/10 mL. Corrections for the blank must be made for each sample:

μg = μg sample - μg blank

where:

μg sample = μg found in sample filter
μg blank = μg found in blank filter

Divide the total weight by the recovery to obtain the corrected μg/sample:

Corrected μg/sample = total weight ÷ recovery

For personal sampling pumps with rotameters only, the following correction should be made:

Corrected volume = $ft(P_1T_2/P_2T_1)^{1/2}$

where:

f = sample flow rate
t = sampling time
P_1 = pressure during calibration of sampling pump (mm Hg)
P_2 = pressure of air sampled (mm Hg)
T_1 = temperature during calibration of sampling pump (°K)
T_2 = temperature of air sampled (°K)

The concentration of the analyte in the air sampled can be expressed in mg/m³ (μg/L = mg/m³):

mg/m³ = corrected μg ÷ volume of air sampled in L

References
1. Cobalt Metal Dust and Fume, S203, Backup Data Report prepared under NIOSH Contract No. 210-76-0123.
2. Memoranda, Kenneth A. Busch (Chief, Statistical Services, DLCD), to Deputy Director, DLCD, January 6, 1975, November 8, 1974, subject: Statistical Protocol for Analysis of Data from Contract CDC-99-74-45.

COPPER DUST OR FUME

ACGIH TLV: 0.2 mg/m³ (fume); 1 mg/m³ (dust)   Cu
OSHA Standard: 0.1 mg/m³

Method	Sampling Duration	Sampling Location	Useful Range (mg/m³)	System Cost ($)	Test Cost ($)	Manufacturer
LS	Cont	Personal	0.01-100	1695	0	GCA
LS	Cont	Portable	0.001-100	2785	0	MDA
LS	Cont	Fixed	0.001-200	3990	0	GCA
BA	1-4 min	Portable	0.02-150	4890	0	GCA

NIOSH METHOD NO. S354

Principle

A known volume of air is drawn through a mixed cellulose ester membrane filter to collect copper fume. The samples are ashed using concentrated nitric acid to destroy the filter and other organic materials in the sample. The copper is taken up in 1% nitric acid. The solutions of samples and standards are aspirated into the oxidizing air acetylene flame of an atomic absorption spectrophotometer (AAS). A hollow cathode lamp for copper is used.

Range and Sensitivity

This method was validated over the range of 0.0548-0.372 mg/m³ using a 480 L sample at an atmospheric temperature of 22° C. and an atmospheric pressure of 761 mm Hg. Under the conditions of sample size (480 L), the working range of the method is estimated to be 0.01-0.3 mg/m³. The method may be extended to higher values by dilution of the solution before AA analysis.

The sensitivity of the AA analysis as reported by the manufacturer is 0.04 µg/mL of aqueous solution. This is the concentration of the solution which produces a signal of 1% absorbance. The detection limit of the AA analysis as reported by the manufacturer is 0.003 µg/mL of aqueous standard solution.

Interferences

Any particulate material other than copper fume containing copper or copper salts will result in higher values being reported as copper fume. Incomplete digestion of the filter or high zinc/copper ratio in the atmosphere may lead to depression of the copper absorbance.

Precision and Accuracy

The coefficient of variation for the total analytical and sampling method in the range of 0.0548-0.372 mg/m^3 was 0.058. This value corresponds to a 0.006 mg/m^3 standard deviation at the OSHA standard level. Statistical information can be found in reference 2. Details of the test procedures can be found in reference 3.

A collection efficiency of 1.00 was determined for the collecting medium, thus no bias was introduced in the sample collection step, and no correction for collection efficiency is necessary. The coefficient of variation is a good measure of the accuracy of the method. Storage stability studies on samples collected from a test atmosphere at a concentration of 0.1181 mg/m^3 indicate that collected samples are stable for at least 7 days.

Advantages and Disadvantages

The sampling device is small, portable, and involves no liquids. Samples collected on filters are analyzed by means of a quick, instrumental method.

Apparatus

Filter unit. The filter unit consists of a 37 mm diameter, 0.8 micron pore size mixed cellulose ester membrane filter supported by a glass fiber backup pad, and an appropriate 37 mm two piece cassette filter holder.

Personal sampling pump. A calibrated personal sampling pump whose flow can be determined to an accuracy of 5% at the recommended flow rate.

Atomic absorption spectrophotometer. Equipped with an air-acetylene burner head.

Copper hollow cathode lamp.

Compressed air.

Acetylene.

Pressure reducing valves. A 2 gauge, 2 stage pressure reducing valve and appropriate hose connections are needed for each compressed gas tank.

Manometer.

Thermometer.

Glassware, borosilicate.

Phillips beakers, 125 mL.

Watchglass covers.

Pipets of convenient sizes.

Volumetric flasks. 10 mL and other appropriate sizes.

Polyethylene bottles. Five 100 mL capacity, for working atomic absorption standards.

Polyethylene bottle. One 1000 mL capacity, for stock atomic absorption standard.

Hot plate. Adjustable thermostatically controlled, capable of reaching 160°C.

Reagents

All reagents must be ACS reagent grade or better.
Water, distilled or deionized.
Nitric acid, concentrated.
1% nitric acid (v/v).
Aqueous standard copper solution (1000 µg/mL). Commercially available.

Procedure

Cleaning of equipment. Before use, all glassware should initially be soaked in a mild detergent solution to remove any residual grease or chemicals. After initial cleaning, glassware must be cleaned with hot concentrated nitric acid and then rinsed thoroughly with tap water and distilled water, in that order, and then dried. For glassware that has previously been subjected to the entire cleaning procedure, a nitric acid rinse will be adequate.

Calibration of personal sampling pumps. Each personal sampling pump must be calibrated with a representative filter cassette in the line. This will minimize errors associated with uncertainties in the sample volume collected.

Collection and shipping of samples. Assemble the filter in the two piece filter cassette holder and close firmly to insure that the center ring seals the edge of the filter. The cellulose membrane filter is held in place by a glass fiber backup pad. Secure the cassette holder together with tape or shrinkable band. Remove the cassette plugs and attach to the personal sampling pump tubing. Clip the cassette to the worker's lapel. The cassette plugs are replaced after sampling.

Air being sampled should not pass through any hose or tubing before entering the filter cassette. A sample size of 480 L is recommended. Sample at a flow rate of 2.0 L/min. The flow rate should be known with an accuracy of 5%. Since the filter may become plugged or overloaded as evidenced by caking, the filter and pump's sampling rate should be checked periodically. When the filter becomes overloaded or when the pump's flow rate cannot be adjusted to 2.0 L/min, terminate sampling. Terminate sampling at the predetermined time and record sample flow rate, collection time, and ambient temperature and pressure. If pressure reading is not available, record the elevation. Also record the type of sampling pump used.

With each batch of ten samples, submit one filter from the same lot of filters which was used for sample collection and which is subjected to exactly the same handling as the samples, except that no air is drawn through it. Label this as a blank. The cassettes in which the samples are collected should be shipped in a suitable container, designed to prevent damage in transit.

Analysis of samples. Transfer each sample to a clean 125 mL Phillips beaker. Treat the sample with 10 mL concentrated nitric acid. Cover each beaker with a watchglass and heat on a hot plate (140-160° C.) in a fume hood until most of the acid has evaporated. Add 3 mL more of concentrated nitric acid and repeat the procedure. Evaporate the sample to 0.5 mL. Keep the beakers covered with a watchglass during the entire ashing procedure. When the beakers are cool, rinse the watchglass with 1% nitric acid into the beaker.

Quantitatively transfer the clear solutions to a 10 mL volumetric flask using 1% nitric acid. Rinse each beaker at least three times with 2 mL portions of 1% nitric acid and quantitatively transfer each rinsing to the solution in the volumetric flask. Dilute all samples to 10 mL with 1% nitric acid.

Aspirate samples and standard solution into an oxidizing air-acetylene flame and record the absorbance at 324.7 nm. The sample absorbance is proportional to the sample concentration and can be determined from the appropriate calibration curve. When very low concentrations are found in the sample, scale expansion can be used to increase instrument response. Note: follow instrument manufacturer's recommendations for specific AAS operating parameters. Appropriate filter blanks must be analyzed in accordance with the total procedure.

Calibration and Standardization

From the 1000 μg/mL stock standard, prepare at least 5 working standards to cover the range from 0.5-15 μg/mL. Prepare all standard solutions in 100 mL polyethylene bottles. Aspirate standards into the air-acetylene flame and record absorbance. Prepare a calibration curve by plotting on linear graph paper the absorbance versus the concentration of each standard in μg/mL. Standards should be run both before and after the series of samples are analyzed to ensure that conditions have not changed.

Calculations

Read the concentration in μg/mL, corresponding to the total absorbance from the calibration curve. The μg in each sample equals:

μg/mL (from curve) × 10 mL = μg Cu/sample

Corrections for the blank must be made for each sample:

μg = μg sample - μg blank

where:

μg sample = μg found in sample filter
μg blank = μg found in blank filter

For personal sampling pumps with rotameters only, the following air volume correction should be made:

Corrected volume = $ft(P_1T_2/P_2T_1)^{1/2}$

where:

f = flow rate sampled
t = sampling time
P_1 = pressure during calibration of sampling pump (mm Hg)
P_2 = pressure of air sampled (mm Hg)
T_1 = temperature during calibration of sampling pump (°K)
T_2 = temperature of air sampled (°K)

The concentration of copper fume in the air sample can be expressed in mg/m³ (μg/L = mg/m³):

mg/m³ = μg ÷ air volume sampled in L

References

1. *Analytical Methods for Flame Spectrophotometry,* Varian Associates, 1972.
2. *Documentation of NIOSH Validation Tests,* National Institute for Occupational Safety and Health, Cincinnati, Ohio (DHEW-NIOSH Publication #77-185), 1977. Available from Superintendent of Documents, U.S. Government Printing Office, Washington, D.C., Order No. 017-033-00231-2.
3. Backup Data Report for Copper Fume, prepared under NIOSH Contract No. 210-76-0123.

CRESOL

Synonyms: methylphenol; o-, m-, and p-cresol
ACGIH TLV: 5 ppm (22 mg/m³) - Skin
OSHA Standard: 5 ppm
NIOSH Recommendation: 10 mg/m³

$CH_3C_6H_4OH$

Method	Sampling Duration	Sampling Location	Useful Range (ppm)	System Cost ($)	Test Cost ($)	Manufacturer
DT	Grab	Portable	0.4-62.5	180	2.25	Bendix
IR	Cont	Portable	1-10	4374	0	Foxboro

NIOSH METHOD NO. S167

Principle

A known volume of air is drawn through a silica gel tube to trap the organic vapors present. The silica gel in the tube is transferred to a small, stoppered sample container, and the analyte is desorbed with acetone. An aliquot of the desorbed sample is injected into a gas chromatograph. The combined areas of the resulting two peaks are determined and compared with areas obtained for standards.

Range and Sensitivity

This method was validated over the range of 10.54-42.2 mg/m³ at an atmospheric temperature and pressure of 22° C. and 760 mm Hg, using a 20 L sample. Under the conditions of sample size (20 L), the probable useful range of this method is 5-60 mg/m³ at a detector sensitivity that gives nearly full detection of the strip chart recorder for a 1 mg sample. The method is capable of measuring much smaller amounts if the desorption efficiency is adequate. Desorption efficiency must be determined over the range used.

The upper limit of the range of the method is dependent on the adsorptive capacity of the silica gel tube. This capacity varies with the concentrations of the analyte and other substances in the air. The first section of the silica gel tube was found to hold at least 1.87 mg of analyte when a test atmosphere containing 42.2 mg/m³ of analyte in air was sampled at 0.185 L/min for 240 min. (The silica gel tube consists of two sections of silica gel separated by a section of urethane foam.) If a particular atmosphere is suspected of containing a large amount of contaminant, a smaller sampling volume should be taken.

Interferences

Silica gel has a high affinity for water, so organic vapors will not be trapped efficiently in the presence of a high relative humidity. This effect may be important even through there is no visual evidence of condensed water in the silica gel tube.

When compounds other than the isomers of cresol are known or suspected to be present in the air, such information, including their suspected identities, should be transmitted with the sample. Since acetone is used to desorb the analyte from the silica gel, it is not possible to measure acetone in the sample.

It must be emphasized that any compound which has the same retention time as the analyte at the operating conditions described in this method is an interference. Retention time data on a single column cannot be considered proof of chemical identity. If the possibility of interference exists, separation conditions (column packing, temperature, etc.) must be changed to circumvent the problem.

Precision and Accuracy

The coefficient of variation for the total analytical and sampling method in the range of 10.54-42.2 mg/m^3 was 0.068. This value corresponds to a 1.5 mg/m^3 standard deviation at the OSHA standard level. Statistical information and details of the validation and experimental test procedures can be found in reference 2.

On the average the concentrations obtained at the OSHA standard level using the overall sampling and analytical method were 4.2% lower than the "true" concentrations for a limited number of laboratory experiments. Any difference between the "found" and "true" concentrations may not represent a bias in the sampling and analytical method, but rather a random variation from the experimentally determined "true" concentration. Therefore, no recovery correction should be applied to the final result.

Advantages and Disadvantages

The sampling device is small, portable, and involves no liquids. Interferences are minimal, and most of those which do occur can be eliminated by altering chromatographic conditions. The tubes are analyzed by means of a quick, instrumental method. The method can also be used for the simultaneous analysis of two or more substances suspected to be present in the same sample by simply changing gas chromatographic conditions from isothermal to a temperature-programmed mode of operation.

One disadvantage of the method is that the amount of sample which can be taken is limited by the number of mg that the tube will hold before overloading. When the sample value obtained for the backup section of the silica gel tube exceeds 25% of that found on the front section, the possibility of sample loss exists. Furthermore, the precision of the method is limited by the reproducibility of the pressure drop across the tubes. This drop will affect the flow rate and cause the volume to be imprecise, because the pump is usually calibrated for one tube only.

Apparatus

Calibrated personal sampling pump. Whose flow can be determined within 5% at the recommended flow rate.

Silica gel tubes. Glass tube with both ends flame sealed, 7 cm long with 6 mm O.D. and 4 mm I.D., containing 2 sections of 20/40 mesh silica gel separated by a 2 mm portion of urethane foam. The adsorbing section contains approximately 150 mg of silica gel, the backup section, approximately 75 mg. A 3 mm portion of urethane foam is placed between the outlet end of the tube and the backup section. A plug of silylated glass wool is placed in front of the adsorbing section. The pressure drop across the tube must be less than one inch of mercury at a flow rate of 1 L/min.

Gas chromatograph. Equipped with a flame-ionization detector.

Column. 10 ft × 1/8 in stainless steel, packed with 10% FFAP on 80/100 mesh, acid washed DMCS Chromosorb W.

Electronic integrator. Or some other suitable method for measuring peak area.

Sample containers. 2 mL, with glass stoppers or Teflon-lined caps. If an automatic sample injector is used, the associated vials may be used.

Syringes. 10 µL, and other convenient sizes for making standards.

Pipets. 0.1 mL delivery type.

Volumetric flasks. Convenient sizes for making standard solutions.

Reagents

Chromatographic quality acetone.

Cresol (all isomers). Prepare a standard mixture of the isomers by adding together 20 g of the ortho, 40 g of the meta, and 30 g of the para isomer and mix.

Prepurified hydrogen.

Filtered compressed air.

Purified nitrogen.

n-Hexane, reagent grade.

Procedure

Cleaning of equipment. All glassware used for the laboratory analysis should be detergent washed and thoroughly rinsed with tap water and distilled water.

Calibration of personal pumps. Each personal pump must be calibrated with a representative silica gel tube in the line. This will minimize errors associated with uncertainties in the sample volume collected.

Collection and shipping of samples. Immediately before sampling, break the ends of the tube to provide an opening at least one-half the internal diameter of the tube (2 mm). The smaller section of silica gel is used as a backup and should be positioned nearest the sampling pump. The silica gel tube should be placed in a vertical direction during sampling to minimize channeling through the silica gel. Air being sampled should not be passed through any hose or tubing before entering the silica gel tube.

A sample size of 20 L is recommended. Sample at a flow of 0.20 L/min or less. The flow rate should be known with an accuracy of at least 5%. The temperature and pressure of the atmosphere being sampled should be recorded. If pressure reading is not available, record the elevation. The silica gel tubes should be capped with the supplied plastic caps immediately after sampling. Under no circumstances should rubber caps be used.

One tube should be handled in the same manner as the sample tubes (break, seal, and transport), except that no air is sampled through this tube. This tube should be labeled as a blank. Capped tubes should be packed tightly and padded before they are shipped to minimize tube breakage during shipping.

Preparation of samples. In preparation for analysis, each silica gel tube is scored with a file in front of the first section of silica gel and broken open. The glass wool and the silica gel in the first (larger) section is transferred to a 2 mL stoppered sample container. The separating section of foam is removed and discarded; the second section is transferred to another stoppered container. These two sections are analyzed separately.

Desorption of samples. Prior to analysis, 1.0 mL of acetone is pipetted into each sample container. Desorption should be done for 30 minutes. Tests indicate that this is adequate if the sample is agitated occasionally during this period. If an automatic sample injector is used, the sample vials should be capped as soon as the solvent is added to minimize volatilization.

GC conditions. The typical operating conditions for the gas chromatograph are:

50 mL/min (60 psig) nitrogen carrier gas flow
65 mL/min (24 psig) hydrogen gas flow to detector
500 mL/min (50 psig) air flow to detector
230° C. injector temperature
250° C. detector temperature
200° C. column temperature

Injection. The first step in the analysis is the injection of the sample into the gas chromatograph. To eliminate difficulties arising from blow back or distillation within the syringe needle, one should employ the solvent flush injection technique. The 10 µL syringe is first flushed with solvent several times to wet the barrel and plunger. Three µL of solvent are drawn into the syringe to increase the accuracy and reproducibility of the injected sample volume. The needle is removed from the solvent, and the plunger is pulled back about 0.2 µL to separate the solvent flush from the sample with a pocket of air to be used as a marker. The needle is then immersed in the sample, and a 5 µL aliquot is withdrawn, taking into consideration the volume of the needle, since the sample in the needle will be completely injected. After the needle is removed from the sample and prior to injection, the plunger is pulled back 1.2 µL to minimize evaporation of the sample from the tip of the needle. Observe that the sample occupies 4.9-5.0 µL in the barrel of the syringe. Duplicate injections of each sample and standard should be made. No more than a 3% difference in area is to be expected. An automatic sample injector can be used if it is shown to give reproducibility at least as good as the solvent flush method.

Measurement of area. Although there are three isomers of cresol, there are only two peaks on the gas chromatogram, because the meta and para isomers have the same retention time. The total area of the two sample peaks is measured by an electronic integrator or some other suitable form of area measurement, and preliminary results are read from a standard curve prepared as discussed below.

Determination of desorption efficiency. The desorption efficiency of a particular compound can vary from one laboratory to another and also from one batch of silica gel to another. Thus, it is necessary to determine at least once the percentage of the specific compound that is removed in the desorption process, provided the same batch of silica gel is used.

Silica gel equivalent to the amount in the first section of the sampling tube (approximately 150 mg) is measured into a 2.5 in, 4 mm I.D. glass tube, flame sealed at one end. This silica gel must be from the same batch as that used in obtaining the samples and can be obtained from unused silica gel tubes. The open end is capped with Parafilm. A standard solution is prepared by placing 1100 mg of the mixture of the isomers of cresol into a 10 mL volumetric flask and making it up to volume with n-hexane. A known amount of the standard solution is injected directly into the silica gel with a microliter syringe, and the tube is capped with more Parafilm. When using an automatic sample injector, the sample injector vials, capped with Teflon-faced septa, may be used in place of the glass tubes.

The amount injected is equivalent to that present in a 20 L air sample at the selected level. Six tubes at each of three levels are prepared in this manner and allowed to stand for at least overnight to assure complete adsorption of the analyte onto the silica gel. These tubes are referred to as the samples. A parallel blank tube should be treated in the same manner except that no sample is added to it. The sample and blank tubes are desorbed and analyzed in exactly the same manner as the sampling tube described above.

Two or three standards are prepared by injecting the same volume of compound into 1.0 mL of acetone with the same syringe used in the preparation of the samples. These are analyzed with the samples. The desorption efficiency (D.E.) is dependent on the amount of analyte collected on the silica gel. The desorption efficiency equals the average weight in mg recovered from the tube divided by weight in mg added to the tube, or:

D.E. = average wt (mg) recovered ÷ wt (mg) added

The desorption efficiency is dependent on the amount of analyte collected on the silica gel. Plot the desorption efficiency versus the weight of analyte found. This curve is used to correct for adsorption losses.

Calibration and Standardization

It is convenient to express concentration of standards in terms of mg/1.0 mL acetone, because samples are desorbed in this amount of acetone. A series of standards varying in concentration over the range of interest, are prepared and analyzed under the same GC conditions and during the same time period as the unknown samples. Curves are established by plotting concentration in mg/1.0 mL versus total peak area. Note: since no internal standard is used in the method, standard solutions must be analyzed at the same time that the sample analysis is done. This will minimize the effect of known day-to-day variations and variations during the same day of the FID response.

Calculations

Read the weight, in mg, corresponding to each combined peak area from the standard curve. No volume corrections are needed, because the standard curve is based on mg/1.0 mL acetone and the volume of sample injected is identical to the volume of the standards injected. Corrections for the blank must be made for each sample:

mg = mg sample - mg blank

where:

mg sample = mg found in front section of sample tube
mg blank = mg found in front section of blank tube

A similar procedure is followed for the backup sections. Add the weights found in the front and backup sections to get the total weight in the sample. Read the desorption efficiency from the curve for the amount found in the front section. Divide the total weight by this desorption efficiency to obtain the corrected mg/sample:

Corrected mg/sample = total weight ÷ D.E.

The concentration of the analyte in the air sampled can be expressed in mg/m^3:

mg/m^3 = (corrected mg)(1000 L/m^3) ÷ (air volume sampled in L)

Another method of expressing concentration is ppm:

ppm = $(mg/m^3)(24.45)(760)(T + 273)$ ÷ $(M.W.)(P)(298)$

where:

P = pressure (mm Hg) of air sampled
T = temperature (°C.) of air sampled
24.45 = molar volume (L/mol) at 25° C. and 760 mm Hg
M.W. = molecular weight (g/mol) of analyte
760 = standard pressure (mm Hg)
298 = standard temperature (°K)

References

1. White, L.D. et al. A convenient optimized method for the analysis of selected solvent vapors in the industrial atmosphere. Amer. Ind. Hyg. Assoc. J. 31: 225, 1970.
2. Documentation of NIOSH Validation Tests, NIOSH Contract No. CDC-99-74-45.
3. Final Report, NIOSH Contract No. HSM-99-71-31. Personal Sampler Pump for Charcoal Tubes, September 15, 1972.

CROTONALDEHYDE

Synonym: 2-butenol
ACGIH TLV: 2 ppm (6 mg/m^3)
OSHA Standard: 2 ppm

$CH_3CH=CHCHO$

Method	Sampling Duration	Sampling Location	Useful Range (ppm)	System Cost ($)	Test Cost ($)	Manufacturer
IR	Cont	Portable	0.5-5	4374	0	Foxboro

NIOSH METHOD NO. P&CAM 285

Principle

A known volume of air is drawn through a midget bubbler containing a buffered solution of hydroxylamine. Crotonaldehyde is derived with hydroxylamine to form crotonaldehyde oxime. The crotonaldehyde oxime derivative is measured by polarography.

Range and Sensitivity

This method was studied over the range of 2.9-23.4 mg/m^3 at an atmospheric temperature and pressure of 21° C. and 766 mm Hg, using a 12 L sample. The upper limit of the method is dependent on the concentration of hydroxylamine. While collection efficiency has not been tested outside of the method range, a 50-fold molar excess of hydroxylamine should be maintained over the total amount of crotonaldehyde sampled.

Interferences

When other compounds are known or suspected to be present in the air, such information, including their suspected identities, should be transmitted with the sample. Other volatile aldehydes such as acrolein, formaldehyde, and benzaldehyde may cause significant interference.

Precision and Accuracy

The coefficient of variation for the total analytical and sampling method in the range of 2.9-11.6 mg/m^3 was 0.061. This corresponds to a 0.4 mg/m^3 standard deviation at 1X OSHA standard. Statistical information can be found in reference 1. Details of the test procedures can be found in reference 2.

The average concentrations obtained from analysis of samples collected from test atmospheres at 0.5X, 1X, 2X, and 4X the OSHA standard were 10.0% lower, 2.1% lower, 3.7% higher, and

0.3% higher, respectively, than the "true" concentrations. The difference between the "found" and "true" concentrations may not represent a bias in the sampling and analytical method, but rather a random variation from the experimentally determined "true" concentration. Therefore, the method has no uncorrected bias. The coefficient of variation is a good measure of the accuracy of the method provided samples are refrigerated after collection. Storage stability studies on samples collected from a test atmosphere at a concentration of 5.80 mg/m^3 indicate that collected samples are stable for at least 7 days at 8° C. Collection efficiency of the midget bubbler was determined to be 0.96 for an average of 24 samples and a correction must be applied.

Advantages and Disadvantages

The sample is not stable unless stored at cold temperatures (8°C.). If samples are analyzed within 4 hours after collection, no refrigeration is required. Collected samples are analyzed by a quick instrumental method.

A disadvantage of the method is the awkwardness in using midget bubblers for collecting personal samples. If the worker's job performance requires much body movement, loss of the collection solution during sampling may occur. The bubblers are more difficult to ship than adsorption tubes or filters due to possible breakage and leakage of the bubblers during shipping.

Apparatus

Glass midget bubbler.

Personal sampling pump. A calibrated personal sampling pump whose flow rate can be determined to an accuracy of 5%. The sampling pump is protected from splashover or solvent condensation by a trap. The trap is a midget bubbler or impinger with the stem broken off which is used to collect spillage. The trap is attached to the pump with a metal holder. The outlet of the trap is connected to the pump by a flexible tubing. Each sampling pump must be calibrated with a representative bubbler and trap in the line to minimize errors associated with uncertainties in the volume sampled.

Thermometer.

Manometer.

Volumetric flasks. 1 L and other convenient sizes for preparing standard solutions.

Pipets. Convenient sizes for preparing stock standard solutions and for measuring the collection medium.

Syringes. Convenient sizes for preparing spiked standard samples.

Polarograph. Capable of differential pulse polarography with reference to a saturated calomel electrode.

China marker.

Teflon tubing or plugs. 15 cm × 7 mm I.D., for sealing the inlet and outlet of the bubbler stem before shipping.

Reagents

All reagents used must be ACS reagent grade or better.

Hydroxylamine hydrochloride.

Sodium hydroxide, 0.1 M.

Formic acid.

Sodium formate.

Hydroxylamine collection solution. Dissolve 0.68 g of hydroxylamine hydrochloride in 1 L of deionized water. Adjust the pH to 5 with 0.1 M sodium hydroxide.

Formic acid-sodium formate buffer. Add 3.4 g sodium formate and 2 mL formic acid into a 500 mL volumetric flask. Bring to volume with deionized water.

Crotonaldehyde, 85%.

Crotonaldehyde stock solutions. Prepare solutions containing 17.19 mg/mL and 1.719 mg/mL of crotonaldehyde in deionized water.

Procedure

Cleaning of equipment. All glassware used for the laboratory analysis should be detergent washed and thoroughly rinsed with tap water and distilled water and dried. All midget bubblers should be heated in an oxidizing atmosphere at a temperature of approximately 580° C. A glass blower's annealing oven is satisfactory for this purpose. This procedure will remove organic contaminants.

Collection and shipping of samples. Rinse each bubbler with 10 mL of the hydroxylamine solution. Pipet 10 mL of hydroxylamine solution buffered at a pH of 5 into each midget bubbler. Mark the liquid level on the bubbler with a china marker. Make sure the bubbler frit is covered.

The outlet of the midget bubbler is attached to the pump's inlet or a trap which may be used to protect the pump during personal sampling. The trap is a midget impinger or bubbler with the stem broken off which is used to collect spillage. The trap is attached to the pump with a metal holder. The outlet of the trap is connected to the pump by flexible tubing.

Air being sampled should not be passed through any hose or tubing before entering the midget bubbler. (Polyvinyl chloride tubing is known to give interferences in the method.) A sample size of 12 L is recommended. Sample at a flow rate of 0.2 L/min or less. Do not sample at a flow rate of less than 0.1 L/min. Turn the pump on and begin sample collection. Periodically check the pump rotameter and adjust the flow as needed. Terminate sampling at the predetermined time and record sample flow rate, collection time, and ambient temperature and pressure. If pressure reading is not available, record the elevation. Also record the type of sampling pump used.

The inlet and outlet of the bubbler stem should be sealed by connecting a piece of Teflon tubing between them or inserting Teflon plugs in the inlet and outlet. Do not seal with rubber or polyvinyl chloride tubing. The standard taper joint of the bubbler should be taped securely to prevent leakage during shipping. Care should be taken to minimize spillage or loss of sample by evaporation at all times. The bubblers should be shipped in a suitable container designed to prevent damage in transit. The samples should be shipped to the laboratory as soon as possible. If samples cannot be analyzed within 4 hours after collection, they must be refrigerated during shipping and until the time of analysis at 8° C.

With each batch or partial batch of five samples, submit one bubbler containing 10 mL of the collection medium prepared from the same stock as that used for sample collection. This bubbler must be subjected to exactly the same handling as the samples except that no air is drawn through it. Label this bubbler as the blank. A sample of the bulk material should be submitted to the laboratory in a glass container with a Teflon-lined cap or equivalent. This sample should not be transported in the same container as the bubblers. Do not ship the material in the trap. If more than 1 mL of material is collected in the trap after sampling, the sample should be considered invalid, otherwise, it can be discarded.

Analysis of samples. Sample analysis should be done in a room with a fairly stable temperature. Each sample and blank is analyzed separately. Remove the bubbler stem and tap the stem lightly against the flask to drain the contents into the bubbler flask. If necessary, bring the liquid volume to the 10 mL mark with distilled water. Add 5 mL of the formic acid-sodium formate buffer. Swirl the bubbler flask to mix the contents well.

Transfer the mixed solution to a clean, dry polarographic receptacle which contains a small amount of mercury (approximately 0.3 mL) on the bottom. Place the receptacle on the polarograph and purge the sample for 3 min with oxygen-free nitrogen at a flow rate of 200 mL/min. Analyze the sample by differential pulse polarography using the following conditions:

Drop time: 1 second
Scan rate: 5 mV/second
Scan from −0.6 V to 1.2 V versus a saturated calomel electrode

Rinse the electrodes with distilled water between samples. Measure the resultant diffusion current from the reduction of the crotonaldehyde oxime. The half-wave potential, $E^{1/2}$, of the crotonaldehyde oxime is −1.03 V versus a saturated calomel electrode. The concentration of crotonaldehyde in the sample is determined by referring to the calibration curve prepared as described below.

Calibration and Standardization

A series of standards, varying in concentration over the range corresponding to approximately 0.1 to 4 times the OSHA standard for the sample under study, is prepared and analyzed under the same polarographic conditions and during the same time period as the unknown samples. Curves are established by plotting concentration in $\mu g/15$ mL versus diffusion current. Standard solutions must be analyzed at the same time that the sample analysis is done. This will minimize the effect of known day-to-day variations. Sample analysis and standards calibration should be done in a room with a fairly stable temperature.

From the stock solutions, appropriate aliquots are withdrawn (less than 100 μL) and added to 10 mL of the hydroxylamine solution. Five mL of formic acid/sodium formate buffer is added. The mixture should be swirled occasionally and after 15 min. Prepare at least 6 working standards to cover the range of 6-280 μg/sample. Prepare at least one blank and analyze it at the same time as the standards. Prepare a standard calibration curve by plotting $\mu g/15$ mL crotonaldehyde versus diffusion current.

Calculations

Read the weight, in μg, corresponding to each sample current from the standard curve. No volume correction is needed, because the standard curve is based on $\mu g/15$ mL of solution. Corrections for the blank must be made for each sample:

$\mu g = \mu g$ sample $- \mu g$ blank

where:

μg sample $= \mu g$ found in sample bubbler
μg blank $= \mu g$ found in blank bubbler

For personal sampling pumps with rotameters only, the following volume correction should be made:

Corrected volume $= ft(P_1 T_2/P_2 T_1)^{1/2}$

where:

f = flow rate sampled
t = sampling time
P_1 = pressure during calibration of sampling pump (mm Hg)
P_2 = pressure of air sampled (mm Hg)
T_1 = temperature during calibration of sampling pump (°K)
T_2 = temperature of air sampled (°K)

The concentration of crotonaldehyde in the air sampled can be expressed in mg/m³:

mg/m³ = μg/L = corrected μg ÷ corrected air volume sampled in L

The collection efficiency of the midget bubblers was determined to be 0.96. The concentration of crotonaldehyde should be corrected for collection efficiency:

Corrected mg/m³ = mg/m³ ÷ 0.96

Another method of expressing concentration is ppm:

ppm = (mg/m³)(24.45)(760)(T + 273) ÷ (70.09)(P)(298)

where:

P = pressure (mm Hg) of air sampled
T = temperature (°C.) of air sampled
24.45 = molar volume (L/mol) at 25° C. and 760 mm Hg
70.09 = molecular weight (g/mol) of crotonaldehyde
760 = standard pressure (mm Hg)
298 = standard temperature (°K)

References

1. *Documentation of NIOSH Validation Tests,* National Institute for Occupational Safety and Health, Cincinnati, Ohio (DHEW-NIOSH-Publication #77-185), 1977. Available from Superintendent of Documents, U.S. Government Printing Office, Washington, D.C., Order No. 017-033-00231-2.
2. Backup Data Report for Crotonaldehyde, prepared under NIOSH Contract No. 210-76-0123.

CUMENE

$C_6H_5CH(CH_3)_2$

Synonym: isopropylbenzene
ACGIH TLV: 50 ppm (245 mg/m³) - Skin
OSHA Standard: 50 ppm

Method	Sampling Duration	Sampling Location	Useful Range (ppm)	System Cost ($)	Test Cost ($)	Manufacturer
DT	Grab	Portable	5-2400	180	2.25	Bendix
IR	Cont	Portable	2-100	4374	0	Foxboro
PI	Cont	Fixed	0.2-20	4950	0	HNU

NIOSH METHOD NO. P&CAM 127

(Use as described for acetone, with 100° C. column temperature)

CYANIDE

Synonyms: potassium cyanide; sodium cyanide CN
ACGIH TLV: 5 mg/m³ as CN - Skin
OSHA Standard: 5 mg/m³
NIOSH Recommendation: 5 mg/m³ ceiling (10 min)

Method	Sampling Duration	Sampling Location	Useful Range (mg/m³)	System Cost ($)	Test Cost ($)	Manufacturer
DT	Grab	Portable	2-25	150	2.90	Nat'l Draeger

NIOSH METHOD NO. S250

Principle
Atmospheric samples are collected by drawing a known volume of air through a cellulose membrane filter and impinger (connected in series) containing 0.1 N sodium hydroxide. The sample containing filters are extracted with 0.1 N sodium hydroxide. The filters and impingers are analyzed separately by direct potentiometry using an ion specific electrode. The millivolt reading is used as a measure of cyanide concentration based on a calibration curve generated by measurement of standard solutions. The samples must be carefully interspersed between calibration standards which give about the same response as the samples in order to obtain reliable results.

Range and Sensitivity
This method was validated over the range of 2.62-9.68 mg/m³ at an atmospheric temperature and pressure of 24° C. and 763 mm Hg, using a 90 L sample. Under the conditions of sample size (90 L), the linear working range of the method is estimated to be 0.5-15 mg/m³. The lower limit of detection for cyanide is noted as 0.1 μg/mL.

Interferences
Gaseous hydrogen cyanide present in the air is an interference. Sulfide ion irreversibly poisons the cyanide ion specific electrode and must be removed if found to be present in the sample. Check for the presence of sulfide ion by touching a drop of sample to a piece of lead acetate paper. The presence of sulfide is indicated by discoloration of the paper.

Sulfide is removed by the addition of a small amount (spatula tip) of powdered cadmium carbonate to the pH 11-13 sample. Swirl to disperse the solid and recheck the liquid by again touching a drop to a piece of lead acetate paper. If sulfide ion has not been removed completely, add more cadmium carbonate. Avoid a large excess of cadmium carbonate and long contact time with the solution. When a drop of liquid no longer discolors a strip of lead acetate paper, remove the solid by filtering the sample through a small plug of glass wool contained in an eye dropper and proceed with the analysis.

It should also be noted that the cyanide electrode will malfunction if other ions like chloride, iodide and bromide, which form insoluble silver salts, are present in sufficient quantity. Several metal ions are also known to complex with cyanide, such as cadmium, zinc, silver, nickel, cuprous, iron, and mercury. Consult the instruction manual for a list of these ions and also the proper procedure to use when such ions are believed to be present.

Precision and Accuracy

The coefficient of variation for the total analytical and sampling method in the range of 2.62-9.68 mg/m^3 was 0.101. This value corresponds to a 0.5 mg/m^3 standard deviation at the OSHA standard level. Statistical information and details of the validation and experimental test procedures can be found in reference 2.

A collection efficiency of 100% was determined for dry particulate cyanide on the cellulose membrane filter and a collection efficiency of 98% was determined for HCN in 0.1 N NaOH; thus, no bias was introduced in the sample collection step based on collection efficiency. Likewise, data on a limited number of analytical filter samples give an average recovery of 97%. In the absence of a significant amount of free HCN in the environment being sampled, the coefficient of variation is a satisfactory measure of both accuracy and precision of the sampling and analytical method.

Advantages and Disadvantages

Advantages are the simplicity, specificity, and speed of the method.

Apparatus

Filter unit. Consists of filter media and a 37 mm, 3 piece cassette filter holder.

Midget impinger. Contains the absorbing solution or reagent.

Personal sampling pump. A calibrated personal samping pump whose flow can be determined to an accuracy of 5% at the recommended flow rate. The pump must be calibrated with a filter and impinger in the line.

Thermometer.

Manometer.

Stopwatch.

Membrane filter. Mixed cellulose ester, 37 mm diameter, 0.8 micron pore size.

Polyethylene scintillation vials. With screw caps, 20 mL.

Cyanide ion specific electrode. Orion 94-06 or equivalent.

Double junction reference electrode. Orion 90-20-00 or equivalent.

pH meter. With readout capacity in increments of 0.1 millivolt.

Magnetic stirrer and stirring bars.

Jars for extraction of filters. 2 oz. ointment jars, squat form with aluminum-lined screw caps.

Pipets. 10 mL and other convenient sizes.

Volumetric flasks. 25 mL and other assorted sizes.

Assorted laboratory glassware.

Reagents

All reagents must be ACS reagent grade or better.

Double distilled water.

Potassium cyanide.

Sodium hydroxide, 0.1 N. Dissolve 2.0 g NaOH in double distilled water and dilute to 500 mL.

Cyanide standard stock solution, 200 μg/mL. Dissolve 0.50 g KCN in 0.1 N NaOH and dilute to 1000 mL with additional 0.1 N NaOH.

Working standards. Prepare at least 6 working standards to cover the concentration range of interest by proper dilution of the stock standard to a total volume of 25 mL. Use 0.1 N NaOH for all dilutions. Prepare these calibration standards fresh daily.

Lead acetate paper.

Cadmium carbonate.

Procedure

Cleaning of equipment. Before use all glassware should initially be soaked in a strong detergent solution to remove any residual grease or chemicals. After initial cleaning, the glassware should be thoroughly rinsed with warm tap water, concentrated nitric acid, tap water, and distilled water, in that order, and then dried.

Calibration of personal pump. Personal sampling pump should be calibrated using integrating volume meter or other means.

Collection and shipping of samples. To collect cyanide salts, a personal sampler pump is used to pull air through a cellulose ester membrane filter connected in series to a midget impinger containing 10 mL of 0.1 N NaOH. The filter holder is held together by tape or a shrinking band. If the middle piece of the filter holder does not fit snugly into the bottom piece of the filter holder, the contaminant will leak around the filter.

A short piece of flexible tubing is used to connect the filter holder to the impinger. A similar piece of flexible tubing, loosely plugged with a piece of glass wool to protect the sampling pump from splashover and condensation, is used to connect the impinger to the sampling pump. The impinger must be maintained in a verticle position during the sampling. Clip the cassette to the worker's lapel.

Air being sampled should not be passed through any hose or tubing before entering the filter cassette. Set the flow rate as accurately as possible using the manufacturer's directions. Record the temperature and pressure of the atmosphere being sampled. If the pressure reading is not available, record the elevations. Position the middle of the rotameter ball of the personal sampling pump to the 1.5 L/min calibration mark as accurately as possible. Since it is possible for the filter to become plugged by heavy particulate loading or by the presence of oil mist or other liquids in the air, the pump rotameter should be observed frequently, and readjusted as needed. If the rotameter cannot be adjusted to correct a problem, terminate the sampling.

Turn on pump to begin sample collection. Care should be taken to measure the flow rate, time and/or volume as accurately as possible. Record atmospheric pressure and temperature. The sample should be taken at a flow rate of 1.5 L/min for 60 min (90 L). After sampling disconnect the filter holder cassettes from the impingers and firmly seal the cassettes with the plugs in both the inlet and outlet. The sample impingers are then handled and shipped separately from the filters. The impinger stems can be removed and cleaned as follows. Tap the stem gently against the inside wall of the impinger bottom to recover as much of the sampling solution as possible and quantitatively transfer the total contents of the impinger bottom into a 20 mL polyethylene scintillation vial with screw cap. Wash the impinger stem with 2 mL of 0.1 N NaOH and add washing into the polyethylene vial. Close cap tightly and secure with plastic tape around edges to avoid sample loss during transit.

Care should be taken to minimize spillage or loss by evaporation at all times. Carefully record sample identity of both filters and impingers and all relevant sample data. With each batch of 10 samples, submit one filter and one impinger solution which is subjected to exactly the same handling as for the samples except that no air is drawn through them. Label these as blanks. Submit one blank filter and one blank impinger solution for every ten samples.

Analysis of samples. Open the cassette filter holder and carefully remove the cellulose membrane filter from the holder and cellulose backup pad with the aid of Millipore filter tweezers and transfer filter to a 2 oz. ointment jar. Pipet 25 mL of 0.1 N NaOH into the jar. Cap and allow to stand for at least 30 min with occasional shaking. Tests indicate that this period is adequate for complete extraction of the cyanide from the filter. Analyze these filter samples within 2 hours after extraction. The impinger sample solutions contained in the polyethylene vials are also made up to a 25 mL volume.

The samples are analyzed by immersing the cyanide ion electrode and reference electrode in the sample solutions and recording the millivolt reading. Both the samples and standards should be stirred while the readings are being taken. The reading should be taken after the meter has stabilized. Follow instrument manufacturer's instruction manual for proper operation and measurement procedures. To obtain the most reliable results, the samples should be carefully interspersed with standards of similar response. Appropriate filter blanks and impinger blanks must be analyzed by the same procedure used for the samples.

Determination of sample recovery. To eliminate any bias in the analytical method for particulate cyanide, it is necessary to determine the recovery of the analyte. The analyte recovery should be determined in duplicate for at least one concentration which corresponds to a weighable amount. If the recovery of the analyte is less than 95%, the appropriate correction factor should be used to calculate the "true" value.

A weighed amount of the analyte, preferably equivalent to the concentration expected in the sample, is added to a representative cellulose membrane filter. The analyte is then recovered from the filter and analyzed as described above. Duplicate determinations should agree within 5%.

For this validation study, an amount of the analyte equivalent to that present in a 90 L sample at the selected level has been used for the recovery studies. Six filters were spiked with weighed amounts of potassium cyanide equivalent to the CN present at 2X OSHA standard level. A parallel blank filter was also treated in the same manner except that no analyte was added to it. All filters were then analyzed as described earlier. The average recovery value obtained was found to be 97%.

The recovery equals the average weight in μg recovered from the filter divided by the weight in μg added to the filter, or:

Recovery = average weight recovered ÷ weight added

Calibration and Standardization

Prepare a series of working standards containing 40-1000 μg of cyanide in 25 mL of 0.1 N sodium hydroxide. These standards should be prepared fresh each time. It is convenient to express concentration of standards in terms of μg/25 mL, because samples are in this volume of solvent. The appropriate calibration standards are alternately analyzed with the samples to determine the response factor. This practice will minimize the effect of observed fluctuations or variations in millivolt readings during any given day. On semilog paper, plot the millivolt readings versus cyanide ion concentrations of the standards. The cyanide ion concentration in μg/25 mL is plotted on the log axis.

Calculations

Determine the weight in μg corresponding to the millivolt response of the sample by using the appropriate response factor or calibration factor for the sample. Separately calculate the amount of particulate cyanide found on the filter and the cyanide (HCN) found in the impinger. Corrections for the blank must be made for each sample filter and impinger:

μg = μg sample − μg blank

where:

μg sample = μg found in sample filter or sample impinger
μg blank = μg found in blank filter or blank impinger

Divide the total weight of CN from filter by the recovery to obtain the corrected μg/sample:

Corrected μg/sample = total weight ÷ recovery

Report the particulate cyanide found and the HCN found as separate values because the OSHA standards for these two species are different. Note that the HCN found may be due to free HCN as well as HCN formed from moisture interaction with the particulate CN during sampling. If HCN is known to be absent in the environment being sampled, then the sum of the CN found on the filter and the impinger solution gives a total measure of particulate cyanide. If air samples were taken under conditions significantly different from standard conditions of 25° C. and 760 mm Hg, a volume correction for the air sampled should be made as follows:

$$Vs = (V)(P)(298) \div (760)(T + 273)$$

where:

Vs = volume of air in L at 25° C. and 760 mm Hg
V = volume of air sampled
P = pressure (mm Hg) of air sampled
T = temperature (°C.) of air sampled
760 = standard pressure (mm Hg)
298 = standard temperature (°K)

The concentration of the analyte in the air sampled can be expressed in mg/m^3 ($\mu g/L = mg/m^3$):

mg/m^3 = corrected μg ÷ air volume sampled in L

References
1. Instruction manual for cyanide ion specific electrode.
2. Documentation of NIOSH Validation Tests, Contract No. CDC-99-74-45.

CYCLOHEXANE

ACGIH TLV: 300 ppm (1050 mg/m³) C_6H_{12}
OSHA Standard: 300 ppm

Method	Sampling Duration	Sampling Location	Useful Range (ppm)	System Cost ($)	Test Cost ($)	Manufacturer
DT	Grab	Portable	100-1500	150	2.60	Nat'l Draeger
DT	Grab	Portable	100-6000	165	1.70	Matheson
DT	Grab	Portable	150-12,000	180	2.25	Bendix
IR	Cont	Portable	6-600	4374	0	Foxboro
PI	Cont	Fixed	20-2000	4950	0	HNU
GC	1 min	Portable	$1-10^6$	10,000	0	Microsensor

NIOSH METHOD NO. P&CAM 127

(Use as described for acetone)

CYCLOHEXANOL

$C_6H_{11}OH$

Synonym: Hexalin
ACGIH TLV: 50 ppm (200 mg/m³)
OSHA Standard: 50 ppm

Method	Sampling Duration	Sampling Location	Useful Range (ppm)	System Cost ($)	Test Cost ($)	Manufacturer
IR	Cont	Portable	1-100	4374	0	Foxboro

NIOSH METHOD NO. S66

(Use as described for allyl alcohol, with 120° C. column temperature)

CYCLOHEXANONE

$C_6H_{10}O$

Synonym: pimelic ketone
ACGIH TLV: 25 ppm (100 mg/m³)
OSHA Standard: 50 ppm

Method	Sampling Duration	Sampling Location	Useful Range (ppm)	System Cost ($)	Test Cost ($)	Manufacturer
DT	Grab	Portable	2-60	180	2.25	Bendix
IR	Cont	Portable	1-100	4374	0	Foxboro
PI	Cont	Fixed	0.2-20	4950	0	HNU

NIOSH METHOD NO. P&CAM 127

(Use as described for acetone, with 110° C. column temperature)

CYCLOHEXENE

Synonym: tetrahydrobenzene C_6H_{10}
ACGIH TLV: 300 ppm (1015 mg/m^3)
OSHA Standard: 300 ppm

Method	Sampling Duration	Sampling Location	Useful Range (ppm)	System Cost ($)	Test Cost ($)	Manufacturer
IR	Cont	Portable	6-600	4374	0	Foxboro
GC	1 min	Portable	1-10^6	10,000	0	Microsensor

NIOSH METHOD NO. P&CAM 127

(Use as described for acetone)

DIACETONE ALCOHOL

Synonym: 4-hydroxy-4-methyl-2-pentanone $CH_3COCH_2C(OH)(CH_3)_2$
ACGIH TLV: 50 ppm (240 mg/m^3)
OSHA Standard: 50 ppm

Method	Sampling Duration	Sampling Location	Useful Range (ppm)	System Cost ($)	Test Cost ($)	Manufacturer
IR	Cont	Portable	1-100	4374	0	Foxboro

NIOSH METHOD NO. S66

(Use as described for allyl alcohol, with 120° C. column temperature)

o-DICHLOROBENZENE

ACGIH TLV: 50 ppm (300 mg/m^3) ceiling
OSHA Standard: 50 ppm ceiling

$C_6H_4Cl_2$

Method	Sampling Duration	Sampling Location	Useful Range (ppm)	System Cost ($)	Test Cost ($)	Manufacturer
DT	Grab	Portable	3-700	180	2.25	Bendix
IR	Cont	Portable	1-100	4374	0	Foxboro
PI	Cont	Fixed	2-200	4950	0	HNU

NIOSH METHOD NO. P&CAM 127

(Use as described for acetone, with 140° C. column temperature)

p-DICHLOROBENZENE

ACGIH TLV: 75 ppm (450 mg/m^3)
OSHA Standard: 75 ppm

$C_6H_4Cl_2$

Method	Sampling Duration	Sampling Location	Useful Range (ppm)	System Cost ($)	Test Cost ($)	Manufacturer
IR	Cont	Portable	1.5-150	4374	0	Foxboro

NIOSH METHOD NO. P&CAM 127

(Use as described for acetone, with 140° C. column temperature)

DICHLORODIFLUOROMETHANE

Synonyms: Freon 12; Refrigerant 12; F-12 CCl_2F_2
ACGIH TLV: 1000 ppm (4950 mg/m^3)
OSHA Standard: 1000 ppm

Method	Sampling Duration	Sampling Location	Useful Range (ppm)	System Cost ($)	Test Cost ($)	Manufacturer
IR	Cont	Portable	20-2000	4374	0	Foxboro
PI	Cont	Fixed	20-2000	4950	0	HNU
GC	1 min	Portable	1-10^6	10,000	0	Microsensor

NIOSH METHOD NO. S111

Principle

A known volume of air is drawn through two tubes in series containing activated coconut charcoal to trap the gaseous dichlorodifluoromethane. Dichlorodifluoromethane is desorbed from the charcoal with methylene chloride, and the sample is analyzed by gas chromatography.

Range and Sensitivity

This method was validated over the range of 2940-10,500 mg/m^3 at an atmospheric temperature of 25° C. and atmospheric pressure of 760 mm Hg, using a 3 L air sample. This maximum sample size is based on the capacity of the charcoal to collect vapors of dichlorodifluoromethane in air at 94% relative humidity. The method may be capable of measuring smaller amounts if the desorption efficiency is adequate. Desorption efficiency must be determined over the range used.

The upper limit range of the method depends on the absorptive capacity of the charcoal. This capacity may vary with the concentrations of dichlorodifluoromethane and other substances in the air. Breakthrough is defined as the time that the effluent concentration from the collection tube (containing 600 mg of charcoal) reaches 5% of the concentration in the test gas mixture. The criterion for acceptance is that the volume of air that has passed through the tube at the time of breakthrough must be greater than 1.5 times the volume of air that would be passed through the tube for a sample at 2X the OSHA standard. At a temperature of 23° C. and relative humidity of 12%, 9730 mg/m^3 was sampled at a flow rate of 0.043 L/min. When the relative humidity was 94%, breakthrough occurred in 96 min when the test concentration was 10,190 mg/m^3 and the flow rate was 0.045 L/min.

Interferences

When interfering compounds are known or suspected to be present in the air, such information, including their suspected identities, should be transmitted with the sample. Any compound that has the same retention time as dichlorodifluoromethane at the operating conditions described in this method is an interference. Retention time data on a single column cannot be considered proof of chemical identity.

Precision and Accuracy

The coefficient of variation for the total analytical and sampling method in the range of

2940-10,500 mg/m^3 was 0.064. This value corresponds to a 317 mg/m^3 standard deviation at the OSHA standard level. Statistical information can be found in reference 2. Details of the test procedures are found in reference 3.

No recovery correction should be applied to the final result. On the average, the concentrations obtained in the laboratory validation study at 0.5X, 1X, and 2X the OSHA standard level were 1.8% lower than the "true" concentrations for 18 samples. Any difference between the "found" and "true" concentrations may not represent a bias in the sampling and analytical method, but rather a random variation from the experimentally determined "true" concentration. Therefore, the method has no bias. The coefficient of variation is a good measure of the accuracy of the method since the recoveries and storage stability were good. Storage stability studies on samples collected from a test atmosphere at a concentration of 5650 mg/m^3 indicate that collected samples are stable for at least 7 days.

Advantages and Disadvantages

The sampling device is small, portable, and involves no liquids. Interferences are minimal, and most of those that occur can be eliminated by altering chromatographic conditions. The tubes are analyzed by means of a quick, instrumental method. One disadvantage of the method is that the amount of sample that can be taken is limited by the number of mg that the tube will hold before overloading. When the amount of dichlorodifluoromethane found on the backup charcoal tube exceeds 6 mg, the probability of sample loss exists. The precision of the method is limited by the reproducibility of the pressure drop across the tubes. This drop will affect the flow rate and cause the volume to be imprecise, because the pump is usually calibrated for one tube only.

Apparatus

Personal sampling pump. A calibrated personal sampling pump whose flow rate can be determined within 5% at the recommended flow rate.

Charcoal tubes. Two charcoal tubes are used in this method. One charcoal tube is used to collect the samples, and a smaller charcoal tube is used as a backup. The charcoal tubes are connected in series with Tygon tubing. The larger charcoal tube is a glass tube, approximately 10 cm long with 8 mm O.D. and 6 mm I.D. It has two sections of 20/40 mesh activated coconut charcoal separated by a 2 mm portion of urethane foam. The front section contains 400 mg of charcoal, the back section 200 mg. A 3 mm portion of urethane foam is placed between the outlet end of the tube and the backup section. A plug of silylated glass wool is placed in front of the adsorbing section. The pressure drop across the tube must be less than 25 mm of mercury at a flow rate of 0.05 L/min. The smaller charcoal tube is 7 cm long with a 6 mm O.D. and a 4 mm I.D. The smaller charcoal tube contains 100 mg of charcoal in the front section and 50 mg in the backup section.

Gas chromatograph. Equipped with a flame-ionization detector.

Column. 4 ft long × 1/4 in O.D. stainless steel, packed with 80/100 mesh Chromosorb 102.

Electronic integrator. Or some other suitable method for measuring peak areas.

Syringes. 10 µL, for injection of samples into the gas chromatograph.

Gas tight syringes. 1, 2, 5, and 10 mL sizes for preparing standards.

Pipets. 20.0 mL delivery pipets.

Serum bottles. 30 mL glass bottles with 20 mm O.D. mouth. Septa: 20 mm rubber septa with Teflon lining. Aluminum tear-away seals to fit serum bottles.

Hand crimper. For sealing septa to serum bottles.

Stopwatch.

Manometer.

Reagents

Methylene chloride, chromatographic quality.
Dichlorodifluoromethane, 99% pure.
Nitrogen, purified.
Hydrogen, prepurified.
Air, filtered, compressed.

Procedure

Cleaning of equipment. All glassware used for the laboratory analysis should be detergent washed and thoroughly rinsed with tap water and distilled water, and dried.

Calibration of sampling pumps. Each personal sampling pump must be calibrated with representative charcoal tubes in the line. This will minimize errors associated with uncertainties in the sample volume collected (reference 4).

Collection and shipping of samples. Immediately before sampling, break the ends of the two charcoal tubes to provide an opening at least one-half the internal diameter of the tube. Connect the larger charcoal tube in series with the smaller charcoal tube by connecting the backup section of the large charcoal tube to the front section of the small charcoal tube with Tygon tubing. The shortest length of tubing compatible with maintaining a leak-free connection should be used. The smaller charcoal tube is used as a backup and should be positioned nearer the pump. Air should flow through the large charcoal tube before entering the small charcoal tube.

Both charcoal tubes should be vertical during sampling to minimize channeling through the charcoal. Air being sampled should not be passed through any hose or tubing before entering the charcoal tube. A maximum sample size of 3 L is recommended. Sample at a flow rate between 0.01 and 0.05 L/min. The flow rate should be known with an accuracy of 5% or better and should be recorded. Also record collection time and the type of sampling pump used. The temperature, pressure, and relative humidity of the atmosphere being sampled should be recorded. If pressure reading is not available, record the elevation.

The charcoal tubes should be separated and capped individually with plastic caps immediately after sampling. Under no circumstances should rubber caps be used. Each set of tubes should be marked to identify the large charcoal tube with the corresponding backup charcoal tube. With each batch of ten samples, submit one set of tubes (a larger charcoal tube and a smaller backup tube) from the same lot of tubes used for sample collection. These tubes must be subjected to exactly the same handling as the samples except that no air is drawn through them. These tubes should be labeled as the blanks. Capped tubes should be packed tightly and padded before they are shipped to minimize tube breakage during shipping.

Analysis of samples. Pipet 20 mL of methylene chloride into the 30 mL serum bottle. Remove the plastic caps from both ends of the larger charcoal tube. Remove the plug of urethane foam from the outlet end of the tube. Transfer the 200 mg portion of charcoal to the serum bottle. Next, remove the glass wool plug from the inlet end of the tube and transfer the 400 mg section of charcoal to the same bottle. Place the Teflon-lined septum over the mouth of the bottle and seal the aluminum seal on with the crimper. It is important that the transfer of the charcoal be carried out as quickly as possible once the methylene chloride has been added to the bottle. The backup charcoal tube should be handled in a similar manner, using a separate serum bottle. The two samples are analyzed separately.

Gently shake the sample. The extract is now ready for analysis. Analyses should be completed within one day after the dichlorodifluoromethane is desorbed. Prepare standards as described below. Fresh standards should be prepared daily.

GC conditions. The typical operating conditions for the gas chromatograph are:

50 mL/min (60 psig) nitrogen carrier gas flow rate
65 mL/min (24 psig) hydrogen gas flow rate to detector
500 mL/min (50 psig) air flow rate to detector
200° C. injector temperature
260° C. detector temperature
110° C. column temperature

A retention time of approximately 2.5 min is to be expected for the analyte under these conditions and using the column recommended. The methylene chloride will elute from the column after the dichlorodifluoromethane.

Injection. The first step in the analysis is the injection of the sample into the gas chromatograph. To eliminate difficulties arising from blow back or distillation within the syringe needle, one should employ the solvent flush injection technique. The 10 μL syringe is first flushed with solvent several times to wet the barrel and plunger. Three μL of solvent are drawn into the syringe to increase the accuracy and reproducibility of the injected sample volume. The needle is removed from the solvent, and the plunger is pulled back about 0.2 μL to separate the solvent flush from the sample with a pocket of air to be used as a marker. The needle is then immersed in the sample, and a 5 μL aliquot is withdrawn, taking into consideration the volume of the needle, since the sample in the needle will be completely injected. After the needle is removed from the sample and prior to injection, the plunger is pulled back 1.2 μL to minimize evaporation of the sample from the tip of the needle. Observe that the sample occupies 4.9-5.0 μL in the barrel of the syringe. Duplicate injections of each sample and standard should be made. No more than a 3% difference in area is to be expected. It is not advisable to use an automatic sample injector for dichlorodifluoromethane in methylene chloride. The area of the sample peak is measured by an electronic integrator or some other suitable form of area measurement, and results are read from a standard curve prepared as discussed below.

Determination of desorption efficiency. The desorption efficiency of a particular compound can vary from one batch of charcoal to another. Thus, it is necessary to determine the fraction of the specific compound that is removed in the desorption process for a particular batch of charcoal.

Known amounts of dichlorodifluoromethane are sorbed from the gas phase on the charcoal in the tubes as described below. The apparatus to be used for spiking charcoal with known amounts of dichlorodifluoromethane consists of a union tee connected to a source of purified nitrogen, a charcoal tube, and a septum for introducing dichlorodifluoromethane from a gas tight syringe. A known amount of dichlorodifluoromethane is injected through the septum where nitrogen at a flow rate of approximately 20 mL/min carries dichlorodifluoromethane into the charcoal tube. The nitrogen is allowed to flow through the tube for an additional 30 sec after discharge of the contents of the syringe through the septum. The charcoal tube is capped with plastic caps.

The amount sorbed is equivalent to that present in a 3 L air sample at the selected level. This requires 1.5, 3, and 6 mL of dichlorodifluoromethane for 0.5X, 1X, and 2X the OSHA standard level, respectively. The atmospheric temperature, pressure, and relative humidity at which the tubes are prepared should be recorded. Six tubes at each of the three levels are prepared in this manner and allowed to stand for at least overnight to assure complete adsorption of dichlorodifluoromethane onto the charcoal in the tube. These tubes are referred to as the samples. A parallel blank tube should be treated in the same manner except that no sample is added to it.

Desorption and analysis experiments are done on the contents of the larger charcoal tube as described above. In preparation of the standards, 20 mL of methylene chloride is added to a 30 mL serum bottle. The bottle is then sealed using a Teflon-lined septum and aluminum seal. An appropriate amount of dichlorodifluoromethane gas is bubbled slowly into the methylene chloride, using a gas tight syringe. The syringe needle should be immersed in the methylene chloride during discharge of dichlorodifluoromethane from the syringe. The bottle should be shaken gently after removal of the syringe needle. Standards should be prepared at the same time that the sample analysis is done, and analyzed with the samples.

The weight (mg) of dichlorodifluoromethane in each standard and sample is calculated by the formula in the next section. The desorption efficiency (D.E.) equals the average weight in mg recovered from the tube divided by the weight in mg added to the tube, or:

D.E. = average weight recovered ÷ weight added

The desorption efficiency is dependent on the amount of dichlorodifluoromethane collected on the charcoal. Plot the desorption efficiency versus weight of dichlorodifluoromethane found. This curve is used to correct for losses, as described later.

Calibration and Standardization

A series of standards, varying in concentration over the range of interest, is prepared as described above and analyzed under the same GC conditions and during the same time period as the unknown samples. Curves are established by plotting concentration in mg/20 mL versus peak area. It is convenient to express concentration of standards in terms of mg/20 mL methylene chloride, because samples are desorbed in this amount of methylene chloride. The number of mg in the standard is calculated as follows:

mg = (V)(M.W.)(298)(P) ÷ (24.45)(T + 273)(760)

where:

V = mL of dichlorodifluoromethane used to prepare the standard
T = temperature (°C.) at which the sample was prepared
P = pressure (mm Hg) at which the sample was prepared
24.45 = molar volume (L/mol) at 25° C. and 760 mm Hg
M.W. = molecular weight of dichlorodifluoromethane
298 = standard temperature (°K)
760 = standard pressure (mm Hg)

Calculations

Read the weight, in mg, corresponding to each peak area from the standard curve. No volume corrections are needed, because the standard curve is based on mg/20 mL methylene chloride and the volume of sample injected is identical to the volume of the standards injected. Corrections for the blank must be made for each sample:

mg = mg sample − mg blank

where:

mg sample = mg found in both sections of sample tube
mg blank = mg found in both sections of blank tube

A similar procedure is followed for the backup tubes. Add the weights found in the front and backup tubes to get the total weight in the sample. Read the desorption efficiency from the curve for the amount found in the first tube. Divide the total weight by this desorption efficiency to obtain the corrected mg/sample:

Corrected mg/sample = total weight ÷ D.E.

For personal sampling pumps with rotameters only, the following correction should be made:

Corrected volume = $ft(P_1T_2/P_2T_1)^{1/2}$

where:

f = flow rate sampled
t = sampling time
P_1 = pressure during calibration of sampling pump (mm Hg)
P_2 = pressure of air sampled (mm Hg)
T_1 = temperature during calibration of sampling pump (°K)
T_2 = temperature of air sampled (°K)

The concentration of the dichlorodifluoromethane in the air sampled can be expressed in mg/m^3:

mg/m^3 = (corrected mg)(1000 L/m^3) ÷ (air volume sampled in L)

Another method of expressing concentration is ppm:

ppm = (mg/m^3)(24.45)(760)(T + 273) ÷ (M.W.)(P)(298)

where:

P = pressure (mm Hg) of air sampled
T = temperature (°C.) of air sampled
24.45 = molar volume (L/mol) at 25° C. and 760 mm Hg
M.W. = molecular weight (g/mol) of analyte
760 = standard pressure (mm Hg)
298 = standard temperature (°K)

References

1. White, L.D. et al. A convenient optimized method for the analysis of selected solvent vapors in the industrial atmosphere. Amer. Ind. Hyg. Assoc. J. 31: 225, 1970.
2. Documentation of NIOSH Validation Tests, NIOSH Contract No. CDC-99-74-45.
3. Backup Data Report for Dichlorodifluoromethane, prepared under NIOSH Contract No. 210-76-0123.
4. Final Report, NIOSH Contract No. HSM-99-71-31. Personal Sampler Pump for Charcoal Tubes, September 15, 1972.

1,1-DICHLOROETHANE

Synonym: ethylidene chloride $CHCl_2CH_3$
ACGIH TLV: 200 ppm (810 mg/m^3)
OSHA Standard: 100 ppm

Method	Sampling Duration	Sampling Location	Useful Range (ppm)	System Cost ($)	Test Cost ($)	Manufacturer
IR	Cont	Portable	2-200	4374	0	Foxboro
PI	Cont	Fixed	2-200	4590	0	HNU
GC	1 min	Portable	1-10^6	10,000	0	Microsensor

NIOSH METHOD NO. P&CAM 127

(Use as described for acetone, with 50° C. column temperature)

1,2-DICHLOROETHYLENE

Synonym: acetylene dichloride $CHCl=CHCl$
ACGIH TLV: 200 ppm (790 mg/m^3)
OSHA Standard: 200 ppm

Method	Sampling Duration	Sampling Location	Useful Range (ppm)	System Cost ($)	Test Cost ($)	Manufacturer
DT	Grab	Portable	10-450	180	2.25	Bendix
IR	Cont	Portable	4-400	4374	0	Foxboro
PI	Cont	Fixed	2-200	4950	0	HNU
GC	1 min	Portable	1-10^6	10,000	0	Microsensor

NIOSH METHOD NO. P&CAM 127

(Use as described for acetone)

DICHLOROETHYL ETHER

Synonym: bis-β-chloroethyl ether $(ClC_2H_4)_2O$
ACGIH TLV: 5 ppm (30 mg/m^3) - Skin
OSHA Standard: 15 ppm ceiling

Method	Sampling Duration	Sampling Location	Useful Range (ppm)	System Cost ($)	Test Cost ($)	Manufacturer
IR	Cont	Portable	0.15-15	4374	0	Foxboro

NIOSH METHOD NO. P&CAM 127

(Use as described for acetone, with 135° C. column temperature)

DICHLOROMONOFLUOROMETHANE

Synonyms: Freon 21; Refrigerant 21; F-21 $CHCl_2F$
ACGIH TLV: 10 ppm (40 mg/m^3)
OSHA Standard: 1000 ppm

Method	Sampling Duration	Sampling Location	Useful Range (ppm)	System Cost ($)	Test Cost ($)	Manufacturer
IR	Cont	Portable	20-2000	4374	0	Foxboro
PI	Cont	Fixed	0.2-20	4950	0	HNU

NIOSH METHOD NO. P&CAM 127

(Use as described for acetone, with 55° C. column temperature)

DICHLOROTETRAFLUOROETHANE

Synonyms: Freon 114; Refrigerant 114; F-114 $CClF_2CClF_2$
ACGIH TLV: 1000 ppm (7000 mg/m^3)
OSHA Standard: 1000 ppm

Method	Sampling Duration	Sampling Location	Useful Range (ppm)	System Cost ($)	Test Cost ($)	Manufacturer
IR	Cont	Portable	20-2000	4374	0	Foxboro
PI	Cont	Fixed	20-2000	4950	0	HNU
GC	1 min	Portable	1-10^6	10,000	0	Microsensor

NIOSH METHOD NO. S111

(Use as described for dichlorodifluoromethane)

DICHLORVOS

Synonyms: DDVP; Vapona $(CH_3O)_2P(S)OCH=CCl_2$
ACGIH TLV: 0.1 ppm (1 mg/m^3) - Skin
OSHA Standard: 0.1 ppm

Method	Sampling Duration	Sampling Location	Useful Range (ppm)	System Cost ($)	Test Cost ($)	Manufacturer
IR	Cont	Portable	0.05-1	4374	0	Foxboro

NIOSH METHOD NO. P&CAM 295

Principle
A known volume of air is drawn through a tube containing XAD-2 to trap the organic vapors present. The sampling tube consists of a front adsorbing section and a backup section. The XAD-2 in each tube is transferred to respective vials and the dichlorvos is desorbed with toluene. An aliquot of this sample solution is injected into a gas chromatograph equipped with a flame photometric detector and phosphorus filter. The area of the resulting peak is determined and compared with areas obtained from the injection of standards.

Range and Sensitivity
This method was tested over the range of 0.382-1.707 mg/m^3 at an atmospheric temperature of 24.0° C., and an atmospheric pressure of 758.1 mm Hg using a 120 L sample volume. The method

may be capable of measuring smaller amounts if the desorption efficiency is adequate. Desorption efficiency must be determined over the range used.

The upper limit of the range of the method is dependent on the absorptive capacity of the XAD-2 resin. This capacity varies with the concentrations of dichlorvos and other substances in the air. Breakthrough is defined as the time that the effluent concentration from the collection tube (containing 100 mg XAD-2) reaches 5% of the concentration in the test gas mixture. When an atmosphere at 90% relative humidity containing an average of 1.73 mg/m^3 of dichlorvos was sampled at 1.092 L/min, no breakthrough was observed after 240 min (capacity = 26.3 L or 0.24 mg). The sample size recommended is less than two-thirds the 5% breakthrough capacity to minimize the probability of overloading the sampling tube. The detection limit for dichlorvos was found to be 0.2 μg/sample.

Interferences

When two or more compounds are known or suspected to be present in the air, such information, including their suspected identities, should be transmitted with the sample. It must be emphasized that any compound which has the same retention time as the analyte at the operating conditions described in this method is an interference. Retention time data on a single column cannot be considered as proof of chemical identity.

Precision and Accuracy

The coefficient of variation for the total analytical and sampling method in the range of 0.382-1.707 mg/m^3 was 0.0535. This value corresponds to a 0.0535 mg/m^3 standard deviation at the 1 mg/m^3 level. Statistical information and details of the experimental test procedures can be found in references 1 and 2.

The accuracy of this method has not been determined. The desorption efficiency was determined to be 0.922 for a collector loading of 0.063 mg. In storage stability studies, the mean of generated samples analyzed after seven days was 94.5% of the mean of samples analyzed immediately after collection. Experiments performed in these studies are described in reference 2.

Advantages and Disadvantages

The sampling device is small, portable, and involves no liquids. Interferences are minimal, and most of those which do occur can be eliminated by altering chromatographic conditions. The collected samples are analyzed by means of a quick, instrumental method.

One disadvantage of the method is that the amount of sample that can be taken is limited by the number of mg that the tube will hold before overloading. When the amount of dichlorvos found on the backup XAD-2 section exceeds 25% of that found on the front section, the probability of sample loss exists.

The precision of the method is affected by the reproducibility of the pressure drop across the tubes. This drop will affect the flow rate and may cause the volume to be imprecise because the pump is usually calibrated for one tube only. The accuracy of the analytical method is unknown.

Apparatus

Sampling pump. A calibrated personal sampling pump suitable for sampling at 1.0 L/min for 120 min. The pump must be accurate to within 5% at the recommended flow rate.

Sampling tubes. Wash the glass tubes with acetone prior to packing with the sorbent. Allow them to air dry. (This will prevent the XAD-2 resin from adhering to the walls of the tube.) The sampling tube consists of a glass tube, flame sealed at both ends, 10 cm long with 8 mm O.D. and 6 mm I.D., packed with two sections of 20/50 mesh XAD-2. The XAD-2 used must be prewashed with toluene, and air dried. The two sections include a front adsorbing section containing 100 mg of XAD-2 and a backup section containing 30 mg. The two sections are separated by a portion of silylated glass wool. A plug of silylated glass wool is placed at each end of the sorbent tube. The pressure drop across the tube must be less than one inch of mercury at a flow rate of 1.0 L/min.

Gas chromatograph. With a flame photometric detector and phosphorus filter.

Column. 6 ft × 1/4 in glass, packed with 1.5% OV-101 stationary phase on 100/120 mesh Supelcoport.

Electronic integrator. Or some other suitable method for measuring peak areas.

Syringes. 10 and 100 µL and other convenient sizes for making standards and for taking sample aliquots.

Pipettes. 2 mL, delivery type.

Volumetric flasks. 10 mL or other convenient sizes for making standard solutions.

Sample vials. 4 mL, screw cap, with Teflon cover.

Reagents

Whenever possible, reagents used should be ACS reagent grade or better.

Dichlorvos.

Toluene.

Tributyl phosphate. 98%, or other suitable internal standard. The appropriate solution of the internal standard is prepared in toluene.

Pre-cleaned resin. XAD-2 resin (20/50 mesh) can be obtained from Rohm and Haas Company. XAD-2 resin is purified by charging an amount into a Soxhlet extractor. Twenty-four hour extractions are then performed successively with water, methanol, and methylene chloride. Resin has been prepared in this manner using charges of about 700 g of resin and 1.5 L of each solvent. The resin is dried in a fluidized bed using nitrogen gas at room temperature from a liquid nitrogen cylinder. The drying process is terminated when essentially no solvent is detected in the effluent. A final quality control check is performed by desorbing a portion of the resin and analyzing the resulting solution by gas chromatography. Residual solvent should be less than 1000 ppm in concentration. Finally, several washings with toluene are recommended to reduce possible interferences to a minimum when the sorbent is desorbed with this solvent. This can be done in a beaker of the appropriate volume. The resin is then air dried in a hood.

Helium, purified.

Hydrogen, prepurified.

Air, filtered, compressed.

Procedure

Cleaning of equipment. All glassware used for the laboratory analysis should be detergent washed and thoroughly rinsed with tap water and distilled water.

Calibration of personal sampling pumps. Each personal sampling pump must be calibrated with a representative resin tube in the line. This will minimize errors associated with uncertainties in the sample volume collected.

Collection and shipping of samples. Immediately before sampling, break the ends of the tube to provide an opening at least one-half the internal diameter of the tube (3 mm). The section containing 30 mg of XAD-2 is used as a backup and should be positioned nearest the sampling pump. The XAD-2 tube should be maintained in a vertical position during sampling to avoid channeling and subsequent premature breakthrough of the analyte. Air being sampled should not be passed through any hose or tubing before entering the front section of the XAD-2 tube.

A sample size of 120 L is recommended. Sample at a known flow rate between 0.5 and 1 L/min. Set the flow rate as accurately as possible using the manufacturer's directions. Record the necessary information to determine flow rate and also record the initial and final sampling time.

Record the temperature and pressure of the atmosphere being sampled. If pressure reading is not available, record th elevation. The XAD-2 tubes should be labeled properly and capped with the supplied plastic caps immediately after sampling.

One XAD-2 tube should be handled in the same manner as the tubes (break, seal, and transport), except for the taking of an air sample. This set of tubes should be labeled as a blank. Submit one blank for every ten samples. At least two unused XAD-2 tubes should accompany the samples. These tubes are used in desorption efficiency studies in conjunction with these samples because desorption efficiency may vary from one batch of XAD-2 to another. Capped XAD-2 tubes should be packed tightly and padded before they are shipped to minimize tube breakage during shipping.

Preparation of samples. In preparation for analysis, each tube is scored with a file and broken open. The glass wool is removed and discarded. The XAD-2 in each tube is transferred to a 5 mL Teflon-lined screw cap sample vial. Each tube is analyzed separately.

Desorption of samples. Prior to analysis, 2.0 mL of toluene is pipetted into each sample vial. Desorption should be done for 30 minutes. Tests indicate that this is adequate if the sample is agitated occasionally during this period. The sample vials should be capped as soon as the solvent is added to minimize volatilization. For the internal standard method, desorb using 2.0 mL of toluene containing a known amount of internal standard.

GC conditions. The typical operating conditions for the gas chromatograph are:

35 mL/min (60 psig) helium carrier gas flow
50 mL/min (25 psig) hydrogen gas flow to detector
200 mL/min (60 psig) air flow to detector
235° C. injector temperature
220° C. flame photometric detector
150° C. column temperature

A retention time of approximately 75 sec is to be expected for the analyte using these conditions and the column recommended. The internal standard elutes in approximately 6 min.

Injection. A 1 μL aliquot of the sample solution is injected into the gas chromatograph. The solvent flush method or other suitable alternative such as an automatic sample injector can be used provided that duplicate injections of a solution agree well. No more than a 3% difference in area is to be expected.

Measurement of area. The signal of the sample peak is measured by an electronic integrator or some other suitable form of measurement such as peak height, and preliminary results are read from a standard curve prepared as discussed below.

Determination of desorption efficiency. The desorption efficiency of a particular compound may vary from one laboratory to another and also from one batch of XAD-2 to another. Thus, it is necessary to determine the percentage of the specific compound that is removed in the desorption process for a particular batch of resin used for sample collection and over the concentration range of interest. The desorption efficiency must be determined over the sample concentration range of interest. In order to determine the range which should be tested, the samples are analyzed first and then the analytical samples are prepared based on the amount of dichlorvos found in the samples.

The analytical samples are prepared as follows: XAD-2 resin, equivalent to the amount in the front section (100 mg), is measured into a 4 mL, screw-cap vial. This resin must be from the same batch used in obtaining the samples. A known amount of a solution of dichlorvos in toluene (spiking solution) is injected directly into the resin by means of a microliter syringe. Adjust the concentration of the spiking solution such that no more than a 10 μL aliquot is used to prepare the analytical samples.

Six analytical samples at each of the three concentration levels (0.5, 1, and 2X the OSHA standard) are prepared by adding an amount of dichlorvos equivalent to a 120 L sample at the selected level. A stock solution containing 30.0 mg of dichlorvos per mL of toluene is prepared. Aliquots (2.0, 4.0 and 8.0 µL) of the solutions are added to the XAD-2 resin vials to produce 0.5, 1 and 2X the OSHA standard level. The analytical samples are allowed to stand overnight to assure complete adsorption of the analyte onto the resin. A parallel blank vial is treated in the same manner except that no sample is added to it.

Desorption and analysis. Desorption and analysis experiments are done on the analytical samples as described above. Calibration standards are prepared by adding the appropriate volume of spiking solution to 2.0 mL of toluene with the same syringe used in the preparation of the samples. Standards should be prepared and analyzed at the same time the sample analysis is done. If the internal standard method is used, prepare calibration standards by using 2.0 mL of toluene containing a known amount of the internal standard. The desorption efficiency (D.E.) equals the average weight in mg recovered from the vial divided by the weight in mg added to the vial, or:

D.E. = (average weight recovered − wt of resin blank) ÷ weight added

The desorption efficiency may be dependent on the amount of dichlorvos collected on the resin. Plot the desorption efficiency versus weight of dichlorvos. This curve is used to correct for adsorption losses.

Calibration and Standards

A series of standards varying in concentration over the range corresponding to 120 L collections at 0.1-3 times the OSHA standard is prepared and analyzed under the same GC conditions and during the same time period as the unknown samples. This is done in order to minimize variations in FPD response. It is convenient to express concentration of standards in terms of mg/2.0 mL since the samples are desorbed in 2.0 mL of toluene. A calibration curve is established by plotting peak area versus concentration in mg/2.0 mL. Prepare a stock standard solution containing about 30 mg/mL of dichlorvos in toluene.

From the above stock solution, appropriate aliquots are added to 2.0 mL of toluene. Prepare at least five standards to cover the range of 0.012-0.360 mg/sample. The range is based on a 120 L air sample. For the internal standard method, use toluene containing a predetermined amount of an internal standard such as tributyl phosphate. The internal standard concentration used for these studies was equivalent to the analyte concentration for a standard solution representing a 120 L collection at 2X the OSHA standard. However, the area of the tributyl phosphate peak at this level was approximately 70% of the area of the dichlorvos due to differences in detector response to the two compounds. The area ratio of the analyte to that of the internal standard is plotted against the analyte concentration in mg/2.0 mL.

Calculations

Read the weight, in mg, corresponding to each peak area from the standard curve. No volume corrections are needed, because the standard curve is based on mg/2.0 mL and the volume of sample injected is identical to the volume of the standards injected. Corrections for the blank must be made for each sample:

mg = mg sample − mg blank

where:

mg sample = mg found in front section of sample vial
mg blank = mg found in front section of blank vial

A similar procedure is followed for the backup sections. Add the weights found in the front and backup sections to determine the total weight in the sample. Read the desorption efficiency from the curve for the amount found in the front section of the tube. Divide the total weight by this desorption efficiency to obtain the corrected mg/sample:

Corrected mg/sample = total weight ÷ D.E.

Determine the volume of air sampled at ambient conditions in L based on the appropriate information, such as flow rate in L/min multiplied by sampling time. If a pump using a rotameter for flow rate control was used for sample collection, a pressure and temperature correction must be made for the indicated flow rate. The expression for this correction is:

Corrected volume = $ft(P_1T_2/P_2T_1)^{1/2}$

where:

    f = sampling flow rate
    t = sampling time
    P_1 = pressure during calibration of sampling pump (mm Hg)
    P_2 = pressure of air sampled (mm Hg)
    T_1 = temperature during calibration of sampling pump (°K)
    T_2 = temperature of air sampled (°K)

The concentration of the analyte in the air sampled can be expressed in mg/m^3, which is numerically equal to $\mu g/L$:

mg/m^3 = (corrected mg)(1000 L/m^3) ÷ air volume sampled in L

Another method of expressing concentration is ppm (corrected to standard conditions of 25° C. and 760 mm Hg):

ppm = (mg/m^3)(24.45)(760)(T + 273) ÷ (220.93)(P)(298)

where:

    P = pressure (mm Hg) of air sampled
    T = temperature (°C.) of air sampled
    24.45 = molar volume (L/mol) at 25° C. and 760 mm Hg
    220.93 = molecular weight of dichlorvos
    298 = standard temperature (°K)

References

1. *Documentation of NIOSH Validation Tests,* National Institute for Occupational Safety and Health, Cincinnati, Ohio (DHEW-NIOSH Publication No. 77-185), 1977. Available from Superintendent of Documents, U.S. Government Printing Office, Washington D.C., Order No. 017-033-00231-2.
2. Backup Data Report No. 295 for Dichlorvos, prepared under NIOSH Contract No. 210-76-0123.

DIETHYLAMINE

ACGIH TLV: 10 ppm (30 mg/m³)
OSHA Standard: 25 ppm (75 mg/m³)

$(C_2H_5)_2NH_2$

Method	Sampling Duration	Sampling Location	Useful Range (ppm)	System Cost ($)	Test Cost ($)	Manufacturer
DT	Grab	Portable	1-60	180	2.25	Bendix
CS	Grab	Portable	0.5-10	2465	0.80	MDA
IR	Cont	Portable	0.5-50	4374	0	Foxboro
PI	Cont	Fixed	0.2-20	4950	0	HNU
GC	1 min	Portable	$1-10^6$	10,000	0	Microsensor

NIOSH METHOD NO. P&CAM 221

(See n-butylamine)

2-DIETHYLAMINOETHANOL

Synonym: DEAE
ACGIH TLV: 10 ppm (50 mg/m³) - Skin
OSHA Standard: 10 ppm

$(C_2H_5)_2NCH_2CH_2OH$

Method	Sampling Duration	Sampling Location	Useful Range (ppm)	System Cost ($)	Test Cost ($)	Manufacturer
IR	Cont	Portable	0.2-20	4374	0	Foxboro

NIOSH METHOD NO. P&CAM 270

Principle

A known volume of air is drawn through a tube containing silica gel to trap the aminoethanol compounds. Concentrated HCl is added to stabilize amines collected. The sorbent sections are transferred to glass tubes and treated with methanol solution. An aliquot is alkalinized with sodium hydroxide. Benzaldehyde is added to derivatize 2-aminoethanol forming benzylidene-aminoethanol. The solution is analyzed by gas chromatography with flame-ionization detection.

Range and Sensitivity

The ranges of the method for the various analytes are at least:

2-aminoethanol	0.1-15 mg/sample
2-diethylaminoethanol	0.1-12 mg/sample
2-dibutylaminoethanol	0.1-17 mg/sample

The lower ends of the ranges given are levels where the desorption and analysis procedure was evaluated and found to be acceptably precise. The upper limits are determined by the capacities of the first section of the sampling tube used to collect the sample. Amounts of individual compounds retained on 150 mg of a 42/60 mesh silica gel in 6 mm I.D. sample tubes at 1 L/min sampling flow rate were found to be those given above as the upper limits.

Interferences

Water vapor does not significantly affect collection efficiency. Large amounts of ammonia or primary amines will reduce the amount of benzaldehyde available for reaction with 2-aminoethanol. Such interferences should be compensated for by adding additional benzaldehyde.

Any compound which has nearly the same retention time on the GC column as one of the analytes is an interference. Retention time data on a single GC column cannot be considered as proof of chemical identity. The presence and identity of possible interfering substances may be determined by the analysis of bulk samples from the process or environment.

Precision and Accuracy

The volume of air sampled can be measured to within 2% if a pump with a calibrated volume indicator and an adequate battery is used. Volumes calculated from initially set flow rates may be less accurate (5-10%) unless changes in flow rate are manually or electronically monitored and compensated. Below the following levels precisions of replicate GC analyses of standards were worse than 9% relative standard deviation:

2-aminoethanol	0.10 mg/sample
2-diethylaminoethanol	0.08 mg/sample
2-dibutylaminoethanol	0.02 mg/sample

At the upper limits of the method analytical precision becomes 2% relative standard deviation or better. The precision (relative standard deviation) of the method based on analyses of replicate 0.1 mg samples injected by syringe into silica gel sections was:

2-aminoethanol	3%
2-diethylaminoethanol	7%
2-dibutylaminoethanol	2%

Replicate samples of vapor at higher sample levels had precisions ranging from 2-7%. The precision of the analysis is dependent upon the precision and sensitivity of the technique used to quantitate the GC peaks of samples and standards. An electronic integrator with baseline correction capability is best for this purpose.

Collection efficiency is 100% until breakthrough of the sampling tube beds occurs. Desorption efficiencies based on analyses of replicate 0.1 mg samples on silica gel and standards in 2 mL eluent were found to be:

2-aminoethanol	0.97 (7%)
2-diethylaminoethanol	0.85 (22%)
2-dibutylaminoethanol	0.93 (3%)

with the relative standard deviations given in parentheses. Measurements at higher sample levels have given desorption efficiencies consistently in the range 90-100%.

This method has given quantitative results equivalent to an independent method involving desorption of 2-aminoethanol vapors in water in a bubbler and titration with standard acid. Using each method, four or five 0.030 m^3 vapor in nitrogen samples were taken at 1 L/min. The following sample amounts and relative standard deviations in parenthesis were determined:

Tube and chromatography	Bubbler and titration
0.31 mg (7.3%)	0.35 mg (3.4%)
1.91 mg (5.0%)	1.74 mg (1.5%)

Advantages and Disadvantages

The method uses a small, portable sampling device involving no liquids. This is an advantage for sampling air in a worker's breathing zone without interfering with normal work activities. Transportation to the analytical laboratory is simplified with the solid sorbent tube. The sorbent tube has a high capacity even at high relative humidities. It can be used for at least 8 hours to measure a workday average concentration, or for shorter times to measure excursion concentrations. Desorption and preparation of samples for analysis involve simple procedures and equipment.

Several amines can be collected and determined simultaneously. The GC analysis distinguishes which are present and at what individual concentrations they occur. Interferences by other amines are much less likely than in colorimetric or titrimetric methods. A major disadvantage is the tendency of amines to oxidize while adsorbed on surfaces exposed to air. This is overcome by addition of concentrated hydrochloric acid to the sorbent sections soon after sampling.

Apparatus

Sorbent. The silica gel used should be the equivalent of Silica Gel D-08, chromatographic grade, activated and fines free, 42/60 mesh, as produced by Coast Engineering Laboratories and sold by Applied Science Laboratories (State College, PA).

Sampling tubes. Glass tubes 7 cm long and 6 mm I.D., tapered and flame sealed at one end, are packed with two 150 mg sections of silica gel. Glass wool plugs are used to separate and enclose the sections. The second end is flame sealed to prevent contamination during storage prior to use. Note: sampling tubes constructed with metal parts or urethane foam plugs should not be used. Concentrated hydrochloric acid will react with these materials.

Personal sampling pump. Battery operated pumps are required, capable of operation at 0.2 L/min for up to 8 hours or at 1.0 L/min for up to 2 hours with a sampling tube in line. The pump is to be calibrated with a representative sorbent tube in line. A wet or dry test meter or bubble meter capable of measuring a flow rate of 0.2 L/min and/or 1.0 L/min to within 2% may be used in setting the pump flow.

Gas chromatograph. With a flame-ionization detector. Temperature programming capability is necessary to determine more than one compound simultaneously.

GC column. 1.8 mm × 2 mm I.D. glass, silanized and packed with 10% (by weight) Carbowax 20M and 2% KOH on 80/100 mesh Chromosorb W AW, or equivalent support.

Strip chart recorder. Compatible with the GC. An electronic digital integrator is desirable.

Test tubes or glass vials. 2 mL, sealed by a septum.

Syringes. 10 µL.

Syringe for dispensing conc. HCl. 50 µL, with glass barrel, fluorocarbon-tipped plunger, and inert (platinum or fluorocarbon) needle.

Pipettes. 0.5 and 2 mL.

Volumetric flasks. 10 mL.

File and forceps.

Reagents

2-Aminoethanol (ethanolamine).

2-Diethylaminoethanol.
2-Dibutylaminoethanol.
Water, doubly distilled and aldehyde-free. Deionized water should not be used as it may contain formaldehyde and other impurities leached from the ion exchange resins.
Concentrated HCl (38%, 12 M).
Benzaldehyde.
Eluent solution, 4:1 methanol-water (by volume).
Alkalinizing solution. 0.20 N NaOH in 4:1 methanol-water (by volume).
Helium, Bureau of Mines Grade A.
Hydrogen, prepurified.
Air, compressed and filtered.

Procedure

Cleaning of equipment. All glassware is washed with detergent solution, rinsed with tap water and distilled water, and dried in an oven.

Collection and shipping of samples. Immediately before beginning the collection of a sample, break each end of the sorbent tube so as to provide openings at least 2 mm in diameter. Attach the tubing from the sampling pump to the backup end of the sampling tube. Sampled air must not pass through any hose or tubing before entering the sorbent tube. With the sorbent tube in a vertical position as much as is practical, sample the air at 1.0 L/min for 15 min-2 hours or at 0.2 L/min for 2-8 hours. Intermediate flow rates and other sampling periods may be used, as appropriate.

Immediately after sampling is completed, add 20 μL of concentrated HCl to each section of silica gel in the tube. Use a 50 μL glass syringe with fluorocarbon tipped plunger in inert needle, so that the acid does not contact and react with stainless steel. The silica gel will turn yellow upon addition of the acid due to iron impurities in the silica gel. Seal the sorbent tubes with polyethylene caps.

Obtain a blank sample by handling one tube in the same manner as the sample tubes (break, add acid, seal, and ship) except that no air is pumped through it. For shipping to the laboratory, pack the tubes tightly to minimize chances of breakage during transit. Tubes should not be subjected to extremes of high temperature or low pressure.

Preparation of samples. Score each tube with a file 5 mm in front of the glass wool plug that precedes the first sorbent section and break the tube there. Transfer this plug and the initial section to a 2 mL test tube or glass vial that can be septum sealed. Likewise, transfer the second plug and sorbent section to another test tube. Label each appropriately for separate analysis.

Desorption. Add 2 mL of eluent solution to each sorbent section. Seal the samples. Shake the mixture occasionally over a period of 2 hours. Tests have shown that complete desorption occurs within 2 hours and samples are stable for a least a day.

Preparation. Transfer a 0.50 mL aliquot of each eluate to another test tube or vial. Add 0.50 mL of alkalinizing solution and mix thoroughly. Verify the basicity of the solution with litmus paper.

Reaction. If 2-aminoethanol is present in the sample, repeat the preparation step with another 0.50 mL aliquot. Add 10 μL of benzaldehyde to the basic solution, mix thoroughly, and allow at least 20 min for reaction.

GC conditions. Typical operation conditions are:

50 mL/min helium gas flow rate
150° C. injection port temperature
250° C. detector temperature
90° C. column temperature for 3 min, heat to 225° C. at 16° C./min, and hold for 6 min

Under these conditions, retention times are:

213 sec for 2-diethylaminoethanol
365 sec for 2-aminoethanol
515 sec for 2-dibutylaminoethanol
866 sec for 2-benzylideneaminoethanol

When only one compound is present, isothermal analysis is best at:

90° C. (retention time, 280 sec) for 2-diethylaminoethanol
150° C. (retention time, 180 sec) for 2-dibutylaminoethanol
225° C. (retention time, 195 sec) for 2-benzylideneaminoethanol

Injection. Inject and analyze 3 µL aliquots of each sample and standard. To eliminate difficulties arising from blow back distillation within the 10 µL syringe, use a solvent flush technique.

GC peak measurement. Determine the areas of the peaks of the compounds of interest from analyses of samples and standards.

Determination of desorption efficiency. Desorption efficiency for a particular compound can vary from one lot of silica gel to another and from one laboratory to another. Also, for a given lot of silica gel the desorption efficiency may vary with the amount of material adsorbed during sampling. Therefore, it is necessary to determine at least once the desorption efficiency for each aminoethanol with each lot of silica gel used for at least two levels within the normal range of sample size. If the desorption efficiencies are significantly different from quantitative recovery, they should be used to correct the measured weight of aminoethanols as described below.

Place 150 mg of silica gel in a 2 mL glass stoppered tube. The silica gel must be from the same lot as that used in collecting the sample; it can be obtained from unused sorbent tubes. With a microliter syringe, inject a known amount of the aminoethanol, either pure or in water solution, directly onto the silica gel. Also inject 20 µL of concentrated HCl. Close the tube with the glass stopper and allow it to stand overnight to insure complete adsorption of the amine. Prepare at least three tubes for each at two different levels. These tubes are referred to as samples. Prepare a blank, in the same manner, omitting the aminoethanol. Analyze the samples and blank as described above. Also analyze three standards prepared by adding identical amounts of the aminoethanol and 20 µL of concentrated HCl to a 2.0 mL of eluent solution. Determine the concentrations of the aminoalcohols in the blank, samples, and standards using calibration curves prepared as described below. The desorption efficiency is calculated by dividing the concentration of amine found in the sample by concentration obtained in the corresponding standard.

Calibration and Standardization

Preparation of a stock solution. Calculate for each compound i the volume V_i (µL) of pure liquid required for preparation of 10 mL of a stock solution, 10 µL of which contains amounts of aminoethanols equal to those collected from volume of air V_s at concentrations X_i:

$$V_i = (X_i)(V_s)(1000) \div d_i$$

where:

x_i = the standard concentration limit, or the anticipated average concentration (mg/m^3), of compound i in air

V = volume (m³) of air sampled
di = density (mg/μL) of pure compound i
1000 = aliquot factor

Add the calculated volumes of each compound to a 10 mL volumetric flask and dilute to the mark with 4:1 methanol-water. Neutral solutions of aminoethanol are subject to oxidation and should be prepared fresh when needed.

Preparation of calibration standards. Add 2.5, 5, 10, 15, and 20 μL of the prepared standards to reaction tubes containing 2 mL eluent solutions and 10 μL concentrated hydrochloric acid. Use a 10 μL syringe to inject through septum seals. Shake these tubes. These standards correspond to 0.25, 0.5, 1, 1.5, and 2 times Xi in Vs, respectively. Other standards may be similarly prepared, if desired. React and analyze standards with samples as described above. Prepare a calibration curve for each amine by plotting peak areas obtained from the analyses of standards against nominal amount (mg) in the calibration standards.

Calculations

Read the weights (mg) of each compound corresponding to each peak area from the appropriate calibration curve. Correct each valve for the weight found in a blank, if any. Add the weights found in the front and backup sections of the sample tube to obtain the total weight of compound in the air volume sampled. Divide the total weight by the desorption efficiency to obtain the corrected sample weight.

Determine the volumes (m³) of air sampled at ambient conditions based on the appropriate information, such as flow rate (L/min) multiplied by sampling time (min) and 10^{-3} m³/L. If a pump using a rotameter for flow rate control was used for sample collection, a pressure and temperature correction must be made for the indicated flow rate. The expression for this correction is:

$$V_s = (10^{-3}) ft(P_1 T_2 / P_2 T_1)^{1/2}$$

where:

f = sample flow rate (L/min)
t = sampling time (min)
P_1 = pressure during calibration of sampling pump (torr)
P_2 = pressure of air samples (torr)
T_1 = temperature during calibration of sampling pump (°K)
T_2 = temperature of air sampled (°K)

Divide the total corrected weight (mg) of each compound by Vs (m³) to obtain the concentration of the compound in the air sampled (mg/m³).

References

1. G.O. Wood, R.G. Anderson, and J.W. Nickols. Sampling and Analysis of Aminoethanols in Air, Report LA-UR-77-1398, 1977, Industrial Hygiene Group, Los Alamos Scientific Laboratory, Los Alamos, New Mexico. (Presented at the 1977 American Industrial Hygiene Conference, May 1977, New Orleans).
2. G.O. Wood, and J.W. Nickols. Development of Air-Monitoring Techniques Using Solid Sorbents, October 1, 1976-December 31, 1977, Progress Report LA-7295-PR, Los Alamos Scientific Laboratory, Los Alamos, New Mexico. Also available from the National Technical Information Services, Springfield, VA, 1978.

DIFLUORODIBROMOMETHANE

Synonym: dibromodifluoromethane CF_2Br_2
ACGIH TLV: 100 ppm (860 mg/m³)
OSHA Standard: 100 ppm

Method	Sampling Duration	Sampling Location	Useful Range (ppm)	System Cost ($)	Test Cost ($)	Manufacturer
IR	Cont	Portable	2-200	4374	0	Foxboro

NIOSH METHOD NO. S107

Principle

A known volume of air is drawn through two charcoal tubes in series to trap the organic vapors present. The charcoal in the tubes is transferred to a small, stoppered sample container and the analyte is desorbed with isopropyl alcohol. An aliquot of the desorbed sample is injected into a gas chromatograph. The area of the resulting peak is determined and compared with areas obtained from the injection of standards.

Range and Sensitivity

This method was validated over the range of 470-1875 mg/m³ at an atmospheric temperature and pressure of 25° C. and 745 mm Hg, using a 10 L sample. Under the conditions of sample size (10 L) the probable range of this method is 90-2580 mg/m³ at a detector sensitivity that gives nearly full deflection on the strip chart recorder for a 19 mg sample. The method is capable of measuring much smaller amounts if the desorption efficiency is adequate. Desorption efficiency must be determined over the range used.

The upper limit of the range of the method is dependent on the adsorptive capacity of the charcoal tube. This capacity varies with the concentrations of the analyte and other substances in the air. The first section of the charcoal tube was found to hold 29.3 mg of the analyte when a test atmosphere of 1875 mg/m³ of the analyte in dry air was sampled at 0.2 L/min for 1 hour, 18 min; breakthrough was observed at this time, i.e., the concentration of the analyte in the effluent was 5% of that in the influent. (The charcoal tube consists of two sections of activated charcoal separated by a section of urethane foam.) If a particular atmosphere is suspected of containing a large amount of contaminant, a smaller sampling volume should be taken.

Interferences

When the amount of water in the air is so great the condensation actually occurs in the tube, organic vapors will not be trapped efficiently. Preliminary experiments with toluene indicate that high humidity severely decreases the breakthrough volume. When two or more compounds are known or suspected to be present in the air, such information, including their suspected identities, should be transmitted with the sample.

It must be emphasized that any compound which has the same retention time as the specific compound under study at the operating conditions described in this method is an interference. Retention time data on a single column cannot be considered as proof of chemical identity. If the possibility of interference exists, separation conditions (column packing, temperature, etc.) must be changed to circumvent the problem.

Precision and Accuracy

The coefficient of variation for the total analytical and sampling method in the range of 470 to 1875 mg/m^3 was 0.090. This value corresponds to a standard deviation of 77.4 mg/m^3 at the OSHA standard level. Statistical information and details of the validation and experimental test procedures can be found in reference 2. The average values obtained using the overall sampling and analytical method were 4.7% higher than the "true" value at the OSHA standard level.

Advantages and Disadvantages

The sampling device is small, portable, and involves no liquids. Interferences are minimal, and most of those which do occur can be eliminated by altering chromatographic conditions. The tubes are analyzed by means of a quick, instrumental method. The method can also be used for the simultaneous analysis of two or more compounds suspected to be present in the same sample by simply changing gas chromatographic conditions from isothermal to a temperature-programmed mode of operation.

One disadvantage of the method is that the amount of sample which can be taken is limited by the number of mg that the tube will hold before overloading. When the sample value obtained for the backup tube exceeds 25% of that found on the front tube, the possibility of sample loss exists. Furthermore, the precision of the method is limited by the reproducibility of the pressure drop across the tubes. This drop will affect the flow rate and cause the volume to be imprecise because the pump is usually calibrated for one tube only.

Apparatus

Calibrated personal sampling pump. Whose flow can be determined accurately (5%) at the recommended flow rate.

Charcoal tubes. Two glass tubes in series, each with both ends flame sealed, 7 cm long with 6 mm O.D. and 4 mm I.D., containing 2 sections of 20/40 mesh activated charcoal separated by a 2 mm portion of urethane foam. The activated charcoal is prepared from coconut shells and is fired at 600° C. prior to packing. The adsorbing section contains 100 mg of charcoal, the backup section 50 mg. A 3 mm portion of urethane foam is placed between the outlet end of the tube and the backup section. A plug of silylated glass wool is placed in front of the adsorbing section. The pressure drop across the tube must be less than one inch of mercury at a flow rate of 1 L/min.

Gas chromatograph. Equipped with a flame-ionization detector.

Column. 10 ft × 1/8 in stainless steel, packed with 10% FFAP on 80/100 Chromosorb W-AW.

Electronic integrator. Or some other suitable method for determining peak size areas.

Two mL glass sample containers. With glass stoppers or Teflon-lined caps. If an automatic sample injector is used, the sample injector vials will also be needed.

Syringes. 10 μL, and other convenient sizes for making standards.

Gastight syringe. 10 mL or other convenient size.

Pipets. 1.0 mL delivery type.

Volumetric flasks. 10 mL or convenient sizes for making standard solutions.

Cold room or cold box. For preparing standards at 0° C. or lower.

Reagents

Eluent. Isopropyl alcohol (chromatographic grade).
Difluorodibromomethane.
Purified nitrogen.
Prepurified hydrogen.
Filtered compressed air.

Procedure

Cleaning of equipment. All glassware used for the laboratory analysis should be detergent washed and thoroughly rinsed with tap water and distilled water.

Calibration of personal pumps. Each personal pump must be calibrated with a representative charcoal tube in the line. This will minimize errors associated with uncertainties in the sample volume collected.

Collection and shipping of samples. Immediately before sampling, break the ends of the sample and backup tubes to provide an opening at least one-half the internal diameter of the tube (3 mm). One of the charcoal tubes is used as a backup and should be positioned nearest the sampling pump. This should be labeled as the backup tube before shipping. The charcoal tubes should be placed in a vertical direction during sampling to minimize channeling through the charcoal.

Air being sampled should not be passed through any hose or tubing before entering the charcoal tubes. A maximum sample size of 10 L is recommended. Sample at a flow of 0.20 L/min or less. The flow rate should be known with an accuracy of at least 5%. The temperature and pressure of the atmosphere being sampled should be recorded. If pressure reading is not available, record the elevation. The charcoal tubes should be capped with the supplied plastic caps immediately after sampling. Under no circumstances should rubber caps be used.

One tube should be handled in the same manner as the sample tube (break, seal, and transport), except that no air is sampled through this tube. This tube should be labeled as a blank. Capped tubes should be packed tightly and padded before they are shipped to minimize tube breakage during shipping. A sample of the suspected compound should be submitted to the laboratory in glass containers with Teflon-lined caps. These liquid bulk samples should not be transported in the same container as the charcoal tubes.

Preparation of samples. In preparation for analysis, each charcoal sampling and backup tube is scored with a file in front of the first section of charcoal and broken open. The glass wool is removed and discarded. The charcoal in the first tube is transferred to a 2 mL vial. The second tube is transferred to another vial. These two tubes are analyzed separately.

Desorption of samples. Prior to analysis, 1.0 mL of isopropyl alcohol is pipetted into each vial. Desorption should be done overnight. Tests indicate that this is adequate. The sample vials should be capped as soon as the solvent is added to minimize volatilization. If an automatic sampler is to be used, the samples can be transferred to automatic sample injector vials after desorption.

GC conditions. The typical operating conditions for the gas chromatograph are:

30 mL/min (80 psig) nitrogen carrier gas flow
30 mL/min (50 psig) hydrogen gas flow to detector
300 mL/min (50 psig) air flow to detector
200° C. injector temperature
300° C. detector temperature
60° C. column temperature

Injection. The first step in the analysis is the injection of the sample into the gas chromatograph. To eliminate difficulties arising from blow back or distillation within the syringe needle, one should employ the solvent flush injection technique. The 10 μL syringe is first flushed with solvent several times to wet the barrel and plunger. Three μL of solvent are drawn into the syringe to increase the accuracy and reproducibility of the injected sample volume. The needle is removed from the solvent, and the plunger is pulled back about 0.2 μL to separate the solvent flush from the sample with a pocket of air to be used as a marker. The needle is then immersed in the sample, and a 5 μL aliquot is withdrawn, taking into consideration the volume of the needle, since the sample in the needle will be completely injected. After the needle is removed from the sample and prior to injection, the plunger is pulled back 1.2 µL to minimize evaporation of the sample from the tip of the needle. Observe that the sample occupies 4.9-5.0 µL in the barrel of the syringe. Duplicate injections of each sample and standard should be made. No more than a 3% difference in area is to be expected. An automatic sample injector can be used if it is shown to give reproducibility at least as good as the solvent flush technique. In this case 2 µL injections are satisfactory.

Measurement of area. The area of the sample peak is measured by an electronic integrator or some other suitable form of area measurement, and preliminary results are read from a standard curve prepared as discussed below.

Determination of desorption efficiency. The desorption efficiency of a particular compound can vary from one laboratory to another and also from one batch of charcoal to another. Thus, it is necessary to determine at least once the percentage of the specific compound that is removed in the desorption process, provided the same batch of charcoal is used.

Activated charcoal equivalent to the amount in the first section of the sampling tube (100 mg) is measured into a 2.0 mL sample container. This charcoal must be from the same batch as that used in obtaining the samples and can be obtained from unused charcoal tubes. A known amount of difluorodibromomethane is injected directly into the activated charcoal with a 10 µL syringe, and the container is capped. The amount injected is equivalent to that present in a 10 L sample at the selected level. This must be done in a cold room below the boiling point of difluorodibromomethane, 0° C. or lower.

At least six tubes at each of three levels (0.5X, 1X, and 2X the standard) are prepared in this manner and allowed to stand for at least overnight to assure complete adsorption of the analyte onto the charcoal. These six tubes are referred to as the samples. A parallel blank tube should be treated in the same manner except that no sample is added to it. The sample and blank tubes are desorbed and analyzed in exactly the same manner as the sampling tube described earlier. The weight of analyte found in each tube is determined from the standard curve. Desorption efficiency is determined by the following equation:

D.E. = average wt (mg) recovered ÷ wt (mg) added

The desorption efficiency is dependent on the amount of analyte collected on the charcoal. Plot the desorption efficiency versus the weight of analyte found. This curve is used to correct for adsorption losses.

Calibration and Standards

It is convenient to express concentration of standards in terms of mg/mL of isopropyl alcohol. A series of standards, varying in concentration over the range of interest, is prepared by injecting the liquid with a syringe in a cold room and analyzed under the same GC conditions and during the same time period as the unknown samples. Curves are established by plotting concentrations in mg/mL versus peak area. Note: standard solutions should be analyzed at the same time that the sample analysis is done. This will minimize the effect of variations of FID response.

Calculations

Read the weights, in mg, corresponding to each peak area from the standard curve. No volume corrections are needed, because the standard curve is based on mg/mL isopropyl alcohol and the volume of sample injected is identical to the volume of the standards injected. Corrections for the blank must be made for each sample:

mg = mg sample − mg blank

where:

mg sample = mg found in front tube of sample
mg blank = mg found in blank tube

A similar procedure is followed for the backup tubes. Add the weights present in the front and backup tubes of the same sample to determine the total weight in the sample. Read the desorption efficiency from the curve for the amount of the analyte found in the front section. Divide the total weight by this desorption efficiency to obtain the corrected mg/sample:

Corrected mg/sample = total weight ÷ D.E.

The concentration of the analyte in the air sampled can be expressed in mg/m^3, which is numerically equal to μg/L of air:

mg/m^3 = (corrected mg)(1000 L/m^3) ÷ air volume sampled in L

Another method of expressing concentration is ppm:

ppm = (mg/m^3)(24.45)(760)(T + 273) ÷ (M.W.)(P)(298)

where:

P = pressure (mm Hg) of air sampled
T = temperature (°C.) of air sampled
24.45 = molar volume (L/mol) at 25° C. and 760 mm Hg
M.W. = molecular weight (g/mol) of analyte
760 = standard pressure (mm Hg)
298 = standard temperature (°K)

References

1. White, L.D. et al. A convenient optimized method for the analysis of selected solvent vapors in the industrial atmosphere. Amer. Ind. Hyg. Assoc. J. 31: 225, 1970.
2. Documentation of NIOSH Validation Tests, NIOSH Contract No. CDC-99-74-45.
3. Final Report, NIOSH Contract No. HSM-99-71-31. Personal Sampler Pump for Charcoal Tubes, September 15, 1972.

DIISOBUTYL KETONE

Synonyms: 2,6-dimethyl-4-heptanone; isovalerone; valerone $((CH_3)_2CHCH_2)_2CO$
ACGIH TLV: 25 ppm (150 mg/m^3)
OSHA Standard: 50 ppm (290 mg/m^3)

Method	Sampling Duration	Sampling Location	Useful Range (ppm)	System Cost ($)	Test Cost ($)	Manufacturer
IR	Cont	Portable	1-100	4374	0	Foxboro

NIOSH METHOD NO. P&CAM 127

(Use as described for acetone, with 71° C. column temperature)

DIISOPROPYLAMINE

ACGIH TLV: 5 ppm (20 mg/m^3) - Skin $(CH_3)_2CHNHCH(CH_3)_2$
OSHA Standard: 5 ppm

Method	Sampling Duration	Sampling Location	Useful Range (ppm)	System Cost ($)	Test Cost ($)	Manufacturer
DT	Grab	Portable	1-60	180	2.25	Bendix
CS	Grab	Portable	0.5-10	2465	0.80	MDA
IR	Cont	Portable	0.2-10	4374	0	Foxboro
PI	Cont	Fixed	0.2-20	4950	0	HNU

NIOSH METHOD NO. P&CAM 221

(See n-butylamine)

DIMETHYLACETAMIDE

Synonym: N,N-dimethylacetamide $CH_3CON(CH_3)_2$
ACGIH TLV: 10 ppm (35 mg/m^3) - Skin
OSHA Standard: 10 ppm

Method	Sampling Duration	Sampling Location	Useful Range (ppm)	System Cost ($)	Test Cost ($)	Manufacturer
DT	Grab	Portable	10-40	150	3.50	Nat'l Draeger
DT	Grab	Portable	1.5-240	180	2.25	Bendix
IR	Cont	Portable	0.6-20	4374	0	Foxboro
PI	Cont	Fixed	0.2-20	4950	0	HNU

NIOSH METHOD NO. S254

Principle
A known volume of air is drawn through a silica gel tube to trap the organic vapors present. The silica gel in the tube is transferred to a small, stoppered sample container, and the analyte is desorbed with methanol. An aliquot of the desorbed sample is injected into a gas chromatograph. The area of the resulting peak is determined and compared with areas obtained for standards.

Range and Sensitivity
This method was validated over the range of 18.17-105.6 mg/m^3 at an atmospheric temperature and pressure of 24° C. and 760 mm Hg, using a 45 L sample. Under the conditions of sample size (45 L) the probable useful range of this method is 10-105 mg/m^3. The method is capable of measuring much smaller amounts if the desorption efficiency is adequate. Desorption efficiency must be determined over the range used.

The upper limit of the range of the method is dependent on the adsorptive capacity of the silica gel tube. This capacity varies with the concentrations of the analyte and other substances in the air. The first section of the silica gel tube was found to hold at least 22.2 mg of analyte when a test atmosphere containing 105.6 mg/m^3 of analyte in air was sampled at 0.876 L/min for 240 min; at that time the concentration of dimethylacetamide in the effluent was less than 5% of that in the influent. (The silica gel tube consists of two sections of silica gel separated by a section of urethane foam.) If a particular atmosphere is suspected of containing a large amount of contaminant, a smaller sampling volume should be taken.

Interferences
Silica gel has a high affinity for water, so organic vapors may not be trapped efficiently in the presence of a high relative humidity. This effect may be important even though there is no visual evidence of condensed water in the silica gel tube. When interfering compounds are known or suspected to be present in the air, such information, including their suspected identities, should be transmitted with the sample.

It must be emphasized that any compound which has the same retention time as the analyte at the operating conditions described in this method is an interference. Retention time data on a single column cannot be considered proof of chemical identity. If the possibility of interference exists, separation conditions (column packing, temperature, etc.) must be changed to circumvent the problem.

Precision and Accuracy

The coefficient of variation for the total analytical and sampling method in the range of 18.17-105.6 mg/m^3 was 0.067. This value corresponds to a 2.3 mg/m^3 standard deviation at the OSHA standard level. Statistical information and details of the validation and experimental test procedures can be found in reference 1.

A collection efficiency of 1.00 was determined for the collection medium, thus, no bias was introduced in the sample collection step and no correction for collection efficiency is necessary. There was also no bias in the sampling and analytical method for which a desorption efficiency correction was made. Thus, coefficient of variation is a satisfactory measure of both accuracy and precision of the sampling and analytical method.

Advantages and Disadvantages

The sampling device is small, portable, and involves no liquids. Interferences are minimal, and most of those which do occur can be eliminated by altering chromatographic conditions. The tubes are analyzed by means of a quick, instrumental method. The method may also be used for the simultaneous analysis of two or more substances suspected to be present in the same sample.

One disadvantage of the method is that the amount of sample which can be taken is limited by the number of mg that the tube will hold before overloading. When the sample value obtained for the backup section of the silica gel tube exceeds 25% of that found in the front section, the possibility of sample loss exists. Furthermore, the precision of the method is limited by the reproducibility of the pressure drop across the tubes. This drop will affect the flow rate and cause the volume to be imprecise, because the pump is usually calibrated for one tube only.

Apparatus

Calibrated personal sampling pump. Whose flow can be determined within 5% at the recommended flow rate

Silica gel tubes. Glass tube with both ends flame sealed, 7 cm long with 6 mm O.D. and 4 mm I.D., containing 2 sections of 20/40 mesh silica gel separated by a 2 mm portion of urethane foam. The adsorbing section contains approximately 150 mg of silica gel, the backup section, approximately 75 mg. A 3 mm portion of urethane foam is placed between the outlet end of the tube and the backup section. A plug of silylated glass wool is placed in front of the adsorbing section. The pressure drop across the tube must be less than one inch of mercury at a flow rate of 1 L/min.

Gas chromatograph. Equipped with a flame-ionization detector.

Column. 6 ft × 1/4 in O.D. stainless steel, packed with 20% UCON 50LB 550X on Chromosorb P 60/80 treated with 20% KOH.

Electronic integrator. Or some other suitable method for measuring peak areas.

Sample containers. 2 mL, with glass stoppers or Teflon-lined caps. If an automatic sample injector is used, the associated vials may be used.

Syringes. 10 μL, and other convenient sizes for making standards.

Pipets. 1.0 mL delivery pipets.

Volumetric flasks. 10 mL or convenient sizes for making standard solutions.

Reagents

Methanol, reagent grade.
Dimethyl acetamide, reagent grade.
Acetone, reagent grade.
Prepurified hydrogen.
Filtered compressed air.
Purified nitrogen.

Procedure

Cleaning of equipment. All glassware used for the laboratory analysis should be detergent washed and thoroughly rinsed with tap water and distilled water.

Calibration of personal pumps. Each personal pump must be calibrated with a representative silica gel tube in the line. This will minimize errors associated with uncertainties in the sample volume collected.

Collection and shipping of samples. Immediately before sampling, break the ends of the tube to provide an opening at least one-half the internal diameter of the tube (2 mm). The smaller section of silica gel is used as a backup and should be positioned nearest the sampling pump. The silica gel tube should be placed in a vertical direction during sampling to minimize channeling through the silica gel. Air being sampled should not be passed through any hose or tubing before entering the silica gel tube.

A sample size of 50 L is recommended. Sample at a flow of 1.0 L/min or less. The flow rate should be known with an accuracy of at least 5%. The temperature and pressure of the atmosphere being sampled should be recorded. If pressure reading is not available, record the elevation. The silica gel tubes should be capped with the supplied plastic caps immediately after sampling. Under no circumstances should rubber caps be used.

With each batch of ten samples, submit one tube from the same lot of tubes which was used for sample collection and which is subjected to exactly the same handling as the samples except that no air is drawn through it. Label this as a blank. Capped tubes should be packed tightly and padded before they are shipped to minimize tube breakage during shipping. A sample of the bulk material should be submitted to the laboratory in a glass container with a Teflon-lined cap. This sample should not be transported in the same container as the silica gel tubes.

Preparation of samples. In preparation for analysis, each silica gel tube is scored with a file in front of the first section of silica gel and broken open. The glass wool is removed and discarded. The silica gel in the first (larger) section is transferred to a 2 mL stoppered sample container. The separating section of foam is removed and discarded; the second section is transferred to another stoppered container. These two sections are analyzed separately.

Desorption of samples. Prior to analysis, 1.0 mL of methanol is pipetted into each sample container. The sample is desorbed for one hour using an ultrasonic bath to aid desorption. Note that the water in the ultrasonic bath can get hot (ca. 50-60° C.) during the desorption period. Therefore, all vials must be tightly capped to minimize evaporation losses. If an automatic sample injector is used, the sample vials should be capped as soon as the solvent is added to minimize volatilization.

GC conditions. The typical operating conditions for the gas chromatograph are:

50 mL/min (60 psig) nitrogen carrier gas flow
65 mL/min (24 psig) hydrogen gas flow to detector
500 mL/min (50 psig) air flow to detector
240° C. injector temperature
320° C. detector temperature
160° C. column temperature

Injection. The first step in the analysis is the injection of the sample into the gas chromatograph. To eliminate difficulties arising from blow back or distillation within the syringe needle, one should employ the solvent flush injection technique. The 10 µL syringe is first flushed with solvent several times to wet the barrel and plunger. Three µL of solvent are drawn into the syringe to increase the accuracy and reproducibility of the injected sample volume. The needle is removed from the solvent, and the plunger is pulled back about 0.2 µL to separate the solvent flush from the sample with a pocket of air to be used as a marker. The needle is then immersed in the sample, and a 5 μL aliquot is withdrawn, taking into consideration the volume of the needle, since the sample in the needle will be completely injected. After the needle is removed from the sample and prior to injection, the plunger is pulled back 1.2 μL to minimize evaporation of the sample from the tip of the needle. Observe that the sample occupies 4.9-5.0 μL in the barrel of the syringe. Duplicate injections of each sample and standard should be made. No more than a 3% difference in area is to be expected. An automatic sample injector can be used if it is shown to give reproducibility at least as good as the solvent flush method.

Measurement of area. The area of the sample peak is measured by an electronic integrator or some other suitable form of area measurement, and preliminary results are read from a standard curve prepared as discussed below.

Determination of desorption efficiency. The desorption efficiency of a particular compound can vary from one laboratory to another and also from one batch of silica gel to another. Thus, it is necessary to determine at least once the percentage of the specific compound that is removed in the desorption process, provided the same batch of charcoal is used.

Silica gel equivalent to the amount in the first section of the sampling tube (approximately 150 mg) is measured into a 64 mm, 4 mm I.D. glass tube, flame sealed at one end. This silica gel must be from the same batch as that used in obtaining the samples and can be obtained from unused silica gel tubes. The open end is capped with Parafilm. A known amount of an acetone solution of dimethyl acetamide containing 472 μg/μL is injected directly into the silica gel with a 10 μL syringe. Cap the silica gel tube with more Parafilm. When using an automatic sample injector, the sample injector vials, capped with Teflon faced septa, may be used in place of the glass tubes.

The amount injected is equivalent to that present in the 50 L air sample at the selected level. Six tubes at each of three levels (0.5X, 1X, and 2X the standard) are prepared in this manner and allowed to stand for at least overnight to assure complete adsorption of the analyte onto the silica gel. These tubes are referred to as the samples. A parallel blank tube should be treated in the same manner except that no sample is added to it. The sample and blank tubes are desorbed and analyzed in exactly the same manner as the sampling tube described above.

Two or three standards are prepared by injecting the same volume of compound into 1.0 mL of methanol with the same syringe used in the preparation of the samples. These are analyzed with the samples. The desorption efficiency equals the average weight in mg recovered from the tube divided by the weight in mg added to the tube, or:

D.E. = average wt (mg) recovered ÷ wt (mg) added

The desorption efficiency is dependent on the amount of analyte collected on the silica gel. Plot the desorption efficiency versus the weight of analyte found. This curve is used to correct for adsorption losses.

Calibration and Standards

It is convenient to express concentration of standards in terms of mg/1.0 mL methanol, because samples are desorbed in this amount of methanol. The density of the analyte is used to calculate mg from μL for easy measurement with a microliter syringe. A series of standards, varying in concentration over the range of interest, is prepared and analyzed under the same GC conditions and during the same time period as the unknown samples. Curves are established by plotting concentration in mg/1.0 mL versus peak area. Note: since no internal standard is used in the method, standard solutions must be analyzed at the same time that the sample analysis is done. This will minimize the effect of known day-to-day variations and variations during the same day of the FID response.

Calculations

Read the weight, in mg, corresponding to each peak area from the standard curve. No volume corrections are needed, because the standard curve is based on mg/1.0 mL methanol and the volume of sample injected is identical to the volume of the standards injected. Corrections for the blank must be made for each sample:

mg = mg sample − mg blank

where:

mg sample = mg found in front section of sample tube
mg blank = mg found in front section of blank tube

A similar procedure is followed for the backup sections. Add the weights found in the front and backup sections to get the total weight in the sample. Read the desorption efficiency from the curve for the amount found in the front section. Divide the total weight by this desorption efficiency to obtain the corrected mg/sample:

Corrected mg/sample = total weight ÷ D.E.

The concentration of the analyte in the air sampled can be expressed in mg/m^3:

mg/m^3 = (corrected mg)(1000 L/m^3) ÷ air volume sampled in L

Another method of expressing concentration is ppm:

ppm = (mg/m^3)(24.45)(760)(T + 273) ÷ (M.W.)(P)(298)

where:

P = pressure (mm Hg) of air sampled
T = temperature (°C.) of air sampled
24.45 = molar volume (L/mol) at 25° C. and 760 mm Hg
M.W. = molecular weight (g/mol) of analyte
760 = standard pressure (mm Hg)
298 = standard temperature (°K)

Reference

1. Documentation of NIOSH Validation Tests. NIOSH Contract No. CDC-99-75-45.

DIMETHYLFORMAMIDE

Synonyms: N,N-dimethylformamide; DMF $HCON(CH_3)_2$
ACGIH TLV: 10 ppm (30 mg/m^3) - Skin
OSHA Standard: 10 ppm

Method	Sampling Duration	Sampling Location	Useful Range (ppm)	System Cost ($)	Test Cost ($)	Manufacturer
DT	Grab	Portable	10-40	150	3.80	Nat'l Draeger
DT	Grab	Portable	0.8-90	180	2.25	Bendix
IR	Cont	Portable	0.2-20	4374	0	Foxboro
PI	Cont	Fixed	0.2-20	4950	0	HNU

NIOSH METHOD NO. S254

(Use as described for dimethylacetamide, with 140° C. column temperature)

DIMETHYL SULFATE

Synonym: methyl sulfate $(CH_3)_2SO_4$
ACGIH TLV: 0.1 ppm (0.5 mg/m^3) - Skin, possible carcinogen
OSHA Standard: 1 ppm (5 mg/m^3)

Method	Sampling Duration	Sampling Location	Useful Range (ppm)	System Cost ($)	Test Cost ($)	Manufacturer
DT	Grab	Portable	0.005-0.05	150	2.70	Nat'l Draeger
IR	Cont	Portable	0.04-2	4374	0	Foxboro

NIOSH METHOD NO. P&CAM 301

Principle
Dimethyl sulfate is collected by passing a known volume of air through a Porapak P sorbent tube. The dimethyl sulfate is desorbed with diethyl ether. An aliquot of the solution is analyzed by gas chromatography using electrolytic conductivity detection in the oxidative mode. The area of the resulting peak is determined and compared with areas obtained by injecting standards.

Range and Sensitivity
A log-log plot of detector response versus amount injected was linear over the range of 4.6-157 ng. Assuming 1 mL of desorbate and a 4 µL injection, this corresponds to a range of 1.1-39

µg/sample. The corresponding ceiling concentrations for a 0.75 L air sample range from 1.5-52 mg/m^3, or 0.3-10 ppm. The corresponding time weighted average concentrations for a 12 L air sample range from 0.1-3.2 mg/m^3, or 0.02-0.6 ppm. It is expected, however, that the capacity of the sampling device and the linearity of the detector response will be sufficient to permit the measurement of concentrations up to 15 mg/m^3, or 3 ppm for 12 L air samples.

Over the range of 4.6-157 ng/injection, the integrated response of the electrolyte conductivity detector ranged from 20-510 µmho-sec, for a sensitivity of approximately 3.2 µmho-sec/ng. The detection limit was not determined, but is estimated to be on the order of 1 ng/injection under the conditions used.

Interferences

Halogen, sulfur, and nitrogen containing compounds which are detected by the electrolyte conductivity detector in the oxidative mode, and have retention times comparable to that of dimethyl sulfate, will interfere with the analysis.

Precision and Accuracy

The pooled relative standard deviation for the total sampling and analytical method for concentrations ranging from 1.8-24.5 mg/m^3, as measured by this method, was 0.07. The results of this method were compared with those of an independent method consisting of collection on Tenax-GC, methylation of p-nitrophenol with the collected dimethyl sulfate, and HPLC of the resulting p-nitroanisole. For concentrations determined to be 1.8, 4.35, and 24.5 µg/L by this method, the independent method found 3.18, 7.18, and 24.4 µg/L, respectively. Since the results are inconsistent and the accuracy of the independent method is unknown, the bias and accuracy of this method remains to be determined. Samples collected at high humidity were stable for at least 1 week.

Advantages and Disadvantages

The combination of chromatography and selective detection reduces the number of potential interferences. The signal of the electrolytic conductivity detector is quite noisy, making quantitation difficult for some integrators.

Apparatus

Solid sorbent tube. Flame sealed, consisting of a glass wool plug followed by a 100 mg section of 50/80 mesh Porapak P, a urethane foam plug, a 50 mg section of Porapak P, and a final urethane foam plug packed in 6 mm outside diameter, standard wall glass tubing. Because of batch-to-batch variations in the Porapak P, it is advisable to pretreat the sorbent by heating overnight in a vacuum oven at 180° C.

Personal sampling pump. Calibrated with a representative sorbent tube in the line and capable of maintaining the flow of air through the sorbent tube at a rate of 50 mL/min with an accuracy of 5%.

Flexible tubing. For connecting the sorbent tube and pump.

Thermometer.

Manometer.

Stopwatch.

Polyethylene caps. Suitable for capping the sorbent tube after sampling.

Vials. 2 mL with Teflon lined septum caps, and other sizes, as needed.

Pipets. 1 mL, and other sizes, as needed.

Syringes. 10 µL, and other sizes, as needed.

Gas chromatograph. Equipped with an electrolytic conductivity detector, operating in the oxidative mode.

Gas chromatographic column. 183 cm × 3.18 mm stainless steel, packed with 5% DEGS on 80/100 mesh Supelcoport, or equivalent. The column is conditioned by heating at 180° C. overnight while carrier gas is flowing at 20 mL/min. After connecting the conditioned column to the detector, prime the column by repeatedly injecting several μg of dimethyl sulfate until consistent analytical results are obtained.

Reagents

Whenever possible, reagents used should be ACS reagent grade or better.
Diethyl ether. Distilled in glass, preserved with 2% ethanol.
Helium, high purity.
Oxygen, high purity.
Water, deionized and filtered.
Dimethyl sulfate.

Procedure

Cleaning of equipment. Detergent wash all non-disposable glassware used for the laboratory analysis and thoroughly rinse with tap water and distilled water.

Collection and shipping of samples. Immediately before sampling, break the ends of the tube to provide an opening at least one-half the internal diameter of the tube. The smaller section of Porapak P is used as a backup. Position it nearest the sampling pump. Place the sorbent tube in a vertical direction during sampling to minimize channeling through the resin. Do not pass the air being sampled through any hose or tubing before it enters the sorbent tubes.

Sample sizes of 12 L for time weighted average and 0.75 L for ceiling determinations are recommended. Sample at 50 mL/min. Higher flows may be used if the pressure drop across the sorbent tube does not exceed the capacity of the pump. The flow should be known with an accuracy of at least 5%. Record the temperature and pressure of the atmosphere being sampled. If the pressure is not available, record the elevation. Cap the sorbent tubes with polyethylene caps immediately after sampling. For every 10 samples, handle one tube in the same manner as the sample tube (break, seal, and transport), except that no air is sampled through this tube. Label this tube as a blank.

Preparation of samples. In preparation for analysis, score each sorbent tube with a file in front of the first section, and break it open. Remove and discard the glass wool. Transfer the first, larger section of Porapak P to a sample container and pipette 1.0 mL of diethyl ether into the container. Immediately crimp seal with a Teflon lined septum cap. (All work with diethyl ether should be performed in a well ventilated hood because of its high flammability.) Remove the separating section of urethane foam and transfer it along with the smaller section of Porapak P to a 2 mL container, add 1.0 mL of diethyl ether, and crimp seal the container. Desorption for 30 min with occasional agitation is adequate. Store the solutions in a spark-proof refrigerator if they are not to be analyzed immediately.

GC conditions. Typical operating conditions for the gas chromatograph are as follows:

20 mL/min helium carrier gas flow
180° C. injector temperature
180° C. manifold temperature
column temperature program: 50° C. isothermal for 2 min, then to 120° C. at 30° C./min, then isothermal at 120° C.
950° C. detector temperature
20 mL/min oxygen flow (detector reactant gas)
1 mL/min conductivity solvent flow (deionized, filtered water)

Injection. Inject 4.0 μL of the sample or standard solution into the gas chromatograph, using an appropriate technique such as solvent flush to achieve reproducibility. Due to the sorbent fines present, special effort must be made to prevent air bubbles from forming in the syringe because of obstruction of the tip of the syringe needle as the sample is being drawn in.

Measurement of area. Measure the area of the sample and standard peaks by an electronic integrator or some other suitable form of area measurement.

Determination of desorption efficiency. The desorption efficiency of a particular compound may vary from one laboratory to another and also from one batch of resin to another. Thus, it is necessary to determine, at least once, the percentage of the specific compound that is removed in the desorption process for each batch of resin that is used.

Measure an amount of Porapak P equivalent to that in the first section of the sampling tube (100 mg) into a 2 mL vial and cap with a Teflon lined septum. Inject a known amount of dimethyl sulfate solution of appropriate concentration in ether directly into the Porapak P with a 10 μL syringe and allow it to stand overnight to assure complete adsorption of the analyte onto the resin. Prepare six such samples for each of three levels covering the concentration range of interest. For a time weighted average determination, appropriate amounts would be those present in 12 L samples at concentrations 0.3, 1, and 3 times the standard. At the same time, prepare standards at each level by injecting the same amounts of dimethyl sulfate into 1.0 mL portions of diethyl ether in capped 2 mL vials with the same syringe used for spiking the Porapak P samples. Prepare the parallel blank by treating the Porapak P in the same manner, except that no sample is added to it. Analyze the sample and blank tubes in the same manner as described above. The desorption efficiency, D, is given by:

$$D = R \div A$$

where:

R = average weight recovered
A = weight added

The desorption efficiency may be dependent on the amount of analyte collected on the resin. Plot the desorption efficiency versus weight of analyte found. Use this curve to correct for adsorption losses.

Calibration and Standardization

CAUTION: Dimethyl sulfate has been found carcinogenic in rats, and is possibly carcinogenic in humans (reference 1). The pure liquid and solutions must be handled in a manner so as to avoid exposure of personnel either through inhalation of vapors or skin contact.

Prepare a 1.33 μg/μL stock solution by diluting 100 μL of dimethyl sulfate to 100 mL with diethyl ether. Prepare at least five standard solutions varying from 6-180 μg/mL by diluting appropriate aliquots of the stock solution with diethyl ether. This range is based on a 12 L air sample and air concentrations ranging from approximately 0.1-3 times the OSHA standard. For other sample volumes or concentrations, adjust the range of the standards accordingly.

Analyze the standard solutions at the same time that the samples are analyzed, using the same analytical procedure. This will minimize the effect of possible day-to-day variations and variations during the same day of detector response. Prepare a calibration curve by plotting the peak areas of the standards versus the corresponding concentrations in μg/mL.

Calculations

Read the weight, in µg/sample, corresponding to each peak area from the standard curve. No volume corrections are needed, because the standard curve is based on µg/1.0 mL diethyl ether and the volume of sample injected is identical to the volume of the standards injected. For each sample, add the weights found in the front and backup sections to find the total weight, W. Note: if a blank is found, the source of the blank must be determined and steps taken to eliminate the problem. Read the desorption efficiency from the curve for the amount found in the front section. Divide the total weight by this desorption efficiency to obtain the µg/sample, Q, corrected for desorption efficiency:

$$Q = W \div D$$

The concentration of the analyte in the air sampled, C, can be expressed in mg/m^3 which is numerically equal to µg/L:

$$C = Q \div V$$

where:

V = volume of air sampled (L)

Another method of expressing concentration is ppm (corrected to standard conditions of 25° C. and 760 mm Hg):

$$ppm = (C)(24.45)(760)(T + 273) \div (M.W.)(P)(298)$$

where:

P = pressure (mm Hg) of air sampled
T = temperature (°C.) of air sampled
24.45 = molar volume (L/mol) at 25° C. and 760 mm Hg
M.W. = molecular weight = 126.1
760 = standard pressure (mm Hg)
298 = standard temperature (°K)

References

1. *IARC Monographs on the Evaluation of the Carcinogenic Risk of Chemicals to Man: Some Aromatic Amines, Hydrazine and Related Substances, N-Nitroso Compounds and Miscellaneous Alkylating Agents,* World Health Organization, Lyon, France, IARC Monographs, 4: 271, 1974.
2. Lunsford, R.A., and P. Fey. Backup Data Report for P&CAM 301, NIOSH, Cincinnati, Ohio, December 29, 1978.

2,4-DINITROTOLUENE

Synonym: 1-methyl-2,4-dinitrobenzene $(NO_2)_2C_6H_5CH_3$
ACGIH TLV: 1.5 mg/m^3 - Skin
OSHA Standard: 1.5 mg/m^3

Method	Sampling Duration	Sampling Location	Useful Range (ppm)	System Cost ($)	Test Cost ($)	Manufacturer
CS	Cont	Fixed	0.02-0.4	4950	Variable	MDA

NIOSH METHOD NO. S215

Principle

A known volume of air is drawn through a mixed cellulose ester membrane filter connected in series to a midget bubbler containing 10 mL of ethylene glycol to collect dinitrotoluene. The filter and bubbler are disconnected. Five mL of methanol are added to the bubbler flask, and the filter is removed from the cassette holder and added to the bubbler flask. The resulting sample is analyzed by high pressure liquid chromatography.

Range and Sensitivity

This method was validated over the range of 0.900-5.02 mg/m^3 at an atmospheric temperature of 20° C. and pressure of 764 mm Hg, using 90 L samples. The upper limit of the range of the method is dependent on the capacity of the mixed cellulose ester membrane filter connected in series to the midget bubbler and the capacity of the midget bubbler.

Interferences

When interfering compounds are known or suspected to be present in the air, such information, including their suspected identities, should be transmitted with the sample. Any compound that has the same retention time as dinitrotoluene at the operating conditions described in this method is an interference. Retention time data on a single column cannot be considered proof of chemical identity.

Precision and Accuracy

The coefficient of variation for the total analytical and sampling method in the range of 0.900-5.02 mg/m^3 was 0.063. This value corresponds to a standard deviation of 0.09 mg/m^3 at the OSHA standard level. Statistical information can be found in reference 1. Details of the test procedures can be found in reference 2.

On the average, the concentrations obtained in the laboratory validation study at 0.5X, 1X, and 2X the OSHA standard level were 4% lower than the "true" concentrations for 17 samples. Any difference between the "found" and "true" concentrations may not represent a bias in the sampling and analytical method, but rather a random variation from the experimentally determined "true" concentration. The coefficient of variation is a good measure of the accuracy of the method since the recoveries, storage stability, and collection efficiency were good and would not contribute to a bias in a determined concentration. Storage stability studies on samples collected from a test atmosphere at a concentration of 1.800 mg/m^3 indicate that collected samples are stable for at least 7 days.

Advantages and Disadvantages

Collected samples are analyzed by means of a quick, instrumental method. A disadvantage of the method is the awkwardness in using midget bubblers for collecting personal samples. If the worker's job performance requires much body movement, loss of the collection solution during sampling may occur.

The precision of the method is limited by the reproducibility of the pressure drop across the filter and bubbler. This drop will affect the flow rate and cause the volume to be imprecise, because the pump is usually calibrated for one filter/bubbler combination only. The bubblers are more difficult to ship than adsorption tubes or filters due to possible breakage and leakage of the bubblers during shipping.

Apparatus

Filter unit. The filter unit consists of a 37 mm diameter cellulose ester membrane filter (Millipore Type AA or equivalent) with a pore size of 0.80 micron, and a 37 mm two piece cassette filter holder supported by a stainless steel screen. It is important that a stainless steel screen be used since other filter supports may retain part of the vapor.

Glass midget bubbler. Containing 10 mL of ethylene glycol.

Personal sampling pump. A calibrated personal sampling pump whose flow can be determined to an accuracy of 5%. The sampling pump is protected from splashover or solvent condensation by a 5 cm long by 6 mm I.D. glass tube loosely packed with a plug of glass wool and inserted between the exit arm of the bubbler and the pump.

Manometer.

Thermometer.

High pressure liquid chromatograph. Equipped with a 254 nm fixed wavelength uv detector and a sample injection valve with a 10 µL external sample loop. The injection valve is fitted with a syringe filter to remove filter fibers which would eventually block the flow to the LC column.

Column. 250 mm × 3 mm I.D. stainless steel, packed with Spherisorb ODS. The superficially porous packing material consists of spherical silica particles with a 5% bonded coating of octadecyl groups. This packing can be obtained from Spectra-Physics (Santa Clara, CA).

Electronic integrator. Or some other suitable method for measuring peak areas.

Tweezers.

Syringes. 10 and 100 µL.

Volumetric flasks. Convenient sizes for preparing standard solutions.

Pipets. Convenient sizes for preparing standard solutions and 5 and 10 mL pipets for measuring the extraction medium.

Teflon tubing or plugs. 15 cm long × 7 mm I.D., for sealing the inlet and outlet of the bubbler stem before shipping.

Reagents

2,4-Dinitrotoluene, reagent grade.

Ethylene glycol, reagent grade.

Methanol, distilled in glass.

2-Propanol, reagent grade.

Water, deionized and distilled.

Procedure

Cleaning of equipment. All glassware used for the laboratory analysis should be detergent washed and thoroughly rinsed with tap water and distilled water, and dried.

Calibration of personal sampling pumps. Each personal sampling pump must be calibrated with a representative filter cassette and bubbler in the line to minimize errors associated with uncertainties in the volume sampled.

Collection and shipping of samples. Assemble the filter in the two piece filter cassette holder and close firmly. The filter is backed up by a stainless steel screen. Secure the cassette holder together with tape or shrinkable band. Pipet 10 mL of ethylene glycol into each midget bubbler, and mark the liquid level. Be sure that the bubbler frit is completely immersed in the ethylene glycol.

Remove the cassette plugs and attach the outlet of the filter cassette to the inlet arm of the midget bubbler using a short piece of flexible tubing. Connect the outlet arm of the midget bubbler with a 5 cm glass splashover tube (6 mm I.D.) containing the glass wool plug, then to the personal sampling pump, using short pieces of flexible tubing. The bubbler must be maintained in a vertical position during sampling. Air being sampled should not pass through any hose or tubing before entering the filter cassette.

A sample size of 90 L is recommended. Sample at a flow rate of 1.5 L/min. The flow rate should be known with an accuracy of 5%. Turn the pump on and begin sample collection. Since it is possible for a filter to become plugged by heavy particulate loading or by the presence of oil mists or other liquids in the air, the pump rotameter should be observed frequently, and the sampling should be terminated by any evidence of a problem.

Terminate sampling at the predetermined time and record sample flow rate, collection time and ambient temperature and pressure. If pressure reading is not available, record the elevation. Also record the type of sampling pump used. After sampling, disconnect the filter and bubbler. The inlet and outlet of the bubbler stem should be sealed by connecting a piece of Teflon tubing between them or inserting Teflon plugs in the inlet and outlet. Do not seal with rubber. The standard taper joint of the bubbler should be taped securely to prevent leakage during shipping.

The filters are shipped in the cassette filter holder. Each filter should be marked to identify it with its corresponding backup bubbler. With each batch of ten samples submit one filter and bubbler containing 10 mL of ethylene glycol from the same lot of filters and bubblers used for sample collection. This filter and bubbler must be subjected to exactly the same handling as the samples except that no air is drawn through them. Label this filter and bubbler as the blank. The cassette filter holders and bubblers should be shipped in a suitable container, designed to prevent damage in transit. The sample should be shipped to the laboratory as soon as possible.

Analysis of samples. Mix the contents in the bubbler by swirling. If the sample volume is less than 10 mL, add ethylene glycol until the volume reaches the 10 mark. If the sample volume is more than 10 mL, determine the volume and make an appropriate volume correction in the calculations as described below. Add 5 mL of methanol to each sample and mix the solution by swirling. Remove the filter from the filter cassette with clean tweezers, and add the filter to the bubbler. Stopper the bubbler and shake. Allow the samples to stand for two hours before analysis.

HPLC conditions. The typical operating conditions for the high pressure liquid chromatograph are:

Column temperature: ambient
Column pressure: 1290 psi
Flow rate: 1.0 mL/min
Mobile phase: 50% methanol/50% water (v/v)
Detector: uv photometer at 254 nm
Capacity ratio: 4.9

Injection. The first step in the analysis is to inject the sample into the liquid chromatograph. The chromatograph is fitted with a sample injection valve and a 10 μL sample loop. Flush this loop thoroughly with the sample (100 μL), and inject the sample.

Measurement of peak area. The area of the sample peak is measured by an electronic integrator or some other suitable form of area measurement, and results are read from a standard curve prepared as discussed below.

Determination of analytical recovery. To eliminate any bias in the analytical method, it is necessary to determine the recovery of dinitrotoluene from the cellulose ester membrane filter. The analytical method recovery should be determined over the concentration range of interest.

An appropriate aliquot of a 3.375 mg/mL solution of dinitrotoluene in 2 propanol is added to a 37 mm MCE filter. The amount added is equal to the amount of dinitrotoluene collected in a 90 L sample at 2X, 1X, and 0.5X the OSHA standard level. Six spiked samples are prepared at each level along with six blanks. The filters are allowed to stand for 2-3 min to allow evaporation of 2 propanol. The filter is transferred to a container containing 10 mL of ethylene glycol and 5 mL methanol. The resulting sample is mixed well and allowed to stand for 1 hour before analysis. Analytical method recovery (A.M.R.) equals weight in mg found on the filter divided by the weight in mg added to the filter, or:

A.M.R. = mg found ÷ mg taken

Calibration and Standards

A series of standards, varying in concentration over the range corresponding to approximately 0.1-3 times the OSHA standard for the sample under study, is prepared and analyzed under the same LC conditions and during the same time period as the unknown samples. Curves are established by plotting concentration in mg/15 mL versus peak area. Note: since no internal standard is used in this method, standard solutions must be analyzed at the same time as the samples. This will minimize the effect of known day-to-day variations and variations during the same day of the uv detector response.

Prepare a 3.375 mg/mL dinitrotoluene stock standard solution by dissolving 33.75 mg dinitrotoluene in 2-propanol and diluting to 10 mL in a volumetric flask. From the above stock solution, appropriate aliquots are withdrawn and added to 10 mL ethylene glycol and 5 mL methanol. Prepare at least 5 working standards to cover the range of 0.013-0.40 mg/15 mL. This range is based on a 90 L sample. Analyze samples as described above. Prepare a standard calibration curve by plotting concentration of dinitrotoluene in mg/15 mL versus peak area.

Calculations

Read the weight, in mg, corresponding to each peak from the standard curve. No volume correction is needed, because the standard curve is based on mg/15 mL of ethylene glycol/methanol and the volume of sample injected is identical to the volume of the standards injected. A correction for the blank must be made for each sample:

mg = mg sample − mg blank

where:

mg sample = mg found in sample filter
mg blank = mg found in blank filter

Corrections for analytical method recovery (A.M.R.) must be made:

Corrected mg/sample = mg ÷ A.M.R.

For personal sampling pumps with rotameters only, the following volume correction should be made:

Corrected volume = $ft(P_1T_2/P_2T_1)^{1/2}$

where:

  f = flow rate sampled
  t = sampling time
  P_1 = pressure during calibration of sampling pump (mm Hg)
  P_2 = pressure of air sampled (mm Hg)
  T_1 = temperature during calibration of sampling pump (°K)
  T_2 = temperature of air sampled (°K)

The concentration of dinitrotoluene in the air sample can be expressed in mg/m³:

mg/m³ = (mg)(1000 L/m³) ÷ corrected air volume sampled in L

References

1. Documentation of NIOSH Validation Tests, Contract No. CDC-99-74-45.
2. Backup Data Report for Dinitrotoluene, prepared under NIOSH Contract No. 210-76-0123.

p-DIOXANE

Synonym: 1,4-dioxane $\quad C_4H_8O_2$
ACGIH TLV: 25 ppm (90 mg/m³) - Skin
OSHA Standard: 100 ppm (360 mg/m³)
NIOSH Recommendation: 1 ppm ceiling (30 min)

Method	Sampling Duration	Sampling Location	Useful Range (ppm)	System Cost ($)	Test Cost ($)	Manufacturer
DT	Grab	Portable	500-25,000	165	1.70	Matheson
DT	Grab	Portable	6000-56,000	180	2.25	Bendix
IR	Cont	Portable	2-200	4374	0	Foxboro
PI	Cont	Fixed	2-200	4950	0	HNU
GC	1 min	Portable	1-10⁶	10,000	0	Microsensor

NIOSH METHOD NO. P&CAM 127

(See acetone)

EPICHLORHYDRIN

Synonym: 1-chloro-2,3-epoxypropane
ACGIH TLV: 2 ppm (10 mg/m^3) - Skin
OSHA Standard: 5 ppm (20 mg/m^3)
NIOSH Recommendation: 0.5 ppm (2 mg/m^3)

$\overline{OCH_2CHCH_2}Cl$

Method	Sampling Duration	Sampling Location	Useful Range (ppm)	System Cost ($)	Test Cost ($)	Manufacturer
DT	Grab	Portable	5-50	150	2.80	Nat'l Draeger
IR	Cont	Portable	0.6-10	4374	0	Foxboro
GC	1 min	Portable	1-10^6	10,000	0	Microsensor

NIOSH METHOD NO. P&CAM 127

(Use as described for acetone, with 120° C. column temperature)

ETHANOLAMINE

Synonym: 2-aminoethanol
ACGIH TLV: 3 ppm (8 mg/m^3)
OSHA Standard: 3 ppm

$NH_2CH_2CH_2OH$

Method	Sampling Duration	Sampling Location	Useful Range (ppm)	System Cost ($)	Test Cost ($)	Manufacturer
IR	Cont	Portable	3-10	4374	0	Foxboro

NIOSH METHOD NO. P&CAM 270

(See diethylaminoethanol)

2-ETHOXYETHANOL

Synonyms: glycol monoethyl ether; cellosolve
ACGIH TLV: 5 ppm (19 mg/m³) - Skin
OSHA Standard: 200 ppm (740 mg/m³)

$C_2H_5OCH_2CH_2OH$

Method	Sampling Duration	Sampling Location	Useful Range (ppm)	System Cost ($)	Test Cost ($)	Manufacturer
IR	Cont	Portable	4-400	4374	0	Foxboro

NIOSH METHOD NO. S76

(Use as described for 2-butoxyethanol, with 140° C. column temperature)

2-ETHOXYETHYL ACETATE

ACGIH TLV: 5 ppm (27 mg/m³) - Skin
OSHA Standard: 100 ppm (540 mg/m³)

$CH_3COOCH_2CH_2OC_2H_5$

Method	Sampling Duration	Sampling Location	Useful Range (ppm)	System Cost ($)	Test Cost ($)	Manufacturer
IR	Cont	Portable	2-200	4374	0	Foxboro

NIOSH METHOD NO. P&CAM 127

(Use as described for acetone, with 100° C. column temperature)

ETHYL ACETATE

ACGIH TLV: 400 ppm (1400 mg/m³)
OSHA Standard: 400 ppm

$CH_3COOC_2H_5$

Method	Sampling Duration	Sampling Location	Useful Range (ppm)	System Cost ($)	Test Cost ($)	Manufacturer
DT	Grab	Portable	200-3000	150	2.50	Nat'l Draeger
DT	Grab	Portable	1000-50,000	165	1.70	Matheson
DT	Grab	Portable	400-15,000	180	2.25	Bendix
DT	8 hr	Personal	125-9000	850	3.10	Nat'l Draeger
IR	Cont	Portable	8-800	4374	0	Foxboro
PI	Cont	Fixed	0.2-20	4950	0	HNU
GC	1 min	Portable	$1-10^6$	10,000	0	Microsensor

NIOSH METHOD NO. P&CAM 127

(Use as described for acetone)

ETHYL ACRYLATE

Synonym: ethyl propenoate
ACGIH TLV: 5 ppm (20 mg/m³) - Skin
OSHA Standard: 25 ppm (100 mg/m³)

$CH_2=CHCOOC_2H_5$

Method	Sampling Duration	Sampling Location	Useful Range (ppm)	System Cost ($)	Test Cost ($)	Manufacturer
DT	Grab	Portable	200-7500	180	2.25	Bendix
IR	Cont	Portable	0.5-50	4374	0	Foxboro
PI	Cont	Fixed	2-200	4950	0	HNU

NIOSH METHOD NO. P&CAM 127

(Use as described for acetone, with 70° C. column temperature)

ETHYL ALCOHOL

Synonym: ethanol C_2H_5OH
ACGIH TLV: 1000 ppm (1900 mg/m³)
OSHA Standard: 1000 ppm

Method	Sampling Duration	Sampling Location	Useful Range (ppm)	System Cost ($)	Test Cost ($)	Manufacturer
DT	Grab	Portable	100-3000	150	2.70	Nat'l Draeger
DT	Grab	Portable	1000-50,000	165	1.70	Matheson
DT	Grab	Portable	200-75,000	180	2.25	Bendix
DT	8 hr	Personal	62.5-8000	850	3.10	Nat'l Draeger
IR	Cont	Portable	20-2000	4374	0	Foxboro
PI	Cont	Fixed	20-2000	4950	0	HNU
GC	1 min	Portable	$1-10^6$	10,000	0	Microsensor

NIOSH METHOD NO. S66

(Use as described for butyl alcohol, with 70° C. column temperature)

ETHYLAMINE

Synonym: aminoethane $C_2H_5NH_2$
ACGIH TLV: 10 ppm (18 mg/m³)
OSHA Standard: 10 ppm

Method	Sampling Duration	Sampling Location	Useful Range (ppm)	System Cost ($)	Test Cost ($)	Manufacturer
IR	Cont	Portable	0.4-20	4374	0	Foxboro

NIOSH METHOD NO. P&CAM 221

(See n-butylamine)

ETHYLBENZENE

Synonym: phenylethane
ACGIH TLV: 100 ppm (435 mg/m³)
OSHA Standard: 100 ppm

$C_2H_5C_6H_5$

Method	Sampling Duration	Sampling Location	Useful Range (ppm)	System Cost ($)	Test Cost ($)	Manufacturer
DT	Grab	Portable	30-600	150	2.60	Nat'l Draeger
DT	Grab	Portable	10-500	165	1.70	Matheson
DT	Grab	Portable	7-700	180	2.25	Bendix
IR	Cont	Portable	2-200	4374	0	Foxboro
PI	Cont	Fixed	2-200	4950	0	HNU
GC	1 min	Portable	$1\text{-}10^6$	10,000	0	Microsensor

NIOSH METHOD NO. P&CAM 127

(Use as described for acetone, with 85° C. column temperature)

ETHYL BROMIDE

Synonym: bromoethane
ACGIH TLV: 200 ppm (890 mg/m³)
OSHA Standard: 200 ppm

C_2H_5Br

Method	Sampling Duration	Sampling Location	Useful Range (ppm)	System Cost ($)	Test Cost ($)	Manufacturer
DT	Grab	Portable	10-90	180	2.25	Bendix
IR	Cont	Portable	4-400	4374	0	Foxboro
PI	Cont	Fixed	2-200	4950	0	HNU
GC	1 min	Portable	$1\text{-}10^6$	10,000	0	Microsensor

NIOSH METHOD NO. S107

(Use as described for difluorodibromomethane)

ETHYL BUTYL KETONE

Synonym: 3-heptanone $C_2H_5COC_4H_9$
ACGIH TLV: 50 ppm (230 mg/m^3)
OSHA Standard: 50 ppm

Method	Sampling Duration	Sampling Location	Useful Range (ppm)	System Cost ($)	Test Cost ($)	Manufacturer
IR	Cont	Portable	1-100	4374	0	Foxboro

NIOSH METHOD NO. S66

(Use as described for n-butyl alcohol, with 120° C. column temperature)

ETHYL CHLORIDE

Synonym: chloroethane C_2H_5Cl
ACGIH TLV: 1000 ppm (2600 mg/m^3)
OSHA Standard: 1000 ppm

Method	Sampling Duration	Sampling Location	Useful Range (ppm)	System Cost ($)	Test Cost ($)	Manufacturer
IR	Cont	Portable	20-2000	4374	0	Foxboro

NIOSH METHOD NO. P&CAM 127

(Use as described for acetone, with 110° C. column temperature)

ETHYLENE CHLOROHYDRIN

Synonym: 2-chloroethanol $ClCH_2CH_2OH$
ACGIH TLV: 1 ppm (3 mg/m^3) ceiling - Skin
OSHA Standard: 5 ppm (16 mg/m^3)

Method	Sampling Duration	Sampling Location	Useful Range (ppm)	System Cost ($)	Test Cost ($)	Manufacturer
IR	Cont	Portable	0.2-10	4374	0	Foxboro

NIOSH METHOD NO. S52

(Use as described for allyl alcohol, with 130° C. column temperature)

ETHYLENEDIAMINE

Synonym: 1,2-ethanediamine $NH_2CH_2CH_2NH_2$
ACGIH TLV: 10 ppm (25 mg/m^3)
OSHA Standard: 10 ppm

Method	Sampling Duration	Sampling Location	Useful Range (ppm)	System Cost ($)	Test Cost ($)	Manufacturer
DT	Grab	Portable	2-120	180	2.25	Bendix
IR	Cont	Portable	0.8-20	4374	0	Foxboro

NIOSH METHOD NO. P&CAM 276

Principle
A known volume of air is drawn through a tube containing silica gel to trap vapors of the compounds. Concentrated HCl is added to stabilize amines collected. The sorbent sections are transferred to glass vials and treated with methanol solution. An aliquot is alkalinized with sodium hydroxide. Benzaldehyde is added to form dibenzylidene derivatives. The solution is analyzed by gas chromatography with flame-ionization detection.

Range and Sensitivity
The ranges of the method for the two analytes are at least:

    ethylenediamine 0.2-7 mg/sample
    diethylenetriamine 0.2-9 mg/sample

Data relating to the lower ends of the range are discussed below. The upper limits are determined in the capacities of the first section of the sampling tube used to collect the sample. Amounts of individual compounds retained on 150 mg of a 42/60 mesh silica gel in 6 mm I.D. sampling tubes at 1 L/min sampling flow rate were found to be those given above as the upper limits.

Interferences

Water vapor does not significantly affect collection efficiency. Large amounts of ammonia or primary amines will reduce the amounts of benzaldehyde available for reaction with the analytes. Such interferences should be compensated for by adding additional benzaldehyde.

Any compound which has nearly the same retention time on the GC column as one of the analytes is an interference. Retention time data on a single GC column cannot be considered as proof of chemical identity. The presence and identity of possible interfering substances may be determined by the analysis of bulk samples from the process or environment.

Precision and Accuracy

The volume of air sampled can be measured to within 2% if a pump with a calibrated volume indicator and an adequate battery is used. Volumes calculated from initially set flow rates may be less accurate (5-10%) unless changes in flow rate are manually or electronically monitored and compensated. Precision (relative standard deviation) of replicate GC analyses of standards was measured to be:

    ethylenediamine 5% at 0.05 mL eluent
    diethylenetriamine 8% at 0.09 mL eluent

These levels corresponds to sample sizes of 0.1 mg and 0.2 mg, respectively. At the upper limits of the method analytical precision becomes 2% relative standard deviation or better.

The precision (relative standard deviation) of the method based on analyses of replicate 0.1 mg samples injected by syringe into silica gel sections or into 2 mL of eluent was 14% for both compounds. The precision of the analysis is dependent upon the precision and sensitivity of the technique used to quantitate the GC peaks of samples and standards. An electronic digital integrator with baseline correction capability is best for this purpose.

Collection efficiency is 100% until breakthrough of the sampling tube bed occurs. Desorption efficiency for both compounds based on analyses of replicate 0.1 mg samples on silica gel and standards in 2 mL eluent were found to be 102% with a relative standard deviation of 20%. Measurements at higher sample levels have given desorption efficiencies consistently in the range 90-100% with better precision.

This method has given quantitative results equivalent to an independent method involving adsorption of ethylenediamine vapors in water in a bubbler and titration with standard acid. Using each method six 0.030 m^3 vapor in nitrogen samples were taken at 1 L/min. The following sample amounts and relative standard deviations in parentheses were determined:

    tube and chromatography 3.30 mg (4%)
    bubbler and titration   3.42 mg (2%)

Advantages and Disadvantages

The method uses a small, portable sampling device involving no liquids. This is an advantage for sampling air in a worker's breathing zone without interfering with normal work activities. Transportation to the analytical laboratory is simplified with the solid sorbent tube. The sorbent tube has a high capacity even at high relative humidities. It can be used for at least 8 hours to measure a workday average concentration, or for shorter times to measure excursion concentrations. Desorption and preparation of samples for analysis involve simple procedures and equipment.

Several amines can be collected and determined simultaneously. The GC analysis distinguishes which are present and at what individual concentrations they occur. Interferences by other amines are much less likely than in colorimetric or titrimetric methods. A major disadvantage is the tendency of amines to oxidize while adsorbed on surfaces exposed to air. This is overcome by addition of concentrated hydrochloric acid to the sorbent sections soon after sampling.

Apparatus

Sorbent. The silica gel should be equivalent to Silica Gel D-08, chromatographic grade, activated and fines free, 42/60 mesh, as produced by Coast Engineering Laboratories and sold by Applied Science Laboratories (State College, PA).

Sampling tubes. Glass tubes 7 cm long and 6 mm I.D., tapered and flame sealed at one end, are packed with two 150 mg sections of silica gel. Glass wool plugs are used to separate and enclose the sections. The second end is flame sealed to prevent contamination during storage prior to use. Polyethylene caps are used to seal tubes after sampling is completed. Note: Sampling tubes constructed with metal parts or urethane foam plugs should not be used. Concentrated hydrochloric acid will react with these materials.

Personal sampling pump. Battery operated pumps are required capable of operation at 0.2 L/min for up to 8 hours or at 1.0 L/min for up to 2 hours with a sampling tube in line. The pump is to be calibrated with a representative sorbent tube in line. A wet or dry test meter or bubble meter capable of measuring a flow rate of 0.2 L/min and/or 1.0 L/min to within 2% may be used in setting the pump flow.

Gas chromatograph. With a flame-ionization detector. Temperature programming capability is necessary to determine more than one compound simultaneously.

Column. 0.6 m × 6 mm I.D. glass, silanized and packed with 10% (by weight) silicone SE-30 on 80/100 mesh Supelcoport or equivalent support.

Strip chart recorder. Compatible with the GC. An electronic digital integrator is desirable.

Test tubes or glass vials. 2 mL sealed by a septum.

Syringes. 10 μL.

Syringe for dispensing conc. HCl. 50 μL, with glass barrel, fluorocarbon-tipped plunger, and inert (platinum or fluorocarbon) needle.

Pipettes. 0.5 and 2 mL.

Volumetric flasks, 10 mL.

File and forceps.

Reagents

All chemicals must be analytical reagent grade.

Ethylenediamine.

Diethylenetriamine.

Water, doubly distilled and aldehyde free. Deionized water should not be used as it may contain formaldehyde and other impurities leached from the ion exchange resins.

Conc. HCl (38%, 12 M).

Benzaldehyde.

Eluent solution. 4:1 methanol-water (by volume).

Alkalinizing solution. 0.20 N NaOH in 4:1 methanol-water (by volume).

Helium, Bureau of Mines Grade A.

Hydrogen, prepurified.

Air, compressed and filtered.

Procedure

Cleaning of equipment. All glassware is washed with detergent solution, rinsed with tap water and distilled water, and dried in an oven.

Collection and shipping of samples. Immediately before beginning the collection of a sample, break each end of the sorbent tube so as to provide openings at least 2 mm in diameter. Attach the tubing from the sampling pump to the backup end of the sampling tube. Sampled air must not pass through any hose or tubing before entering the sorbent tube. With the sorbent tube in a vertical position as much as is practical, sample the air at 1.0 L/min for 15 min to 2 hours or at 0.2 L/min for 2-8 hours. Intermediate flow rates and other sampling periods may be used, as appropriate.

Immediately after sampling is completed, add 20 μL of conc. HCl to each section of silica gel in the tube. Use a 50 μL glass syringe with fluorocarbon-tipped plunger and inert needle, so that the acid does not contact and react with stainless steel. The silica gel will turn yellow upon addition of the acid due to iron impurities in the silica gel. Seal the sorbent tubes with polyethylene caps. Obtain the blank sample by handling one tube in the same manner as the sample tubes (break, add acid, seal, and ship) except that no air is pumped through it. For shipping to the laboratory, pack the tubes tightly to minimize chances of breakage during transit. Tubes should not be subjected to extremes of high temperature or low pressure.

Preparation of samples. Score each tube with a file 5 mm in front of the glass wool plug that precedes the first sorbent section and break the tube there. Transfer this plug and the initial section to a 2 mL test tube or glass vial that can be septum sealed. Likewise, transfer the second plug and sorbent section to another test tube. Label each appropriately for separate analysis.

Desorption. Add 2 mL of eluent solution to each sorbent section. Seal the samples. Shake the mixtures occasionally over a period of 2 hours. Tests have shown that complete desorption occurs within 2 hours and samples are stable for at least a day.

Preparation and reaction. Transfer a 0.50 mL aliquot of each eluate to another test tube or vial. Add 0.50 mL of alkalinizing solution and mix thoroughly. Verify the basicity of the solution with litmus paper. Add 10 μL of benzaldehyde to the basic solution, mix thoroughly, and allow at least 20 min for reaction.

GC conditions:

30 mL/min helium gas flow rate
200° C. injection port temperature
250° C. detector temperature
200° C. column temperature for 1,2-(dibenzylideneamino)ethane
235° C. column temperature for 2,2'-(dibenzylideneamino)diethylamine

Under these conditions, retention times are 300 sec for 1,2-(dibenzylideneamino)ethane and 345 sec for 2,2'-(dibenzylideneamino)diethylamine.

Injection. Inject and analyze 3 μL aliquots of each sample and standard. To eliminate difficulties arising from blowback or distillation within the 10 μL syringe, use a solvent flush technique.

GC peak measurement. Determine the areas of the peaks of the compounds of interest from analyses of samples and standards.

Determination of desorption efficiency. Desorption efficiency for a particular compound can vary from one lot of silica to another and from one laboratory to another. Also, for a given lot of silica gel the desorption efficiency may vary with the amount of material adsorbed during sampling. Therefore, it is necessary to determine at least once the desorption efficiency for each analyte with each lot of silica gel used for at least two levels within the normal range of sample size. If the desorption efficiencies are significantly different from quantitative recovery, they should be used to correct the measured weight of analyte as described below.

Place 150 mg of silica gel in a 2 mL glass stoppered tube. The silica gel must be from the same lot as that used in collecting the sample; it can be obtained from unused sorbent tubes. With a microliter syringe, inject a known amount of amine, either pure or in water solution, directly onto the silica gel. Also inject 20 µL of conc. HCl. Close the tube with the glass stopper and allow it to stand overnight to insure complete adsorption of the amine. Prepare at least three tubes for each at two different levels. These tubes are referred to as samples. Prepare a blank in the same manner, omitting the amine. Analyze the samples and blank as described above. Also analyze three standards prepared by adding identical amounts of the amine and 20 µL of conc. HCl to 2.0 mL of eluent solution. Determine the concentrations of the analytes in the blank, samples, and standards using calibration curves prepared as described below. The desorption efficiency is calculated by dividing the concentration of amine found in the sample by the concentration obtained for the corresponding standard.

Calibration and Standardization

Preparation of a stock solution. Calculate for each compound i the volume Vi (µL) of pure liquid required for preparation of 10 mL of a stock solution, 10 µL of which contains amounts of analytes equal to those collected from volume of air Vs at concentrations Xi:

$$Vi = (Xi)(Vs)(1000) \div di$$

where:

xi = the standard concentration limit, or the anticipated average concentration, (mg/m^3) of compound i in air
Vs = volume (m^3) of air sampled
di = density (mg/µL) of a pure compound i
1000 = aliquot factor

Add the calculated volumes of each compound to a 10 mL volumetric flask and dilute to the mark with 4:1 methanol-water. Neutral solutions of amines are subject to oxidation and should be prepared fresh when needed.

Preparation of calibration standards. Add 2.5, 5, 10, 15, and 20 µL of the prepared standard to reaction tubes containing 2 mL eluent solution and 40 µL concentrated hydrochloric acid. Use a 10 µL syringe to inject through septum seals. Shake these tubes. These standards correspond to 0.25, 0.5, 1, 1.5, and 2 times Xi in Vs, respectively. Other standards may be similarly prepared, if desired. React and analyze standards with samples according to the directions in the previous section. Prepare a calibration curve for each amine by plotting peak areas obtained from the analyses of standards against nominal amount (mg) in the calibration standards.

Calculations

Read the weights (mg) of each compound corresponding to each peak area from the appropriate calibration curve. Correct each value for the weight found in a blank, if any. Add the weights found in the front and backup sections of the sample tube to obtain the total weight of compound in the air volume sampled. Divide the total weight by the desorption efficiency to obtain the corrected sample weight.

Determine the volumes (m^3) of air sampled at ambient conditions based on the appropriate information, such as flow rate (L/min) multiplied by sampling tube (min) and 10^{-3} m^3/L. If a pump using a rotameter for flow rate control was used for sample collection, a pressure and temperature correction must be made for the indicated flow rate. The expression for this correction is:

$$V_s = 10^{-3} ft(P_1 T_2 / P_2 T_1)^{1/2}$$

where:

    f = sample flow rate (L/min)
    t = sampling time (min)
    P_1 = pressure during calibration of sampling pump (torr)
    P_2 = pressure of air sampled (torr)
    T_1 = temperature during calibration of sampling pump (°K)
    T_2 = temperature of air sampled (°K)

Divide the total corrected weight (mg) of each compound by V_s (m^3) to obtain the concentration of the compound in the air sampled (mg/m^3).

References
1. G.O. Wood, R.G. Anderson, and J.W. Nickols. Sampling and Analysis of Aminoethanols in Air, Report LA-UR-77-1398, 1977, Industrial Hygiene Group, Los Alamos Scientific Laboratory, Los Alamos, New Mexico. Presented at the 1977 American Industrial Hygiene Conference, May 1977, New Orleans.
2. G.O. Wood and J.W. Nickols. Development of Air-Monitoring Techniques Using Solid Sorbents, October 1, 1976-December 31, 1977, Progress Report LA-7295-PR, 1978, Los Alamos Scientific Laboratory, Los Alamos, New Mexico. Also available from the National Technical Information Services, Springfield, VA.

ETHYLENE DIBROMIDE

Synonyms: 1,2-dibromoethane; EDB   BrCH$_2$CH$_2$Br
ACGIH TLV: Suspected carcinogen - Skin
OSHA Standard: 20 ppm (155 mg/m^3)
NIOSH Recommendation: 0.13 ppm (1 mg/m^3) ceiling (15 min)

Method	Sampling Duration	Sampling Location	Useful Range (ppm)	System Cost ($)	Test Cost ($)	Manufacturer
DT	Grab	Portable	6-80	180	2.25	Bendix
IR	Cont	Portable	0.4-40	4374	0	Foxboro
PI	Cont	Fixed	0.2-20	4950	0	HNU
GC	1 min	Portable	1-10^6	10,000	0	Microsensor

NIOSH METHOD NO. P&CAM 260

Principle
A known volume of air is drawn through a charcoal tube to trap the 1,2-dibromoethane vapor present. The charcoal in the tube is transferred to a 10 mL volumetric flask and the analyte is desorbed with 10.0 mL of 99:1 benzene-methanol (v/v). An aliquot of the desorbed sample is subjected to gas chromatographic analysis using an electron-capture detector.

Range and Sensitivity

The range of an electron capture detector most useful for the quantitation of 1,2-dibromoethane depends upon the type of detector and chromatograph used. For a non-linearized 63-Ni electron capture detector in a Tracor MT 220 Gas Chromatograph, the most useful range for quantitation was 20-400 pg of 1,2-dibromoethane/aliquot injected. This range corresponds to 40-80 ng of 1,2-dibromoethane/sample solution.

For a 25 L air sample, this method is applicable to air concentrations which range from 0.002-8.0 mg/m^3. The upper limit of the method depends upon the capacity of the charcoal tube and involves dilution of the sample solution. The detection limit was approximately 4 pg/injection or 8 ng/sample solution.

Interferences

Compounds which are detected by the electron capture detector and have retention times approximately that of 1,2-dibromoethane will interfere with the analysis. When the amount of water in the air is so great that condensation actually occurs in the charcoal tube, vapors of 1,2-dibromoethane may not be trapped efficiently. When interfering compounds are known or suspected to be present in the air, such information including their suspected identities should be transmitted with the sample.

Precision and Accuracy

The precision, or relative standard deviation, versus the sample size for the analytical method is as follows: 0.032 for 2400 ng; 0.062 for 200 ng; and 0.079 for 40 ng. The precision and accuracy are affected by drift in detector response. Satisfactory precision can be realized by use of an appropriate internal standard. Recoveries of analyte through desorption decrease with increase storage times at room temperature. These decreases are more pronounced at lower levels of analyte.

Storage of the charcoal tube samples at -25° C. permits satisfactory recoveries. Dilution of the sample solution is a potential source of error. Solutions containing 1,2-dibromoethane in concentrations above the useful range of the detector must be diluted. The capacity of the charcoal tube with respect to 1,2-dibromoethane is limited. When the sample loading in the backup section of the tube exceeds 10% of that found in the front section, the possibility of sample loss exists.

The precision of the method is dependent upon the reproducibility of the pressure drop across the tubes. This drop will affect the flow rate and cause the volume to be imprecise because the pump is usually calibrated for one tube only.

Advantages and Disadvantages

The sampling device is small, portable, and involves no liquids. The highly selective nature of an electron-capture detector dramatically reduces the potential number of interferences. The use of an internal standard is required in order to attain good precision. Recoveries of 1,2-dibromoethane from charcoal are decreased upon storage of the charcoal tube samples at room temperature, particularly at lower levels of analyte.

Apparatus

Personal sampling pump. The flow rate of which can be accurately determined at 200 mL/min or less with the charcoal tube in line.

Charcoal tubes. Glass tube with both ends flame sealed, 7 cm long with 6 mm O.D. and 4 mm I.D., containing 2 sections of 20/40 mesh activated charcoal separated by a 2 mm portion of urethane foam. The activated charcoal is prepared from coconut shells and is fired at 600° C.

prior to packing. The adsorbing section contains 100 mg of charcoal, the backup section 50 mg. A 3 mm portion of urethane foam is placed between the outlet end of the tube and the backup section. A plug of silylated glass wool is placed in front of the adsorbing section. The pressure drop across the tube must be less than 25 torr at a flow rate of 200 mL/min.

Gas chromatograph. Equipped with an electron-capture detector. A glass tube filled with 20/40 mesh activated charcoal should be attached to the exit port in order to trap the 1,2-dibromoethane in the effluent gas stream. A glass tube with a 2 cm I.D. and a length of 25 cm was found to have little or no effect upon the retention time of 1,2-bromoethane.

Column. 1.8 m × 4 mm I.D. constructed from borosilicate glass and packed with 3% OV-210 on 80/100 Gas Chrom Q.

Syringes. 10 µL and other convenient sizes for preparing standard solutions.
Glass vials. 2 mL.
Pipets. 10.0 mL and 1.0 mL.
Volumetric flasks. 10.0 mL.
Parafilm.

Reagents

Benzene, pesticide quality.
Methanol, pesticide quality.
Benzene-methanol, 99:1 (v/v).
1,2-Dibromoethane of known purity.
1,1,2,2-Tetrachloroethane or 1,2-dibromopropane. Or other appropriate internal standard.

Procedure

Cleaning of equipment. The syringes are rinsed well, first with methanol and then with benzene. All other glassware should be cleaned with soap and water, then rinsed with the following sequence: distilled water, pesticide quality methanol, pesticide quality benzene.

Collection and shipping of samples. Immediately before sampling, the ends of the charcoal tubes are broken to provide an opening at least one-half the internal diameter of the tube. The smaller section of charcoal is used as a backup and should be positioned nearest the sampling pump. The charcoal tube is be placed in a vertical direction during sampling in order to minimize channeling through the charcoal. Air being sampled should not be passed through any hose or tubing before entering the charcoal tube.

The sampling time, volume and flow rate are measured. The sample is taken at a flow rate of 200 mL/min or less. The total volume sampled should be no more than 25 L. The temperature, pressure and relative humidity of the atmosphere being sampled are recorded. To obtain a blank sample, a charcoal tube is handled in exactly the same manner as each sample tube except that no air is drawn through it. The charcoal tubes are capped with the supplied plastic caps immediately after sampling.

Low levels of 1,2-dibromoethane cannot be stored on charcoal at ambient temperatures for long periods of time. Therefore, if the analysis cannot be performed within 16-24 hours after sampling has been completed, the samples must be stored at -25° C. or below. Refrigerated samples may be stored for 2 weeks. For shipment to the laboratory, the samples are packed firmly in an insulated container cooled with dry ice. If appropriate, a sample of the bulk material in a glass container with a Teflon-lined cap is prepared and shipped to the laboratory in a separate container.

Preparation of samples. Each charcoal tube is scored with a file in front of the first section of charcoal and broken open. The glass wool is removed and discarded. The charcoal in the first (larger) section is transferred to a 10 mL volumetric flask. The separating section of polyurethane foam is removed and discarded. The charcoal in the second (smaller) section is transferred to a separate 10 mL volumetric flask. These two sections are analyzed separately.

Desorption of samples. Prior to analysis, 10.0 mL of 99:1 benzene-methanol (v/v) is added to each 10 mL volumetric flask containing a section of charcoal. Each flask is stoppered and allowed to stand with occasional agitation over a 1 hour period.

GC conditions:

35 mL/min nitrogen carrier gas flow
175° C. inlet temperature
315° C. detector temperature
50° C. column temperature
DC at 90-95% standing current 63-Ni detector operating mode

Injection. By means of the solvent flush injection technique, 5 μL of sample solutions are injected into the gas chromatograph. After each injection, the syringe is rinsed thoroughly, first with methanol and then with benzene. Otherwise, traces of 1,2-dibromoethane will remain in the syringe and contaminate the solvent flush solution. Periodically, an injection of solvent flush solution only (99:1 benzene-methanol, v/v) should be made to check the cleanliness of the syringe and the purity of the solvent flush solution.

A preliminary analysis of the samples is run in order to determine whether the sample solutions are too concentrated for the useful range of the detector, and which region of the chromatogram may be a suitable location for an internal standard peak. If necessary, an appropriate volume of each original sample solution is diluted to 10.0 mL with 99:1 benzene-methanol (v/v). These diluted solutions should be reanalyzed in order to test whether the amount of 1,2-dibromoethane injected is now within the useful range of the detector, and aid in the selection of a suitable internal standard.

The use of an internal standard is required in order to compensate for changes in detector sensitivity with time. The approximate retention times of 1,2-dibromoethane and two potential internal standards (under the GC conditions described in this method) are:

Compound	Retention time (min)	B.P. (°C.)
1,2-dibromoethane	2.2	131
1,2-dibromopropane	2.9	140
1,1,2,2-tetrachloroethane	4.1	146

The selection of an internal standard is based on three criteria: the absence of components in the sample which would interfere with the internal standard; the absence of adsorption of the internal standard by the charcoal while in 99:1 benzene-methanol solution (if the internal standard is other than 1,2-dibromopropane or 1,1,2,2-tetrachloroethane, then it must be tested); and the proximity of the internal standard peak to the 1,2-dibromoethane peak.

A stock solution of the selected internal standard in 99:1 benzene-methanol (v/v) is prepared such that 5.0 μL of this stock solution contains approximately 400 ng of internal standard. To each diluted sample solution and each standard solution, 5.0 μL of this internal standard stock solution is added. Aliquots (5 μL) of these solutions are injected into the gas chromatograph for quantitative analysis.

Determination of recovery. Activated charcoal equivalent to the amount in the front (larger) section of the sampling tube is poured into a 2 mL glass vial. This charcoal must be from the same batch as that used in obtaining the samples and can be obtained from unused charcoal tubes. A known amount of 1,2-dibromoethane in a 5 μL aliquot of benzene solution is added directly to the charcoal with a μL syringe. The vial is sealed immediately with Parafilm.

Six samples at each of three different levels are prepared in the above manner and allowed to stand 16-24 hours in order to assure complete adsorption of the 1,2-dibromoethane onto the charcoal. A parallel blank vial is treated in exactly the same manner except that no sample is added to it. The samples and the blank are treated with 10.0 mL of 99:1 benzene-methanol (v/v) in the manner described above. Three control solutions are prepared by injecting the same quantity of 1,2-dibromoethane into 10.0 mL portions of 99:1 benzene-methanol (v/v) with the same syringe used in the preparation of the samples.

The samples, the blank, and the control solutions are analyzed on the same day. Recovery (R) is equal to the difference between the average quantity of 1,2-dibromoethane recovered from the samples and the quantity recovered from the blank divided by the average quantity of 1,2-dibromoethane measured in the control solutions. The desorption efficiency obtained is dependent upon the amount of analyte collected on the charcoal. Therefore, a plot of desorption efficiency versus quantity of analyte measured is constructed. This curve is used to correct for adsorption losses.

Calibration and Standardization

A series of standards varying in concentration over the range of interest is prepared. The range of interest may be in the vicinity of 40-800 ng/10.0 mL of 99:1 benzene-methanol (v/v). These standards are analyzed during the same time period as the samples. In view of the low concentrations involved, the volume of a solution prepared from X number of ng of 1,2-dibromoethane and 10.0 mL of solvent is very nearly 10.0 mL and can be assumed to be 10.0 mL for the purpose of this analytical method. A calibration curve is constructed daily by plotting quantity of 1,2-dibromoethane/10.0 mL of solution versus the ratio of 1,2-dibromoethane peak area to internal standard peak area.

Calculations

Determine the quantity of 1,2-dibromoethane, W (ng), in 10.0 mL of diluted sample solution from the calibration curve. Determine the total quantity of 1,2-dibromoethane, Q, in the air sampled:

$$Q = (WF/R)_1 + (WF/R)_2$$

where:

  F = dilution factor (equals 10.0 mL ÷ volume of original sample solution)
  R = recovery
  1 refers to the front section of the sampling tube
  2 refers to the back section of the sampling tube

In the event the blank gives rise to a peak with the same retention time as that of 1,2-dibromoethane, the analyst must determine or compensate for it. Determine the concentration, C ($\mu g/m^3$), of 1,2-dibromoethane in air:

$$C = Q \div V$$

where:

  V = the volume of air sampled in L

ETHYLENE DICHLORIDE

Synonym: 1,2-dichloroethane $\quad\quad\quad\quad\quad\quad\quad\quad\quad\quad\quad\quad\quad\quad$ ClCH$_2$CH$_2$Cl
ACGIH TLV: 10 ppm (40 mg/m^3)
OSHA Standard: 50 ppm (200 mg/m^3)
NIOSH Recommendation: 1 ppm (4 mg/m^3)

Method	Sampling Duration	Sampling Location	Useful Range (ppm)	System Cost ($)	Test Cost ($)	Manufacturer
IR	Cont	Portable	1-100	4374	0	Foxboro
PI	Cont	Fixed	0.2-20	4950	0	HNU
GC	1 min	Portable	1-10^6	10,000	0	Microsensor

NIOSH METHOD NO. P&CAM 127

(See acetone)

ETHYLENEIMINE

Synonyms: ethylenimine; dihydroazirine; aziridine $\quad\quad\quad\quad\quad\quad$ NHCH$_2$CH$_2$
ACGIH TLV: 0.5 ppm (1 mg/m^3) - Skin
OSHA Standard: 0.5 ppm

Method	Sampling Duration	Sampling Location	Useful Range (ppm)	System Cost ($)	Test Cost ($)	Manufacturer
DT	Grab	Portable	1-60	180	2.25	Bendix

NIOSH METHOD NO. P&CAM 300

Principle

A known volume of air is drawn through a bubbler containing an aqueous buffered solution of 1,2-naphthoquinone-4-sulfonate (Folin's reagent) which reacts with ethylenimine to produce

4-(1-aziridinyl)-1,2-naphthoquinone. The reaction product is extracted with chloroform. An aliquot of the extract is injected into a high performance liquid chromatograph (HPLC). Peak areas are electronically determined and compared with a calibration curve obtained from injections of standard solutions of 4-(1-aziridinyl)-1,2-naphthoquinone.

Range and Sensitivity

This method was evaluated over the range 1.2-795 μg/sample (24-15,900 μg/m^3 for 50 L air sample) when using 15 mL of reagent and 10 μL injections into the HPLC. It can probably be extended down to 0.7 μg/sample as a lower limit. Tests to determine breakthrough of ethylenimine indicated less than 0.1% in backup bubblers when challenge concentrations of 204 μg/L were sampled for 26 min and 20.6 μg/L for 270 min. Air flows were maintained at approximately 0.15 L/min. The limit of detection is 0.3 μg/sample.

Interferences

In solutions containing ethylenimine at 17 ng/μL, the following compounds were found not to interfere when their concentrations were 2 ng/μL; methylamine, diethylamine, butylamine, ethanolamine, dihexylamine, dicyclohexylamine, benzylamine, dibenzylamine, and aniline. Propylenimine and 2-bromoethylamine interfere. The latter compound and its chloro analogue can be expected to interfere since they are probably converted to ethylenimine in the alkaline reagent solution.

Any compound collected with ethylenimine and having or forming a derivative which has the same retention time as the analyte will interfere with the analysis. This type of interference often can be eliminated by changing the mobile phase composition, solvent programming, or changing the chromatographic column. Compounds, such as ammonia, which consume Folin's reagent can be considered interferences if they prevent ethylenimine from reacting with the regent. The ammonia derivative is not extracted into the chloroform.

Precision and Accuracy

The precision of the overall method is 7% relative standard deviation for samples collected from dry air at room temperature. This value is based on the collection and analysis of 52 samples of laboratory generated atmospheres containing ethylenimine at the levels 0.16-21 mg/m^3 and sampling at approximately 0.2 L/min. The range of volumes sampled was 2.6-39 L such that the range of sample sizes was 1-800 μg.

Folin's reagent spiked with 1.67 and 16.7 μg of ethylenimine and stored at 23° C. yielded recoveries of 86-99% after 7 days and 56-60% after 28 days. Samples stored at 5° C. yielded recoveries of 94-95% after 7 days and 86-94% after 28 days. It is recommended that samples be refrigerated until analyses can be conducted.

The method was evaluated with up to 800 μg of ethylenimine. Although Folin's reagent is present in sufficient excess to react with larger amounts of ethylenimine, it is not known whether the reaction will be quantitative. The accuracy of the method has not been determined.

Advantages and Disadvantages

The samples, collected in bubblers, are analyzed by means of a quick, instrumental method. The derivative formed between ethylenimine and Folin's reagent is indicative of the intact imine rather than its hydrolysis product. A disadvantage of the method is the awkwardness in using bubblers for collecting personal samples. If the employee's job performance requires much body movement, loss of the collection solution during sampling may occur. This method has not been field tested.

Apparatus

Personal air sampling pump. Capable of being maintained for 4 hours at a flow rate of 0.2 L/min with the sampler in line. Each pump should be calibrated with a representative sampler in line to minimize errors in volume measurements. A bubble flow meter or other suitable flow measuring device may be used. The pump should be protected from splashover or solvent condensation by a 5 cm long × 6 mm I.D. glass tube loosely packed with a plug of glass wool and inserted between the exit arm of the bubbler and the pump.

Spillproof bubblers. With fritted glass stems (I-1143-87 body and I-1143-89 bubbler, Ace Glass, Vineland, NJ, or equivalent) containing the collection solution (reagent) described below. The bubblers should be wrapped in aluminum foil to prevent photochemical decomposition of the analyte.

High performance liquid chromatograph. Equipped with an ultraviolet detector (254 nm) and injection valve.

Column. Lichrosorb DIOL or equivalent, 10 micron, 4.6 mm I.D. × 25 cm.

Potentiometric strip chart recorder.

Electronic integrator.

Analytical balance.

Desiccator.

Glass syringes, 10 μL.

Pipettes. Convenient sizes for the preparation of standard solutions.

Volumetric flasks. Low actinic glass, of convenient volumes for the preparation of standard solutions. Ten mL flasks are required for the sample extracts.

Pipette bulb.

Separatory funnels. With Teflon stopcocks, 60 mL and 1 L.

Funnels. 25 mm I.D. × 40 mm stem and 75 mm I.D. × 75 mm stem.

Beakers, 50 mL and 2 L.

Aluminum foil.

Filter paper. Whatman 42, 12.5 cm diameter.

Reagents

Chloroform. With 1% ethanol preservative, UV grade, distilled in glass.

Hexane, UV grade, distilled in glass.

2-Propanol, UV grade, distilled in glass.

Methyl alcohol, UV grade, distilled in glass.

Acetone, UV grade, distilled in glass.

Ethylenimine.

1,2-Naphthoquinone-4-sulfonic acid, sodium salt (Pfaltz and Bauer).

0.1 N Sodium hydroxide.

0.1 M Potassium dihydrogen phosphate.

Buffer solution, pH 7.7. Mix 100 mL of 0.1 M KH_2PO_4 with 93.4 mL of 0.1 N NaOH.

Folin's reagent. Dissolve 0.40 g of 1.2-naphthoquinone-4-sulfonic acid sodium salt in 100 mL of buffer solution. Dilute to 500 mL with distilled water in a volumetric flask. Wrap the flask with aluminum foil and store in a refrigerator. Discard the reagent in 5 days.

4-(1-Aziridinyl)-1,2-naphthoquinone. Wrap a 1 L separatory funnel with aluminim foil and add 2 g of the sodium salt of 1,2-naphthoquinone-4-sulfonic acid dissolved in 250 mL of distilled water. Add 25 mL of 0.5 M trisodium phosphate, shake and check that the pH is between 10.5 and 11.5. Add 0.3 mL ethylenimine and shake intermittently for 10 min. CAUTION: ethylenimine has been identified as a carcinogen. Appropriate precautions must be taken in handling the compound to avoid personnel exposure and area contamination.

Extract the 4-(1-aziridinyl)-1,2-naphthoquinone formed with six 200 mL portions of chloroform. Place the combined extracts in a 2 L beaker wrapped in aluminum foil. Cover the beaker with aluminum foil in which 3 holes have been made. Evaporate the chloroform with a nitrogen purge. Transfer the dry residue to a 50 mL beaker wrapped with aluminum foil. Add 35 mL of methyl alcohol and 1 mL of chloroform to the residue and stir briefly. Not all the residue will dissolve.

Place the beaker in ice water bath for 10 min and then filter the precipitate through Whatman 42 filter paper. Rinse the precipitate in the filter with 4 mL of chilled methyl alcohol and discard the filtrates. Dry the precipitate with a nitrogen purge, transfer it to a brown glass bottle and purge again. Dry the compound overnight in a desiccator containing Drierite. The melting point of the compound is 173-175° C. The compound is to be used for making standard solutions for calibration purposes. The 4-(1-aziridinyl)-1,2-naphthoquinone should be stored in a freezer until standard solutions are to be prepared.

Procedure

Cleaning of equipment. All glassware used for the laboratory analysis is washed with detergent and rinsed with tap water, distilled water, and methyl alcohol.

Collection and shipping of samples. Pour 15 mL of Folin's reagent into each midget bubbler. Connect the 5 cm glass tube containing the glass wool plug to the personal sampling pump and then to the bubbler using short sections of flexible tubing. The air being sampled should not pass through any tubing or other equipment before entering the bubbler. The bubbler should be maintained in a vertical position during sampling. The atmosphere is sampled at a flow rate of 0.2 L/min for 4 hours. The flow rate and sampling time, or the volume of sampled air, must be measured as accurately as possible.

The temperature, pressure, and if possible, humidity of the atmosphere being sampled are measured and recorded, along with other pertinent data such an location, date, and time of sampling. After sampling, use a pipet bulb to force air through the bubbler stem to dislodge the derivatized product and recover as much of the sampling solution as possible. Wash the stem with 2 mL of chloroform, adding the wash to the bubbler. The bubbler is sealed with a hard, non-reactive stopper (preferably Teflon or glass). Do not seal with rubber. The stoppers on the bubbler should be tightly sealed.

Whenever possible, hand delivery of the samples is recommended. Otherwise, transfer liquid to a 20 mL scintillation vial with polyseal cap (Kimble No. 74515 or equivalent) and seal with a 24-20 mm shrink band (Walter H. Jelly Co., Franklin Park, IL, or equivalent). Care should be taken to minimize spillage or loss by evaporation. Refrigerate samples if analysis cannot be conducted within a day. Blank samples are obtained by handling random bubblers in the same manner as in sampling (fill, seal, and transport) except that no air is drawn through them. If samples of bulk material associated with the process under investigation are to be shipped to the laboratory, they should not be placed in the same container as the air samplers of blanks.

Analysis of samples. Transfer the collection solution from each bubbler into individual 60 mL separatory funnels. Rinse the bubbler with 4 mL of chloroform and add the rinse to the separatory funnel. Cap and moderately shake the funnel for about 15 seconds. Allow the liquid phases to separate. With the aid of a small funnel, collect the chloroform extract in a 10 mL actinic glass volumetric flask. Repeat extraction and collection procedures with another 4 mL chloroform extract. Dilute the volumetric flask to 10 mL with chloroform and shake.

Liquid chromatographic conditions (23° C., 585 mm Hg):

Column:	Lichrosorb DIOL (25 cm × 4.6 mm I.D.)
Mobile phase:	Hexane/chloroform (with 1% ethanol)/2-propanol, 59.5/40/0.5 (v/v)
Flow rate:	1.3 mL/min, 300 psi
Detector:	UV(254 nm), 0.04 AUFS
Injection volume:	10 µL
Efficiency:	1936 theoretical plates
Capacity ratio:	0.9
Retention volume:	6.3 mL

Injection. Flush the 10 µL syringe first with acetone (to insure its cleanliness and to wet the barrel and plunger) and then with the sample solution. Draw 10 µL of sample solution into the syringe and inject the aliquot into the liquid chromatograph. Measure the peak area and read the amount of analyte from the calibration curve.

Efficiency of sample recovery. The recovery of the ethylenimine derivative can vary from one laboratory to another. The fraction recovery may also vary with the amount of ethylenimine present. Determinations should therefore, be made for at least three levels in the range of anticipated sample sizes. Recovery should be at least 90%.

To each bubbler containing 15 mL of Folin's reagent, add a known amount of an ethylenimine chloroform solution. Five bubblers are thus prepared at each of three different levels in the range of interest. Place the stem caps on the bubblers immediately and shake intermittently for 5 min. Treat a parallel blank in the same manner adding only an equal volume of chloroform. Allow them to stand overnight in a refrigerator. Analyze the samples as described above. The sample recovery (R) is given by the average amount found, corrected for the blank if necessary, divided by the amount added.

Calibration and Standardization

For accuracy in preparation of standards, it is recommended that one standard be prepared in a relatively large volume and at a high concentration. The initial standard is prepared by weighing a selected amount of 4-(1-aziridinyl)-1,2-naphthoquinone (e.g., 125 mg) into a 250 mL volumetric flask (low actinic glass) and adding chloroform to the calibration mark. Dilutions encompassing the concentration range of interest down to 0.3 ng/µL are then made from the stock solution. Solutions should be stored in a refrigerator when not in use. Prepare fresh standards every 5 days.

The standard solutions should be analyzed under the same liquid chromatographic conditions and during the same time period as the samples. This will minimize inaccuracy due to day-to-day variations in the setting of internal conditions. Calibration curves for 4-(1-aziridinyl)-1,2-naphthoquinone are prepared by plotting the peak areas for the standards against their concentration in µg/10 mL.

Calculations

In the event that the blank produces a peak with the same retention time as the analyte, the analyst should determine the source of the interference and eliminate or compensate for it. Read the weight (W_1) in µg of 4-(1-aziridinyl)-1,2-naphthoquinone present in the sample from the calibration curve. The total weight (Wt) is determined by:

$$Wt = W_1 \div R$$

where R is the recovery factor. The concentration of ethylenimine in air (C, $\mu g/m^3$) is given by:

$$C = (Wt)(1000)(0.216) \div V$$

where:

V = volume of air sampled in L
1000 = factor to convert from L to m^3
0.216 = gravimetric factor to convert the weight of 4-(1-aziridinyl)-1,2-naphthoquinone to that of the ethylenimine collected

The concentration can also be expressed in ppm:

ppm = (μg/m^3)(24.45)(760)(T + 273) ÷ (1000)(43.08)(P)(298)

where:

1000 = factor to convert μg/m^3 to mg/m^3
24.45 = molar volume (L/mol) at 25° C. and 760 mm Hg
43.08 = molecular weight (g/mol) of ethylenimine
760 = standard pressure (mm Hg)
P = pressure (mm Hg) of air sampled
T = temperature (°C.) of air sampled
298 = standard temperature (°K)

Reference

1. R. Morales, J.F. Stampfer, Jr., R.E. Hermes, E.E. Campbell and H.J. Ettinger. Development of a Sampling and Analytical Method for Ethylenimine, January 1, 1978 - October 31, 1978, Los Alamos Scientific Laboratory Progress Report LA-7978-PR, National Technical Information Service, Springfield, Virginia, 1979.

ETHYLENE OXIDE

Synonyms: 1,2-epoxyethane; oxirane (CH$_2$)$_2$O
ACGIH TLV: 1 ppm (2 mg/m^3) - Suspected carcinogen
OSHA Standard: 50 ppm (90 mg/m^3)
NIOSH Recommendation: 75 ppm ceiling (15 min)

Method	Sampling Duration	Sampling Location	Useful Range (ppm)	System Cost ($)	Test Cost ($)	Manufacturer
DT	Grab	Portable	1-500	150	2.80	Nat'l Draeger
DT	Grab	Portable	5-35,000	165	1.70	Matheson
DT	Grab	Portable	2-30,000	180	2.25	Bendix
IR	Cont	Portable	1-100	4374	0	Foxboro
PI	Cont	Fixed	0.5-50	4950	0	HNU
CM	Cont	Port/Fixed	0.01-25	4995	Variable	CEA
GC	1 min	Portable	1-10^6	10,000	0	Microsensor

NIOSH METHOD NO. S286

Principle
A known volume of air is drawn through a large charcoal tube to trap the organic vapors present. The sampling tube consists of two separate tubes, a front adsorbing tube and a backup tube. This sampling arrangement is necessary to prevent sample migration effects upon storage. The charcoal in each tube is transferred to a 5 mL, screw capped sample container and the analyte is desorbed with carbon disulfide. An aliquot of the desorbed sample is injected into a gas chromatograph. The area of the resulting peak is determined and compared with areas obtained from injection of standards.

Range and Sensitivity
This method was validated over the range of 41-176 mg/m^3 at an atmospheric temperature and pressure of 26° C. and 761 mm Hg, using a 5 L sample. Under the conditions of recommended sample size (5 L), the probable useful range of this method is 20-270 mg/m^3 at a detector sensitivity that gives nearly full deflection on the strip chart recorder for a 1.4 mg sample. This method is capable of measuring much smaller amounts if the desorption efficiency is adequate. Desorption efficiency must be determined over the range used.

The upper limit of the range of the method is dependent on the adsorptive capacity of the front charcoal tube. This capacity varies with the concentrations of analyte and other substances in the air. The front adsorbing charcoal tube was found to hold 2.1 mg of analyte when a test atmosphere containing 280 mg/m^3 of analyte in air was sampled at 0.19 L/min for 40 min; i.e., at that time, the concentration of analyte in the effluent was 5% of that in the influent. (The charcoal tube series consists of two separate large tubes; the first tube contains 400 mg of charcoal and the second tube used as the backup tube contains 200 mg of charcoal. The charcoal is held in place by glass wool plugs at the tube ends.) If a particular atmosphere is suspected of containing a large amount of contaminant, a smaller sampling volume should be taken.

Interferences
When the amount of water in the air is so great that condensation actually occurs in the tube, organic vapors will not be trapped efficiently. Preliminary experiments using toluene indicate that high humidity severely decreases the breakthrough volume. When two or more compounds are known or suspected to be present in the air, such information, including their suspected identities, should be transmitted with the sample.

It must be emphasized that any compound which has the same retention time as the analyte at the operation conditions described in this method is an interference. Retention time data on a single column cannot be considered as proof of chemical identity. If the possibility of interference exists, separation conditions (column packing, temperature, etc.) must be changed to circumvent the problem.

Precision and Accuracy
The coefficient of variation for the total analytical and sampling method in the range of 41-176 mg/m^3 is 0.103. This value corresponds to a 9.3 mg/m^3 standard deviation at the OSHA standard level. Statistical information and details of the validation and experimental test procedures can be found in reference 2.

On the average, the concentrations obtained at the OSHA standard level using the overall sampling and analytical method were 0.9% lower than the "true" concentrations for a limited number of laboratory experiments. Any difference between the "found" and "true" concentrations may not represent a bias in the sampling and analytical method, but rather a random variation from the experimentally determined "true" concentration. Therefore, no recovery correction should be applied to the final result.

Advantages and Disadvantages

The sampling device is small, portable, and involves no liquids. Interferences are minimal, and most of those which do occur can be eliminated by altering chromatographic conditions. The tubes are analyzed by means of a quick, instrumental method. The method can also be used for the simultaneous analysis of two or more substances suspected to be present in the same sample by simply changing gas chromatographic conditions from isothermal to a temperature-programmed mode of operation.

One disadvantage of the method is that the amount of sample which can be taken is limited by the number of mg that the tube will hold before overloading. When the sample value obtained for the backup charcoal tube exceeds 25% of that found on the front tube, the possibility of sample loss exists. Furthermore, the precision of the method is limited by the reproducibility of the pressure drop across the tubes. This drop will affect the flow rate and cause the volume to be imprecise, because the pump is usually calibrated for one tube only.

Apparatus

Calibrated personal sampling pump. Whose flow can be determined within 5% at the recommended flow rate.

Charcoal tubes. The sampling tube series consists of two separate large charcoal tubes. The tubes are glass tubes with both ends flame sealed, 10 cm long with 8 mm O.D. and 6 mm I.D. Each tube contains the appropriate amount of 20/40 mesh activated coconut charcoal. The front tube contains 400 mg of charcoal; the backup tube 200 mg. A plug of silylated glass wool is placed at each end of the charcoal tubes. The pressure drop across the tubes must be less than one inch of mercury at a flow rate of 1 L/min.

The sampling tubes can be prepared by modifying commercially available large charcoal tubes as follows. To prepare the front tube, break off exit end of a large tube and using pointed tweezers remove glass wool plug and the backup charcoal section; push in snugly remaining plug and charcoal to minimize channeling. To prepare the backup tube, similarly break off inlet end of a second large tube and remove retaining plug and the front charcoal. The front tube is connected to the backup tube with a minimal piece of tygon or rubber tubing. This sampling tube scheme is necessary to minimize any sample migration effects upon storage.

Gas chromatograph. Equipped with a flame-ionization detector.

Column. 10 ft × 1/8 in stainless steel, packed with Porapak QS.

Electronic integrator. Or some other suitable method for measuring peak areas.

Sample containers. 5 mL, with Teflon-lined screw caps, such as the vials distributed by SKC, Pittsburgh, PA. Note: glass-stoppered containers are not adequate because of significant sample losses during desorption.

Syringes. 10 µL, and other convenient sizes for making standards.

Pipets. 2.0 mL graduated or delivery.

Volumetric flasks. 10 mL or convenient sizes for making standard solutions.

Reagents

Chromatographic quality carbon disulfide.
Ethylene oxide, 99.5%. Available as a liquid or gas.
Purified nitrogen.
Prepurified hydrogen.
Filtered compressed air.

Procedure

Cleaning of equipment. All glassware used for the laboratory analysis should be detergent washed and thoroughly rinsed with tap water and distilled water.

Calibration of personal pumps. Each personal pump must be calibrated with a representative tube series in the line assembled. This will minimize errors associated with uncertainties in the sample volume collected.

Collection and shipping of samples. Immediately before sampling, break the ends of the two charcoal tubes to provide an opening at least one-half the internal diameter of the tube (3 mm). Prepare and assemble as described above. Connect the front 400 mg tube to the 200 mg backup tube with a short piece of tubing. The tube containing 200 mg of charcoal is used as a backup and should be positioned nearest the sampling pump. The charcoal tube series should be placed in a vertical direction during sampling to minimize channeling through the charcoal.

Air being sampled should not be passed through any hose or tubing before entering the charcoal tube. A sample size of 5 L is recommended. Sample at a rate of 0.05 L/min. The flow rate should be known with an accuracy of at least 5%. The temperature and pressure of the atmosphere being sampled should be recorded. If pressure reading is not available, record the elevation. The front charcoal tube should be separated from the backup tube immediately after sampling. Both tubes must be labeled appropriately and capped with the supplied plastic caps. Under no circumstances should rubber caps be used.

One set of tubes should be handled in the same manner as the sample tubes (break, seal, and transport) except that no air is sampled through these tubes. These tubes should be labeled as blanks. Capped charcoal tubes should be packed tightly and padded before they are shipped to minimize tube breakage during shipping.

Preparation of samples. In preparation for analysis, each charcoal tube is scored with a file and broken open. The glass wool is removed and discarded. The charcoal in the front tube is transferred to a 5 mL sample container with Teflon-lined screw cap. Similarly, the charcoal in the backup tube is transferred to another container. These two charcoal tubes are analyzed separately.

Desorption of samples. Prior to analysis, 2.0 mL of carbon disulfide is pipetted into each sample container. Cap the vials tightly immediately after the addition of carbon disulfide. (All work with carbon disulfide should be performed in a hood because of its high toxicity.) Desorption should be done for 30 minutes. Tests indicate that this is adequate if the sample is agitated occasionally during this period. If an automatic sample injector is used, carefully transfer portions of the desorbed sample solution to the automatic sample injector vials and cap tightly. For the internal standard method, desorb using 2.0 mL of carbon disulfide containing a known amount of the chosen internal standard.

GC conditions. The typical operating conditions for the gas chromatograph are:

30 mL/min (60 psig) nitrogen carrier gas flow
30 mL/min (25 psig) hydrogen gas flow to detector
300 mL/min (60 psig) air flow to detector
155° C. injector temperature
200° C. detector temperature
150° C. column temperature

Injection. The first step in the analysis is the injection of the sample into the gas chromatograph. To eliminate difficulties arising from blow back or distillation within the syringe needle, one should employ the solvent flush injection technique. The 10 μL syringe is first flushed with solvent several times to wet the barrel and plunger. Three μL of solvent are drawn into the syringe to increase the accuracy and reproducibility of the injected sample volume. The needle is

removed from the solvent, and the plunger is pulled back about 0.2 µL to separate the solvent flush from the sample with a pocket of air to be used as a marker. The needle is then immersed in the sample, and a 5 µL aliquot is withdrawn, taking into consideration the volume of the needle, since the sample in the needle will be completely injected. After the needle is removed from the sample and prior to injection, the plunger is pulled back 1.2 µL to minimize evaporation of the sample from the tip of the needle. Observe that the sample occupies 4.9-5.0 µL in the barrel of the syringe. Duplicate injections of each sample and standard should be made. No more than a 3% difference in area is to be expected. An automatic sample injector can be used if it is shown to give reproducibility at least as good as the solvent flush technique.

Measurement of area. The area of the sample peak is measured by an electronic integrator or some other suitable form of area measurement, and preliminary results are read from a standard curve prepared as discussed below.

Determination of desorption efficiency. The desorption efficiency of a particular compound can vary from one laboratory to another and also from one batch of charcoal to another. Thus, it is necessary to determine at least once the percentage of the specific compound that is removed in the desorption process, provided the same batch of charcoal is used.

Activated charcoal equivalent to the amount in the front tube (400 mg) is measured into a 10 cm long, 6 mm I.D. glass tube, flame sealed at one end. This charcoal must be from the same batch as that used in obtaining the samples and can be obtained from unused charcoal tubes. The open end is capped with Parafilm. A known amount of a concentrated solution of the analyte in carbon disulfide is injected directly into the activated charcoal with a microliter syringe, and the tube is capped with more Parafilm. Alternatively, the 5 mL tubes or vials used for desorption may be used in place of the glass tubes.

This solution of ethylene oxide in carbon disulfide can be prepared by bubbling ethylene oxide gas into carbon disulfide in a fritted glass bubbler. The ethylene oxide concentration of the solution is determined by comparison with gas standards. The solution can also be prepared by using ethylene oxide liquid. Note that the syringe used must be cooled; otherwise the liquid ethylene oxide will evaporate very rapidly. This solution must be calibrated as above.

Six tubes at each of three concentration levels (0.5X, 1X, and 2X the standard) are prepared by adding an amount of analyte equivalent to that present in a 5 L sample at the selected level. The tubes are allowed to stand for at least overnight to assure complete adsorption of the analyte onto the charcoal. These tubes are referred to as the samples. A parallel blank tube should be treated in the same manner except that no sample is added to it. The sample and blank tubes are desorbed and analyzed in exactly the same manner as the sampling tube described earlier.

Two or three standards are prepared by injecting the same volume of ethylene oxide solution in carbon disulfide into 2.0 mL of carbon disulfide with the same syringe used in the preparation of the spiked samples. These are analyzed with the samples. The desorption efficiency (D.E.) equals the average weight in mg recovered from the tube divided by the weight in mg added to the tube or:

D.E. = average wt (mg) recovered ÷ wt (mg) added

The desorption efficiency is dependent on the amount of analyte collected on the charcoal. Plot the desorption efficiency versus the weight of analyte found. The curve is used to correct for adsorption losses.

Calibration and Standardization

It is convenient to express concentration of standards in terms of mg/2.0 mL carbon disulfide, because samples are desorbed in this amount of carbon disulfide. The calibrated solution of ethylene oxide in carbon disulfide as described above can be used to prepare calibration solutions. In the preparation of standard solutions, observe the necessary precautions noted earlier. A series of standards, varying in concentration over the range of interest, is prepared and analyzed under the same GC conditions and during the same time period as the unknown sample. Curves are established by plotting concentration in mg/2.0 mL versus peak area. Standard solutions should be analyzed at the same time the sample analysis is done. This will minimize the effect of variations in FID response.

Calculations

Read the weight, in mg, corresponding to each peak area from the standard curve. No volume corrections are needed, because the standard curve is based on mg/2.0 mL carbon disulfide and the volume of sample injected is identical to the volume of the standards injected. Corrections for the blank must be made for each sample:

mg = mg sample − mg blank

where:

mg sample = mg found in front (400 mg) sample tube
mg blank = mg found in front (400 mg) blank tube

A similar procedure is followed for the backup tube. Add the amounts present in the front and backup tubes to determine the total weight in the sample. Read the desorption efficiency from the curve for the amount found in the front tube. Divide the total weight by this desorption efficiency to obtain the corrected mg/sample:

Corrected mg/sample = total weight ÷ D.E.

The concentration of the analyte in the air sampled can be expressed in mg/m^3:

mg/m^3 = (corrected mg)(1000 L/m^3) ÷ air volume sampled in L

Another method of expressing concentration is ppm:

ppm = (mg/m^3)(24.45)(760)(T + 273) ÷ (M.W.)(P)(298)

where:

P = pressure (mm Hg) of air sampled
T = temperature (°C.) of air sampled
24.45 = molar volume (L/mol) at 25° C. and 760 mm Hg
M.W. = molecular weight (g/mol) of analyte
760 = standard pressure (mm Hg)
298 = standard temperature (°K)

References

1. White, L.D. et al. A convenient optimized method for the analysis of selected solvent vapors in the industrial atmosphere. Amer. Ind. Hyg. Assoc. J. 31: 225, 1970.
2. Documentation of NIOSH Validation Tests, NIOSH Contract No. CDC-99-74-45.
3. Final Report, NIOSH Contract No. HSM-99-71-31. Personal Sampler Pump for Charcoal Tubes, September 15, 1972.

ETHYL ETHER

Synonyms: diethyl ether; ether $C_2H_5OC_2H_5$
ACGIH TLV: 400 ppm (1200 mg/m^3)
OSHA Standard: 400 ppm

Method	Sampling Duration	Sampling Location	Useful Range (ppm)	System Cost ($)	Test Cost ($)	Manufacturer
DT	Grab	Portable	100-4000	150	2.60	Nat'l Draeger
DT	Grab	Portable	400-14,000	165	1.70	Matheson
DT	Grab	Portable	400-10,000	180	2.25	Bendix
IR	Cont	Portable	8-800	4374	0	Foxboro
PI	Cont	Fixed	20-2000	4950	0	HNU
GC	1 min	Portable	1-10^6	10,000	0	Microsensor

NIOSH METHOD NO. S80

Principle

A known volume of air is drawn through a charcoal tube to trap the organic vapors present. The charcoal in the tube is transferred to a small, stoppered sample container, and the analyte is desorbed with ethyl acetate. An aliquot of the desorbed sample is injected into a gas chromatograph. The area of the resulting peak is determined and compared with areas obtained from the injection of standards.

Range and Sensitivity

This method was validated over the range of 606-2400 mg/m^3 at an atmospheric temperature and pressure of 22° C. and 766 mm Hg, using a 3 L sample. Under the conditions of sample size (3 L), the probable useful range of this method is 120-3630 mg/m^3 at a detector sensitivity that gives nearly full deflection on the strip chart recorder for 11 mg sample. The method is capable of measuring much smaller amounts if the desorption efficiency is adequate. Desorption efficiency must be determined over the range used.

The upper limit of the range of the method is dependent on the adsorptive capacity of the charcoal tube. This capacity varies with the concentrations of ethyl ether and other substances in the air. The first section of the charcoal tube was found to hold 15 mg of ethyl ether when a test atmosphere containing 2470 mg/m^3 of ethyl ether in air was sampled at 0.185 L/min for 33 min; breakthrough was observed at this time, i.e., the concentration of analyte in the effluent was 5% of that in the influent. (The charcoal tube consists of two sections of activated charcoal separated by a section urethane foam.) If a particular atmosphere is suspected of containing a large amount of contaminant, a smaller sampling volume should be taken.

Interferences

When the amount of water in the air is so great that condensation actually occurs in the tube, organic vapors will not be trapped efficiently. Preliminary experiments using toluene indicate that high humidity severely decreases the breakthrough volume.

When two or more compounds are known or suspected to be present in the air, such information, including their suspected identities, should be transmitted with the sample. Since ethyl acetate is used rather than carbon disulfide to desorb the ethyl ether from the charcoal it would not be possible to measure ethyl acetate in the sample with this desorbing solvent. If it is suspected that ethyl acetate is present, a separate sample should be collected for ethyl acetate analysis.

It must be emphasized that any compound which has the same retention time as the analyte at the operating conditions described in this method is an interference. Retention time data on a single column cannot be considered proof of chemical identity. If the possibility of interference exists, separation conditions (column packing, temperature, etc.) must be changed to circumvent the problem.

Precision and Accuracy

The coefficient of variation for the total analytical and sampling method in the range of 606-2400 mg/m^3 was 0.053. This value corresponds to a 64 mg/m^3 standard deviation at the OSHA standard level. Statistical information and details of the validation and experimental test procedures can be found in reference 2. On the average the values obtained using the overall sampling and analytical method were 5.1% higher than the "true" values at the OSHA standard level.

Advantages and Disadvantages

The sampling device is small, portable, and involves no liquids. Interferences are minimal, and most of those which do occur can be eliminated by altering chromatographic conditions. The tubes are analyzed by means of a quick, instrumental method. The method can also be used for the simultaneous analysis of two or more substances suspected to be present in the same sample by simply changing gas chromatographic conditions from isothermal to a temperature-programmed mode of operation.

One disadvantage of the method is that the amount of sample which can be taken is limited by the number of mg that the tube will hold before overloading. When the sample value obtained for the backup section of the charcoal tube exceeds 25% of that found in the front section, the possibility of sample loss exists. Furthermore, the precision of the method is limited by the reproducibility of the pressure drop across the tubes. This drop will affect the flow rate and cause the volume to be imprecise, because the pump is usually calibrated for one tube only.

Apparatus

Calibrated personal sampling pump. Whose flow can be determined within 5% at the recommended flow rate.

Charcoal tubes. Glass tube with both ends flame sealed, 7 cm long with 6 mm O.D. and 4 mm I.D., containing 2 sections of 20/40 mesh activated charcoal separated by a 2 mm portion of urethane foam. The activated charcoal is prepared from coconut shells and is fired at 600° C. prior to packing. The adsorbing section contains 100 mg of charcoal, the backup section 50 mg. A 3 mm portion of urethane foam is placed between the outlet end of the tube and the backup section. A plug of silylated glass wool is placed in front of the adsorbing section. The pressure drop across the tube must be less than one inch of mercury at a flow rate of 1 L/min.

Gas chromatograph. Equipped with a flame ionization detector.

Column. 4 ft × 1/4 in stainless steel, packed with 50/80 mesh Porapak Q.

Electronic integrator. Or some other suitable method for measuring peak area.

Sample containers. 1 mL, with glass stoppers or Teflon-lined caps.

Syringes. 10 µL, and other convenient sizes for making standards.

Pipets. 0.5 mL delivery pipets or 1.0 mL type graduated in 0.1 mL increments.

Volumetric flasks. 10 mL or convenient sizes for making standard solutions.

Reagents
Ethyl acetate, reagent grade.
Ethyl ether, reagent grade.
Purified nitrogen.
Prepurified hydrogen.
Filtered compressed air.

Procedure
Cleaning of equipment. All glassware used for the laboratory analysis should be detergent washed and thoroughly rinsed with tap water and distilled water.

Calibration of personal pumps. Each personal pump must be calibrated with a representative charcoal tube in the line. This will minimize errors associated with uncertainties in the sample volume collected.

Collection and shipping of samples. Immediately before sampling, break the ends of the tube to provide an opening at least one-half the internal diameter of the tube (2 mm). The smaller section of charcoal is used as a backup and should be positioned nearest the sampling pump. The charcoal tube should be placed in a vertical direction during sampling to minimize channeling through the charcoal. Air being sampled should not be passed through any hose or tubing before entering the charcoal tube.

A maximum sample size of 3 L is recommended. Sample at a flow of 0.20 L/min or less. The flow rate should be known with an accuracy of at least 5%. The temperature and pressure of the atmosphere being sampled should be recorded. If pressure reading is not available, record the elevation. The charcoal tubes should be capped with the supplied plastic caps immediately after sampling. Under no circumstances should rubber caps be used. One tube should be handled in the same manner as the sample tubes (break, seal, and transport), except that no air is sampled through this tube. This tube should be labeled as a blank. Capped tubes should be packed tightly and padded before they are shipped to minimize tube breakage during shipping.

Preparation of samples. In preparation for analysis, each charcoal tube is scored with a file in front of the first section of charcoal and broken open. The glass wool is removed and discarded. The charcoal in the first (larger) section is transferred to a 1 mL stoppered sample container. The separating section of foam is removed and discarded; the second section is transferred to another stoppered container. These two sections are analyzed separately.

Desorption of samples. Prior to analysis, 0.5 mL of ethyl acetate is pipetted into each sample container. Desorption should be done for 30 minutes. Tests indicate that this is adequate if the sample is agitated occasionally during this period.

GC conditions. The typical operating conditions for the gas chromatograph are:

50 mL/min (60 psig) nitrogen carrier gas flow
65 mL/min (24 psig) hydrogen gas flow to detector
500 mL/min (50 psig) air flow to detector
195° C. injector temperature
250° C. detector temperature
175° C. column temperature

Injection. The first step in the analysis is the injection of the sample into the gas chromatograph. To eliminate difficulties arising from blow back or distillation within the syringe needle, one should employ the solvent flush injection technique. The 10 μL syringe is first flushed with solvent several times to wet the barrel and plunger. Three μL of solvent are drawn into the syringe to increase the accuracy and reproducibility of the injected sample volume. The needle is

removed from the solvent, and the plunger is pulled back about 0.2 µL to separate the solvent flush from the sample with a pocket of air to be used as a marker. The needle is then immersed in the sample, and a 5 µL aliquot is withdrawn, taking into consideration the volume of the needle, since the sample in the needle will be completely injected. After the needle is removed from the sample and prior to injection, the plunger is pulled back 1.2 µL to minimize evaporation of the sample from the tip of the needle. Observe that the sample occupies 4.9-5.0 µL in the barrel of the syringe. Duplicate injections of each sample and standard should be made. No more than a 3% difference in area is to be expected.

Measurement of area. The area of the sample peak is measured by an electronic integrator or some other suitable form of area measurement, and preliminary results are read from a standard curve prepared as discussed below.

Determination of desorption efficiency. The desorption efficiency of a particular compound can vary from one laboratory to another and also from one batch of charcoal to another. Thus, it is necessary to determine at least once the percentage of the specific compound that is removed in the desorption process, provided the same batch of charcoal is used.

Activated charcoal equivalent to the amount in the first section of the sampling tube (100 mg) is measured into a 2.5 in, 4 mm I.D. glass tube, flame sealed at one end. This charcoal must be from the same batch as that used in obtaining the samples and can be obtained from unused charcoal tubes. The open end is capped with Parafilm. A known amount of the analyte is injected directly into the activated charcoal with a microliter syringe, and the tube is capped with more Parafilm. The amount injected is equivalent to that present in a 3 L air sample at the selected level.

At least six tubes at each of three levels (0.5X, 1X, and 2X the standard) are prepared in this manner and allowed to stand for at least overnight to assure complete adsorption of the analyte onto the charcoal. These tubes are referred to as the samples. A parallel blank tube should be treated in the same manner except that no sample is added to it. The sample and blank tubes are desorbed and analyzed in exactly the same manner as the sampling tube described above.

Two or three standards are prepared by injecting the same volume of compound into 0.5 mL of ethyl acetate with the same syringe used in the preparation of the samples. These are analyzed with the samples. The desorption efficiency (D.E.) equals the average weight in mg recovered from the tube divided by the weight in mg added to the tube, or:

D.E. = average wt (mg) recovered ÷ wt (mg) added

The desorption efficiency is dependent on the amount of analyte collected on the charcoal. Plot the desorption efficiency versus the weight of analyte found. The curve is used to correct for adsorption losses.

Calibration and Standardization

It is convenient to express concentration of standards in terms of mg/0.5 mL ethyl acetate, because samples are desorbed in this amount of ethyl acetate. The density of the analyte is used to convert mg into µL for easy measurement with a microliter syringe. A series of standards, varying in concentration over the range of interest, is prepared and analyzed under the same GC conditions and during the same time period as the unknown samples. Curves are established by plotting concentration in mg/0.5 mL versus peak area. Note: since no internal standard is used in the method, standard solutions must be analyzed at the same time that the sample analysis is done. This will minimize the effect of known day-to-day variations and variations during the same day of the FID response.

Calculations

Read the weight, in mg, corresponding to each peak area from the standard curve. No volume corrections are needed, because the standard curve is based on mg/0.5 mL ethyl acetate and the volume of sample injected is identical to the volume of the standards injected. Corrections for the blank must be made for each sample:

mg = mg sample − mg blank

where:

mg sample = mg found in front section of sample tube
mg blank = mg found in front section of blank tube

A similar procedure is followed for the backup sections. Add the weights found in the front and backup sections to get the total weight in the sample. Read the desorption efficiency from the curve for the amount found in the front section. Divide the total weight by this desorption efficiency to obtain the corrected mg/sample:

Corrected mg/sample = total weight ÷ D.E.

The concentration of the analyte in the air sampled can be expressed in mg/m^3:

mg/m^3 = (corrected mg)(1000 L/m^3) ÷ air volume sampled in L

Another method of expressing concentration is ppm:

ppm = $(mg/m^3)(24.45)(760)(T + 273) \div (M.W.)(P)(298)$

where:

P = pressure (mm Hg) of air sampled
T = temperature (°C.) of air sampled
24.45 = molar volume (L/mol) at 25° C. and 760 mm Hg
M.W. = molecular weight (g/mol) of analyte
760 = standard pressure (mm Hg)
298 = standard temperature (°K)

References

1. White, L.D. et al. A convenient optimized method for the analysis of selected solvent vapors in the industrial atmosphere. Amer. Ind. Hyg. Assoc. J. 31: 225, 1970.
2. Documentation of NIOSH Validation Tests, NIOSH Contract No. CDC-99-74-45.
3. Final Report, NIOSH Contract HSM-99-71-31. Personal Sampler Pump for Charcoal Tubes, September 15, 1972.

ETHYL FORMATE

ACGIH TLV: 100 ppm (300 mg/m^3)
OSHA Standard: 100 ppm

$HCOOC_2H_5$

Method	Sampling Duration	Sampling Location	Useful Range (ppm)	System Cost ($)	Test Cost ($)	Manufacturer
IR	Cont	Portable	2-200	4374	0	Foxboro

NIOSH METHOD NO. P&CAM 127

(Use as described for acetone, with 65° C. column temperature)

ETHYL SILICATE

Synonym: ethyl orthosilicate
ACGIH TLV: 10 ppm (85 mg/m^3)
OSHA Standard: 100 ppm (850 mg/m^3)

$(C_2H_5)_4SiO_4$

Method	Sampling Duration	Sampling Location	Useful Range (ppm)	System Cost ($)	Test Cost ($)	Manufacturer
IR	Cont	Portable	2-200	4374	0	Foxboro

NIOSH METHOD NO. S264

Principle

A known volume of air is drawn through a tube containing XAD-2 resin to adsorb the ethyl silicate present. The XAD-2 resin in the tube is transferred to a small, stoppered sample container and the analyte is desorbed with carbon disulfide. An aliquot of the desorbed sample is injected into a gas chromatograph. The area of the resulting peak is determined and compared with areas obtained from injection of standards.

Range and Sensitivity

This method was validated over the range of 377-1620 mg/m^3 at an atmospheric temperature and pressure of 21° C. and 771 mm Hg, using a 9 L sample. Under the conditions of sample size (9 L), the probable useful range of this method is 85-1700 mg/m^3 at a detector sensitivity that gives nearly full deflection on the strip chart recorder for a 15 mg sample. This method is capable of measuring much smaller amounts if the desorption efficiency is adequate. Desorption efficiency must be determined over the range used. When higher concentrations of ethyl silicate are expected, a smaller volume should be collected.

The upper limit of the range of the method is dependent on the adsorptive capacity of the XAD-2 resin tube. This capacity varies with the concentrations of analyte and other substances in the air. Breakthrough studies indicated that the first section of the XAD-2 resin tube was found to hold 24 mg of analyte when a test atmosphere containing 1640 mg/m^3 of analyte in air was sampled at 0.80 L/min for 18 min; i.e., at that time, the concentration of analyte in the effluent was 5% of that in the influent.

The collection efficiency of ethyl silicate on the XAD-2 resin tube may be dependent on sample flow rate. A separate study gave a collection efficiency of 98% for 9 L samples collected at 0.067 L/min for concentration levels at 2 times the OSHA standard. The resin tube consists of two sections of XAD-2 resin separated by a section of silicated glass wool. If a particular atmosphere is suspected of containing a large amount of contaminant, a smaller sampling volume should be taken.

Interferences

The adsorptive capacity of XAD-2 resin is not severely affected by water vapor. Breakthrough volume will not be substantially affected by high relative humidity. Note, however, that although ethyl silicate is not soluble in water, it is slowly decomposed by water (Merck Index). When two or more compounds are known or suspected to be present in the air, such information, including their suspected identities, should be transmitted with the sample.

It must be emphasized that any compound which has the same retention time as the analyte at the operating conditions described in this method is an interference. Retention time data on a single column cannot be considered as proof of chemical identity. If the possibility of interference exists, separation conditions (column packing, temperature, etc.) must be changed to circumvent the problem.

Precision and Accuracy

The coefficient of variation for the total analytical and sampling method in the range of 377-1620 mg/m^3 was 0.056. This value corresponds to a 48 mg/m^3 standard deviation at the OSHA standard level. Statistical information and details of the validation and experimental test procedures can be found in reference 1.

On the average, the concentration obtained at the OSHA standard level using the overall sampling and analytical method were 7.1% lower than the "true" concentrations for a limited number of laboratory experiments. Any difference between the "found" and "true" concentrations may not represent a bias in the sampling and analytical method, but rather a random variation from the experimentally determined "true" concentration. Therefore, no recovery correction should be applied to the final result. These data are based on validation experiments using the internal standard method.

Advantages and Disadvantages

The sampling device is small, portable, and involves no liquids. Interferences are minimal, and most of those which do occur can be eliminated by altering chromatographic conditions. The tubes are analyzed by means of a quick, instrumental method. One disadvantage of the method is that the amount of sample which can be taken is limited by the number of mg that the tube will hold before overloading. When the sample value obtained for the backup section of the resin tube exceeds 25% of that found in the front section, the possibility of sample loss exists. Furthermore, the precision of the method is limited by the reproducibility of the pressure drop across the tubes. This drop will affect the flow rate and cause the volume to be imprecise, because the pump is usually calibrated for one tube only.

Apparatus

Calibrated personal sampling pump. Whose flow can be determined within 5% at the recommended flow rate.

Resin tubes. Glass tube with both ends flame sealed, 7 cm long with 6 mm O.D. and 4 mm I.D., containing 2 sections of 20/50 mesh XAD-2 resin, prepared as described below. The adsorbing section contains 100 mg of resin, the backup section 50 mg. A small wad of silylated glass wool is placed between the outlet end of the tube and the backup section; a plug of silylated glass wool is also placed in front of the adsorbing section and at the end of the backup section. The pressure drop across the tube must be less than one inch of mercury at a flow rate of 1 L/min.

Gas chromatograph. Equipped with a flame-ionization detector.

Column. 10 ft × 1/8 in stainless steel, packed with 10% OV-101 stationary phase on 100/120 mesh Supelcoport.

Electronic integrator. Or some other suitable method for measuring peak area.

Sample containers. 5 mL, with glass stoppers or Teflon-lined caps.

Syringes. 10 μL, and other convenient sizes for making standards.

Pipets. 2.0 mL delivery type.

Volumetric flasks. 10 mL or convenient sizes for making standard solutions.

Glass rods. 3 mm, or cylindrical wooden medical applicators.

Reagents

Chromatographic quality carbon disulfide.
Ethyl silicate, reagent grade.
Nonane, or other suitable internal standard.
Purified nitrogen.
Prepurified hydrogen.
Filtered compressed air.
Pre-cleaned resin. XAD-2 resin (20/50 mesh) can be obtained from the Rohm and Haas Company. XAD-2 resin is purified by charging an amount into a standard Soxhlet extractor. Larger batches may be prepared by using a Giant extractor. Overnight (24 hr) extractions are then performed successively with water, methyl alcohol, diethyl ether and, finally, n-pentane. Distilled in glass solvents are used in all cases. Resin has been prepared successfully in this manner using charges of about 700 g of resin and 1.5 L of each solvent. The resin is dried by maintaining it under vacuum (1-10 torr) and mild heat for about 24 hours.

Procedure

Cleaning of equipment. All glassware used for the laboratory analysis should be detergent washed and thoroughly rinsed with tap water and distilled water.

Calibration of personal pumps. Each personal pump must be calibrated with a representative XAD-2 resin tube in the line. This will minimize errors associated with uncertainties in the sample volume collected.

Collection and shipping of samples. Immediately before sampling, break the ends of the tube to provide an opening at least one-half the internal diameter of the tube (2 mm). The smaller section of XAD-2 resin is used as a backup and should be positioned nearest the sampling pump. The XAD-2 resin tube should be placed in a vertical direction during sampling to minimize channeling through the charcoal.

Air being sampled should not be passed through any hose or tubing before entering the coated XAD-2 resin tube. A maximum sample size of 9 L is recommended. Sample at a rate of 0.05

L/min only. The flow rate should be known with an accuracy of at least 5%. The temperature and pressure of the atmosphere being sampled should be recorded. If pressure reading is not available, record the elevation.

The XAD-2 resin tubes should be capped with plastic caps immediately after sampling. Under no circumstances should rubber caps be used. One tube for every 10 samples should be handled in the same manner as the sample tube (break, seal, and transport), except that no air is sampled through this tube. This tube should be labeled as a blank. Capped XAD-2 resin tubes should be packed tightly and padded before they are shipped to minimize tube breakage during shipping.

Preparation of samples. In preparation for analysis, each sample tube is scored with a file in front of the backup and the front section and broken open at both ends to give wider tube opening for easier transfer of the resin bed. Remove the glass wool plug next to the backup section (exit end) and discard. Transfer the resin from the backup section into a 5 mL vial. Then remove the glass wool plug at the front end of the 100 mg section and transfer to a separate 5 mL vial. Carefully transfer the resin into the same 5 mL vial. If the resin particles tend to hang on to the inner walls of the glass tube instead of flowing freely, use the remaining glass wool plug to push out the rest of the resin particles. A short piece of 3 mm glass rod or a wooden medical applicator can be used for this purpose. In the event that difficulties are encountered in the transfer operation due to electrostatic charges which make the resin particles cling to the glass walls, it is suggested that the tubes be flushed with a gentle stream of air or nitrogen saturated with water for 5-10 seconds.

Desorption of samples. Prior to analysis, 2.0 mL of carbon disulfide is pipetted into each sample container. (All work with carbon disulfide should be performed in a hood because of its high toxicity.) Desorption should be done for 30 minutes. Tests indicate that this is adequate if the sample is agitated occasionally during this period. The sample vials should be capped as soon as the solvent is added to minimize volatilization. For the internal standard method, desorb using 2.0 mL of carbon disulfide containing a known amount of the chosen internal standard. If an automatic sample injector is used, transfer at least 1 mL aliquot of the desorbed sample solution to the automatic sample injector vial.

GC conditions. The typical operating conditions for the gas chromatograph are:

30 mL/min (60 psig) nitrogen carrier gas flow
30 mL/min (25 psig) hydrogen gas flow to detector
300 mL/min (60 psig) air flow to detector
225° C. injector temperature
250° C. detector temperature
100° C. column temperature

Injection. The first step in the analysis is the injection of the sample into the gas chromatograph. To eliminate difficulties arising from blow back or distillation within the syringe needle, one should employ the solvent flush injection technique. The 10 μL syringe is first flushed with solvent several times to wet the barrel and plunger. Three μL of solvent are drawn into the syringe to increase the accuracy and reproducibility of the injected sample volume. The needle is removed from the solvent, and the plunger is pulled back about 0.2 μL to separate the solvent flush from the sample with a pocket of air to be used as a marker. The needle is then immersed in the sample, and a 5 μL aliquot is withdrawn, taking into consideration the volume of the needle, since the sample in the needle will be completely injected. After the needle is removed from the sample and prior to injection, the plunger is pulled back 1.2 μL to minimize evaporation of the sample from the tip of the needle. Observe that the sample occupies 4.9-5.0 μL in the barrel of the syringe. Duplicate injections of each sample and standard should be made. No more than a 3%

difference in area is to be expected. An automatic sample injector can be used if it is shown to give reproducibility at least as good as the solvent flush technique.

Measurement of area. The area of the sample peak is measured by an electronic integrator or some other suitable form of area measurement, and preliminary results are read from a standard curve prepared as discussed below.

Determination of desorption efficiency. The desorption efficiency of a particular compound can vary from one laboratory to another and also from one batch of XAD-2 resin to another. Thus, it is necessary to determine at least once the percentage of the specific compound that is removed in the desorption process, provided the same batch of XAD-2 resin is used.

An amount of the XAD-2 resin equivalent to that present in the first section of the sampling tube (100 mg) is measured into a 2.5 in, 4 mm I.D. glass tube, flame sealed at one end. This resin must be from the same batch as that used in obtaining the samples and can be obtained from unused XAD-2 resin tubes. The open end is capped with Parafilm. A known amount of ethyl silicate is injected directly onto the resin with a microliter syringe, and the tube is capped with more Parafilm. Alternatively, a 5 mL vial capped with Teflon faced septum may be used in place of the glass tube.

Six tubes at each of three concentration levels (0.5X, 1X, and 2X the OSHA standard) are prepared by adding an amount of analyte equivalent to that present in a 9 L sample at the selected level. The tubes are allowed to stand for at least overnight to assure complete adsorption of the analyte onto the XAD-2 resin. These tubes are referred to as the samples. A parallel blank tube should be treated in the same manner except that no sample is added to it. The sample and blank tubes are desorbed and analyzed in exactly the same manner as the sampling tube described earlier.

Two or three standards are prepared by injecting the same volume of compound into 2.0 mL of carbon disulfide with the same syringe used in the preparation of the samples. These are analyzed with the samples. The internal standard method is used, prepare calibration standards by using 2.0 mL of carbon disulfide containing a known amount of the internal standard. The desorption efficiency (D.E.) equals the average weight in mg recovered from the tube divided by the weight in mg added to the tube, or:

D.E. = average wt (mg) recovered ÷ wt (mg) added

The desorption efficiency is dependent on the amount of analyte collected on the resin. Plot the desorption efficiency versus the weight of analyte found. The curve is used to correct for adsorption losses.

Calibration and Standardization

It is convenient to express concentration of standards in terms of mg/2.0 mL of carbon disulfide, because samples are desorbed in this amount of carbon disulfide. The density of the ethyl silicate is used to convert mg into μL for easy measurement with a microliter syringe. A series of standards, varying in concentration over the range of interest, is prepared and analyzed under the same GC conditions and during the same time period as the unknown sample. Curves are established by plotting concentration in mg/2.0 mL versus peak area.

For the internal standard method, use carbon disulfide containing a predetermined amount of the internal standard. The internal standard concentration used was approximately 70% of the concentration at 2 times the standard. The analyte concentration in mg/2.0 mL is plotted versus the ratio of the area of the analyte to that of the internal standard. Note: whether the external standard or internal standard method is used, standard solutions should be analyzed at the same time the sample analysis is done. This will minimize the effect of variations in FID response.

Calculations

Read the weight, in mg, corresponding to each peak area from the standard curve. No volume corrections are needed, because the standard curve is based on mg/2.0 mL carbon disulfide and the volume of sample injected is identical to the volume of the standards injected. Corrections for the blank must be made for each sample:

mg = mg sample − mg blank

where:

    mg sample = mg found in front section of sample tube
    mg blank = mg found in front section of blank tube

A similar procedure is followed for the backup sections. Add the amounts present in the front and backup sections of the sample tube to determine the total weight in the sample. Read the desorption efficiency from the curve for the amount found in the front section. Divide the total weight by this desorption efficiency to obtain the corrected mg/sample:

Corrected mg/sample = total weight ÷ D.E.

At the recommended sampling rate of 0.05 L/min and a sample volume of 9.0 L, no correction for collection efficiency is necessary. The collection efficiency was determined to be at least 98% under these conditions. The concentration of the analyte in the air sampled can be expressed in mg/m^3:

mg/m^3 = (corrected mg)(1000 L/m^3) ÷ air volume sampled in L

Another method of expressing concentration is ppm:

ppm = (mg/m^3)(24.45)(760)(T + 273) ÷ (M.W.)(P)(298)

where:

    P = pressure (mm Hg) of air sampled
    T = temperature (°C.) of air sampled
    24.45 = molar volume (L/mol) at 25° C. and 760 mm Hg
    M.W. = molecular weight (g/mol) of analyte
    760 = standard pressure (mm Hg)
    298 = standard temperature (°K)

References

1. Documentation of NIOSH Validation Tests, NIOSH Contract No. CDC-99-74-45.
2. Final Report, NIOSH Contract HSM-99-71-31. Personal Sampler Pump for Charcoal Tubes, September 15, 1972.

FLUOROTRICHLOROMETHANE

Synonyms: trichlorofluoromethane; Freon 11; Refrigerant 11; F-11 $CFCl_3$
ACGIH TLV: 1000 ppm (5600 mg/m^3) ceiling
OSHA Standard: 1000 ppm

Method	Sampling Duration	Sampling Location	Useful Range (ppm)	System Cost ($)	Test Cost ($)	Manufacturer
IR	Cont	Portable	20-2000	4374	0	Foxboro
PI	Cont	Fixed	20-2000	4950	0	HNU
GC	1 min	Portable	1-10^6	10,000	0	Microsensor

NIOSH METHOD NO. P&CAM 127

(Use as described for acetone, with 65° C. column temperature)

FORMALDEHYDE

ACGIH TLV: 1 ppm (1.5 mg/m^3) ceiling - Suspected carcinogen   HCHO
OSHA Standard: 3 ppm 2 ppm
NIOSH Recommendation: 1 ppm ceiling (30 min)

Method	Sampling Duration	Sampling Location	Useful Range (ppm)	System Cost ($)	Test Cost ($)	Manufacturer
DT	Grab	Portable	0.5-40	150	2.90	Nat'l Draeger
DT	Grab	Portable	1-35	165	3.80	Matheson
DT	Grab	Portable	0.2-20	180	2.25	Bendix
CM	Cont	Port/fixed	0.1-99	730	0	MDA
IR	Cont	Portable	0.4-10	4374	0	Foxboro
CM	Cont	Port/fixed	0.002-10	4995	Variable	CEA
GC	1 min	Portable	1-10^6	10,000	0	Microsensor

NIOSH METHOD NO. P&CAM 235

Principle

Formaldehyde is collected from air on alumina. The sample is stabilized immediately after collection by desorption into a 1% solution of methanol in water. The formaldehyde in this solution is then determined by the chromotropic acid-sulfuric acid spectrophotometric method (references 1-3).

Range and Sensitivity

The linear range of the spectrophotometric analysis extends from a detector limit of about 1 μg to an amount near 20 μg, or about 0.1-2 μg/mL in the final solution. This corresponds to approximately 0.4-0.9 mg/m^3 (0.3-0.7 ppm) of formaldehyde in a 6 L sample of air. The absorbance of the sample solution at the detection limit is about 0.05 units greater than that of the reagent blank. The capacity of the sorbent tube at high humidity is at least 312 μg. This corresponds to a concentration of 52 mg/m^3 when 6 L of air are sampled. At lower humidities the capacity of the sorbent tube is higher.

Interferences

The interferences discussed below have been evaluated for the chromotropic acid procedure described in reference 3, which involves the collection of formaldehyde in an impinger-bubbler containing 0.1% chromotropic acid in concentrated sulfuric acid. These interferences have not been evaluated for the overall sorbent collection and analysis procedure described here.

The chromotropic acid procedure for formaldehyde suffers very little interference from other aldehydes. Saturated aldehydes give less than 0.01% positive interference. Ethanol and higher molecular weight alcohols and olefins in mixtures with formaldehyde are negative interferences. However, concentrations of alcohols in air are usually much lower than formaldehyde concentrations; therefore, alcohols are not serious interferents.

Phenols cause 10-20% negative interference when present at an 8:1 excess over formaldehyde. They are, however, ordinarily present at lower concentrations than formaldehyde and, therefore, are not serious interferents. Ethylene and propylene in a 10:1 excess over formaldehyde in air cause 5-10% negative interference and 2-methyl-1,3-butadiene in a 15:1 excess over formaldehyde in air causes a 15% negative interference. Aromatic hydrocarbons also constitute a negative interference (reference 3). It has recently been found that cyclohexanone causes a bleaching of the final color.

Precision and Accuracy

The precision of the overall sampling and analytical procedure has been determined at two concentrations of formaldehyde in air, 3.5 and 3.9 mg/m^3. The analysis of two sets of seven consecutive samples, one set exposed at each concentration, indicated a coefficient of variation for all results of 0.06. The accuracy of the sorbent tube sampling method relative to impinger sampling has been measured to be 101% at 2.7 mg/m^3 and 107% at 6.6 mg/m^3.

Advantages and Disadvantages

The sampling device contains a solid sorbent rather than a liquid sorbing solution. This results in easier sample handling without danger of spillage. The sorbent tube is smaller and less subject to accidental breakage than impingers. Eluted alumina sections are transported to the laboratory in small, easily packed containers. Formaldehyde is stabilized in the 1% methanol eluent. Eluted samples can be shipped and stored for at least one month without loss.

A significant loss of formaldehyde absorbed on alumina occurs after an hour. Therefore, the sampling period is limited to 30 min and immediate elution is required. The absorbance of the solution increases slowly on standing. An increase of 3% in absorbance was noted after the solution stood for one day and an increase of 10% was noted after the solution stood for 8 days (reference 3).

Apparatus

Sorbent tubes. The sorbent tubes (see figure) consist of Pyrex glass tubes (8 mm I.D.) packed with two separate sections of 8/14 mesh activated alumina (Fisher Scientific Co., or equivalent).

The weight of alumina in each section is 1.65 g. The sections are enclosed and separated by 100 mesh stainless steel screens held in place by Teflon rings. Other nonreactive plugs of negligible pressure drop may be used. The ends of each tube are flame-sealed after packing to prevent contamination prior to sampling. Polyethylene caps should be provided to seal the tubes after sampling has been completed.

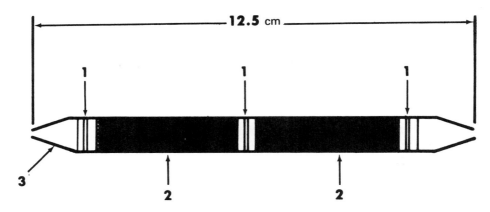

1. Stainless steel-screen plugs, 100 mesh, and Teflon supporting rings
2. Alumina section, 1.64g (2ml), 8/14 mesh
3. Glass tube (8mm i.d. and 12.5 cm long) with tapered ends.

Alumina Sampling Tube For Formaldehyde

Personal sampling pump. The personal sampling pump should be capable of operation at a constant flow rate of 200 mL/min for 30 min. A sample volume indicator is desirable. The pump should be calibrated with a representative sorbent tube in the line. A wet or dry test meter or a glass rotameter capable of measuring flow rate of 200 mL/min to within 5% may be used in calibration.

Thermometer, manometer and stopwatch.

Small glass bottles or vials. With tightly fitted caps, that are sufficiently large to accomodate 1.65 g (2 mL) of 8/14 mesh alumina and 10 mL of the 1% methanol solution.

Spectrophotometer or colorimeter. An instrument capable of measuring absorbance at 580 nm.

Assorted laboratory glassware.

Laboratory centrifuge. Capable of handling tubes containing 10.1 mL of solution.

Reagents

Chromotropic acid reagent. Dissolve in distilled water 0.10 g of 4,5-dihydroxy-2,7-naphthalenedisulfonic acid, disodium salt (Eastman Kodak Company, Rochester, New York, Cat. No. P230, or equivalent). Dilute the solution to 10 mL with water. Filter, if necessary, and store in a brown bottle. Prepare a fresh solution weekly. If the solution darkens, discard it.

Concentrated sulfuric acid.

Formaldehyde standard solution A (1 mg/mL). Dilute about 3 mL of 37% formalin solution to 1 L with distilled water. This solution must be standardized as described below. The solution is stable for at least 3 months. Alternatively, sodium formaldehyde bisulfite (Eastman Kodak Company Cat. No. P6450, or equivalent) can be used as a primary standard (reference 4). Dissolve 4.4703 g in distilled water and dilute to 1 L.

Formaldehyde standard solution B (10 µg/mL). Dilute 1 mL of standard solution A to 100 mL with distilled water. Make up fresh daily.

Iodine, 0.1 N (approximate). Dissolve 25 g of potassium iodide in about 25 mL of water, add 12.7 g of iodine, and dilute to 1 L.

Iodine, 0.01 N. Dilute 100 mL of the 0.1 N iodine solution to 1 L. Standardize against sodium thiosulfate.

Starch solution, 1%. Make a paste of 1 g of soluble starch and 2 mL water and slowly add the paste to 100 mL of boiling water. Cool, add several mL of chloroform as a preservative, and store in a stoppered bottle. Discard when a mold growth is noticeable.

Buffer solution. Dissolve 80 g of anhydrous sodium carbonate in about 500 mL of water. Slowly add 20 mL of glacial acetic acid and dilute to 1 L.

Sodium bisulfite, 1%. Dissolve 1 g of sodium bisulfite in 100 mL of water. It is best to prepare a fresh solution weekly.

Methanol, 1%. Dilute 10 mL of reagent grade methanol to 1 L with doubly distilled water.

Procedure

Cleaning of equipment. Care must be exercised to ensure the absence of contaminants such as organic materials that can be charred by concentrated sulfuric acid. Soak glassware for 1 hr in a 1:1 mixture of nitric and sulfuric acids and follow by thorough rinsing with deionized water to remove all possible organic contaminants.

Collection and shipping of samples. Immediately before sampling, break the ends of a tube to provide openings of at least 2 mm diameter. Connect either end of the tube to the sampling pump with flexible tubing and mark this end as B (for backup) with a permanent marker. Do not sample air through any hose or tubing before it enters the sorbent tube. Attach the sampling tube vertically to the clothing of the test subject. Also attach the pump in a convenient position. Record the time and the volume indicator reading before sampling. Sample at a flow rate of 200 mL/min for exactly 30 min. Record the final volume indicator reading and the time. Measure and record the temperature and pressure of the atmosphere being sampled.

Immediately after sampling, break the tube open and transfer each section of alumina without loss to small bottles containing 10 mL of the 1% methanol solution. Immediate elution is required because of losses of formaldehyde that occur. If the sorbent is stored dry. The methanol prevents polymerization of formaldehyde in solution. Tightly seal and label the bottles. Pack all samples tightly in a suitable container to avoid breakage and transport them to the laboratory for analysis. Handle at least one unexposed tube in the same manner as the sample tubes (break open, transfer both alumina sections to a bottle containing 10 mL of 1% methanol, seal and transport), but sample no air through this tube. Label this bottle as a blank.

Analysis of samples. If the analysis is to be conducted soon after sampling, allow at least 1 hr for desorption of formaldehyde from the alumina. Gently agitate the mixtures several times. Pipette a 4 mL aliquot from each of the sample solutions into glass stoppered test tubes. A reagent blank containing 4 mL of the 1% methanol solution must also be run. If the formaldehyde content of the aliquot is anticipated to exceed the limit of the method, dilute a smaller aliquot to 4 mL with the 1% methanol solution.

Add 0.1 mL of the 1% chromotropic acid reagent to the solution and mix. Pipette slowly and cautiously 6 mL of concentrated sulfuric acid into the solution. The solution becomes extremely hot during the addition of the sulfuric acid. If the acid is not added slowly, some loss of sample could occur because of spattering. Allow the sample solution to cool to room temperature.

Separate the alumina from the solution by centrifugation prior to making the absorbance measuremnt. Determine the absorbance of the solution at 580 nm in a spectrophotometer with a 1 cm cell. No change in absorbance has been noted over a 3 hour period after color development. Subtract the absorbance of the reagent blank. Determine the formaldehyde content of the solution from the calibration curve.

Determination of desorption efficiency. Add 1.65 g of alumina (from the same batch as that used in the sampling tubes) to a 2 mL glass stoppered tube. With a microliter syringe, evenly distribute 18 μL of standard solution A (1 μg/μL) in the alumina in the tube. Note the exact time. Seal the tube and gently shake the alumina. This amount of formaldehyde (18 μg) corresponds to what would be collected in a 6 L air sample at a concentration of 3 mg/m^3 (2 ppm), OSHA standard. Smaller or larger amounts may be used as desired.

Prepare at least three samples in this manner for each amount of formaldehyde tested. After exactly 30 min from the time of addition of the standard transfer each portion of alumina to 10 mL of the 1% methanol solution and extract for at least 1 hr. Prepare a desorption blank in the same manner, omitting the formaldehyde. For each amount of formaldehyde tested prepare three standards by injecting with the same syringe the same volume of standard solution A into 10 mL of the 1% methanol solution.

Analyze these samples, blanks, and standards as described above. The desorption efficiency D.E. is the amount of formaldehyde found in a sample (after correction for the desorption blank) divided by the amount found for the corresponding standards:

D.E. = (μg in sample $-$ μg in blank) $\div$ μg in standard

An average desorption efficiency of 0.85 has been found for amounts of formaldehyde added to alumina in the range of 10-37 μg.

Calibration and Standards

Pipette 1 mL of formaldehyde standard solution A into an iodine flask. Into another flask pipette 1 mL of distilled water; this is the blank. Add 10 mL of 1% sodium bisulfite and 1 mL of 1% starch solution to each flask. Titrate with 0.1 N iodine to a dark blue color. Destroy the excess iodine with 0.05 N sodium thiosulfate. Add 0.01 N iodine until a faint blue end point is reached. The excess inorganic bisulfite is now completely oxidized to sulfate, and the solution is ready for the assay of the formaldehyde bisulfite addition product.

Chill the flask in an ice bath and add 25 mL of chilled buffer solution. Using a microburette, titrate the liberated sulfite with 0.01 N iodine to a faint blue end point. The amount of iodine added in this step must be accurately measured and recorded. Since 1 mL of 0.01 N I_2 solution is equivalent to 0.15 mg of formaldehyde, the concentration on Cs of the standard formaldehyde solution (in mg/mL) is given by the equation:

Cs = (V)(N)(15)

where:

V = volume (in mL) of 0.01 N iodine solution used to titrate the liberated sulfite, corrected for the blank

N = the exact normality of the 0.01 N iodine solution

Pipet 0, 0.1, 0.3, 0.5, 0.7, 1.0, and 2.0 mL of formaldehyde standard solution B into glass-stoppered test tubes. Dilute each standard to 4 mL with distilled water. Develop the color as described in the analysis procedure. Plot absorbance against μg of formaldehyde in the solution. The amount of formaldehyde in each standard is based on the standardization value of solution A.

Calculations

From the calibration curve, read the amount (in μg) of formaldehyde corresponding to the absorbance of the sample solution. This is the amount of formaldehyde in the 4 mL sample aliquot. Multiply this value by 2.5. The result is the weight of formaldehyde desorbed from the alumina. Correct this value for the desorption efficiency. Add the corrected amounts of formaldehyde found in the front and backup sections of the same sorbent tube. Correct this sum for the amount of formaldehyde found in the corresponding unexposed, or blank, sorbent tube to obtain the total weight of formaldehyde in the air sample. These calculations are summarized in the following equation:

$$W = (2.5)(W_1 + W_2 - W_3) \div D.E.$$

where:

W = the total amount (in μg) of formaldehyde in the sample
$D.E.$ = desorption efficiency
W_1 = the amount (in μg) of formaldehyde found in the extract of the first sorbent section
W_2 = the amount (in μg) found in the extract of the backup sorbent section
W_3 = the total amount (in μg) found in the blank tube

The concentration C of formaldehyde in air may be expressed in mg/m³, which is numerically equivalent to μg/L:

$$C(mg/m^3) = W(\mu g) \div V(L)$$

where:

V = volume of air sampled

The concentration may also be expressed in terms of ppm by volume of standard conditions of 25° C. and 760 mm Hg:

$$C(ppm) = (W)(24.45)(760)(T + 273) \div (V)(M.W.)(P)(298)$$

where:

24.45 = the molar volume at standard conditions of 25° C. and 760 mm Hg
$M.W.$ = the molecular weight of formaldehyde (30.03)
P = pressure in mm Hg of air sampled
T = temperature in °C. of air sampled

References

1. Feigl, F. *Spot Tests in Organic Analysis,* 7th ed, American Elsevier Publishing Company, New York, 1966, p 434.
2. Eegriwe, E. Reaktionen und Reagenzien zum Nachweis Organischer Verbindungen IV. Z. Anal. Chem. 110: 22, 1937.
3. Sleva, S.F. *Selected Methods for the Measurement of Air Pollutants,* Public Health Service Publications No. 999-AP-11, H-1, 1965.

4. Altshuller, A.P., L.J. Leng, and A.F. Wartburg. Source and atmospheric analyses for formaldehyde by chromotropic acid procedure. Int. J. Air Water Poll. 6: 381, 1962.

5. Campbell, E.E., G.O. Wood, and R.G. Anderson. Development of Air Sampling Techniques, Los Alamos Scientific Laboratory, Progress Report LA-5973-PR, July, 1975.

6. Intersociety Committee. *Methods of Air Sampling and Analysis,* American Public Health Association, Washington, D.C., 1972, pp. 194-198.

7. MacDonald, W.E. Formaldehyde in air - a specific field test. Am. Ind. Hyg. Assoc. Quarterly 15: 217, 1954.

8. Treadwell, F.P., and W.T. Hall. *Analytical Chemistry,* Vol. II, 9th English ed, John Wiley & Sons, Inc., New York, 1951, pp 588, 590.

9. Wood, G.O. and R.G. Anderson. Development of Air-Monitoring Techniques Using Solid Sorbents, Los Alamos Scientific Laboratory Progress Report LA-6216-PR, February 1976.

10. Wood, G.O., and R.G. Anderson. Air Sampling of Formaldehyde with a Solid Sorbent Tube, presented at the American Industrial Hygiene Conference, Minneapolis, Minnesota, June, 1975.

FORMIC ACID

Synonym: methanoic acid
ACGIH TLV: 5 ppm (9 mg/m^3)
OSHA Standard: 5 ppm

HCOOH

Method	Sampling Duration	Sampling Location	Useful Range (ppm)	System Cost ($)	Test Cost ($)	Manufacturer
DT	Grab	Portable	1-15	150	2.60	Nat'l Draeger
DT	Grab	Portable	1-80	180	2.25	Bendix
IR	Cont	Portable	0.1-10	4374	0	Foxboro
GC	1 min	Portable	1-10^6	10,000	0	Microsensor

NIOSH METHOD NO. P&CAM 232

Principle

Formic acid in air is trapped in a solution of 0.1 N NaOH in a midget impinger. A derivatizing solution consisting of ethanol and sulfuric acid is added to an aliquot of the absorbing solution. The formic acid is converted to ethyl formate by heating to 55° C. for 2 hr. A partitioning of the volatile ethyl formate takes place between the solution and the headspace gas above it. The headspace gas is analyzed for ethyl formate by gas chromatography. The amount of ethyl formate found is proportional to the amount of formic acid in the sample.

Range and Sensitivity

For 15 mL of absorbing solution, the useful range extents from 38-750 μg of formic acid, which corresponds to 3.8-75 mg/m^3 in a 10 L sample of air. The range may be extended to lower amounts by using a smaller volume of absorbing solution. Recovery data indicate that the linear range may extend from the detection limit of 7.5 μg/15 mL to 750 μg/15 mL.

Interferences

Low molecular weight acids and alcohols may interfere in the derivatization step if present in large amounts. These interferences can be reduced by following the analysis procedure carefully. Since water is a product of the derivatization reaction, the ester will not be produced quantitatively in the aqueous reaction mixture. However, the amount of ethyl formate found has been shown to be proportional to the amount of formic acid collected and, also, has been shown to be reproducible (reference 1). Any compound having the same retention times as ethyl formate in the GC procedure will interfere with the analysis. This type of interference can often be eliminated by changing the operating conditions on the chromatograph.

Precision and Accuracy

The average coefficient of variation was 0.11 for the analysis of 24 spiked samples in the range of 38-750 μg/sample. An average recovery of 105% was obtained in the analyses of eight 15 mL samples of 0.1 N NaOH spiked with 15-600 μg of formic acid. The accuracy of the method has not been determined.

Advantages and Disadvantages

The method is a straight forward technique with three steps. It requires no separation or distillation steps. Derivatization is required.

Apparatus

Gas chromatograph. Equipped with dual flame-ionization detectors.

Stainless steel column. 20 ft × 0.125 in, packed with 10% Carbowax 20 M on 80/100 mesh Chromosorb W, or an equivalent column.

Hamilton gas syringe. 1 mL or equivalent.

Reaction vials, 10 mL.

Teflon-coated septums. With tear-away seals for the reaction vials.

Volumetric pipettes, 2 mL.

Eppendorf pipettes. 5, 10, 25, 50, and 100 μL.

Midget impinger, 15 mL.

Calibrated personal sampling pump. Capable of sampling at a flow rate of 2.5 L/min for periods of up to 40 min. The pump should be calibrated with a midget impinger containing 15 mL of the absorbing solution in the sampling train. A dry or wet test meter or a glass rotameter that will measure the appropriate flow rate within 5% may be used for the calibration.

Constant temperature water bath. Capable of maintaining 55° C.

Reagents

Absorbing solution, 0.1 N NaOH. Dissolve 4.0 g of sodium hydroxide pellets, ACS reagent grade, in distilled water, and dilute to 1 L with distilled water.

Absolute ethanol (anhydrous), uv spectrophotometric grade.

Sulfuric acid, concentrated, ACS reagent grade.

Ethyl formate, 99%.

Derivatizing solution. To a 100 mL graduated cylinder add 70 mL of absolute ethanol. To the ethanol slowly add 20 mL of concentrated sulfuric acid. Allow the solution to cool to room temperature. Add ethanol until the volume is 100 mL and mix.

Sodium formate solution. In a 100 mL volumetric flask, dissolve 0.1478 g of sodium formate, ACS reagent grade, in 0.1 N NaOH and fill to the mark with 0.1 N NaOH. This solution is equivalent to a formic acid concentration of 1000 μg/mL.

Helium, Bureau of Mines Grade A, or equivalent.

Prepurified hydrogen.

Filtered compressed air.

Procedure

Cleaning of equipment. Wash all glassware in detergent solution and rinse with tap water. Soak the glassware in chromic acid cleaning solution (saturated solution of sodium dichromate in concentrated sulfuric acid) and rinse with tap water and distilled water.

Collection and shipping of samples. Pour 15 mL of 0.1 N NaOH in a midget impinger and connect the impinger to a vacuum pump with a short piece of flexible tubing. Turn on the pump to begin sample collection. Measure the flow rate and time, or volume, as accurately as possible and record the temperature and pressure of the atmosphere being sampled. Sample at a flow rate of 2.5 L/min. Collect a sample of 45-100 L. A sample of 100 L of air will allow measurement of at least 0.9 mg/m^3, one tenth the OSHA standard.

After sampling, remove the impinger stem. Tap the stem gently against the inside wall of the impinger to recover as much of the solution as possible. Then seal the impinger with a nonreactive stopper. Care should be taken to minimize spillage or loss by evaporation. Whenever possible, hand deliver the samples to the laboratory. Otherwise, ship the samples in the special impinger shipping cases designed by NIOSH. Handle at least one impinger in the same manner as the other samples (fill, seal, and transport), but sample no air through this impinger. Label this sample as a blank.

Analysis of samples. Withdraw a 2 mL aliquot from the impinger and place it in a 10 mL reaction vial. Add 2 mL of derivatizing solution to the vial and seal it immediately with a Teflon coated septum. Place the seal vial in a constant temperature water bath at 55° C. for 2 hr. After 2 hr, pierce the septum with the needle of the gas syringe. Flush the syringe three times with the headspace gas.

Injection. Inject 1 mL of the headspace gas into the gas chromatograph. The operating conditions for the gas chromatographic analysis are as follows:

30 mL/min helium carrier gas flow rate
65 mL/min hydrogen gas flow rate to the detector
500 mL/min air flow rate to the detector
100° C. injector temperature
150° C. detector temperature
70° C. column temperature

Measure the height of the ethyl formate peak. The retention time of ethyl formate is determined by injecting ethyl formate vapor into the gas chromatograph before the sample in injected.

Calibration and Standards

To 2 mL of distilled water in each of five 10 mL reaction vials, add 0.005, 0.010, 0.025, 0.050, or 0.100 mL of sodium formate solution to prepare standards equivalent to 5, 10, 25, 50, or 100 μg/vial of formic acid necessary for the calibration curve. Standards of higher concentration may also be prepared. Analyze the standards by the procedure described earlier. Plot the height of the ethyl formate peak as a function of the equivalent amount (in μg) of formic acid in each vial.

Calculations

From the standard curve, read the weight (in μg) of formic acid corresponding to the peak height for the headspace sample. If the blank impinger solution produces a peak with the same GC retention time as the ethyl formate peak, subtract the calculated amount of formic acid from the sample value. To calculate the total amount of formic acid collected in the sample, use the following equation:

$$W = (Wc)(I) \div A$$

where:

W = amount of formic acid (μg) collected in the impinger
Wc = amount of formic acid (μg) read from the calibration curve and corrected for the blank
I = volume of the impinger sample (15 mL)
A = volume of the aliquot taken (2 mL)

The concentration of formic acid in air may be expressed in mg/m³:

$$mg/m^3 = W \div Vs$$

where:

Vs = volume of air sampled in L

The concentration may also be expressed in terms of ppm by volume:

$$ppm = (mg/m^3)(24.45)(760)(T + 273) \div (M.W.)(P)(298)$$

where:

24.45 = molar volume (L/mol) at 25° C. and 760 mm Hg
M.W. = molecular weight
P = pressure (mm Hg) of air sampled
T = temperature (°C.) of air sampled

References

1. Mraz, M., and V. Sedwec. Determination of toxic substances and their metabolites in biological fluids by gas chromatography. VIII. Formic acid in urine. Collect. Czec. Chem. Comm. 38: 3426, 1973.
2. Smallwood, A.W. Analysis of Formic Acid in Air Samples, presented at American Industrial Hygiene Conference, Atlanta, Georgia, May 1976.

FURFURAL

Synonyms: 2-furancarbonal; 2-furaldehyde; furfuraldehyde C_4H_3OCHO
ACGIH TLV: 2 ppm (8 mg/m^3) - Skin
OSHA Standard: 5 ppm (20 mg/m^3)

Method	Sampling Duration	Sampling Location	Useful Range (ppm)	System Cost ($)	Test Cost ($)	Manufacturer
DT	Grab	Portable	2-60	180	2.25	Bendix
IR	Cont	Portable	0.6-10	4374	0	Foxboro

NIOSH METHOD NO. S17

Principle

A known volume of air is drawn through a midget bubbler containing a buffered solution of Girard T reagent. Furfural is derivatized with the Girard T reagent. The furfural Girard T derivative is analyzed by high pressure liquid chromatography (HPLC) at a uv detection of 312 nm.

Range and Sensitivity

This method was validated over the range of 10.1-40 mg/m^3 at an atmospheric temperature of 20° C. and pressure of 757 mm Hg, using a 120 L sample. The upper limit of the range of the method is dependent on the concentration of Girard T reagent.

Under the instrumental conditions used in the study, a sensitivity of approximately 0.24 absorbance units/μg furfural was obtained. The HPLC flow cell path length was 0.8 cm. The limit of detection is established to be 0.003 μg/injection or 20 μg furfural/bubbler sample. The detection limit may be extended by analyzing the sample without dilution and by attenuating the liquid chromatograph output less.

Interferences

When interfering compounds are known or suspected to be present in the air, such information, including their suspected identities, should be transmitted with the sample. Other volatile aldehydes or ketones such as propionaldehyde, acrolein, or acetone may cause significant interferences or compete with furfural for reaction with Girard T reagent. Chromatographic conditions can be adjusted to separate the various substances.

Precision and Accuracy

The coefficient of variation for the total analytical and sampling method in the range of 10.1-40 mg/m^3 was 0.054. This value corresponds to a 1.1 mg/m^3 standard deviation at the OSHA standard level. Statistical information can be found in reference 1. Details of the test procedures are found in reference 2.

In validation experiments, this method was found to be capable of coming within 25% of the "true value" on the average of 95% of the time over the validation range. The concentrations obtained at 0.5, 1, and 2 times the OSHA environmental limit averaged 5.2% higher than the dynamically generated test concentrations (n=18). The analytical method recovery was determined to be 1.010 for a collector loading of 1.222 mg. Storage stability studies on samples collected from a test atmosphere at a concentration of 20.1 mg/m^3 indicated that collected samples are stable for at least 7 days if protected from exposure to light. The mean of samples analyzed after 7 days were within 1.7% of the mean of samples analyzed immediately after collection. Collection efficiency of the midget bubbler containing Girard T reagent was determined to be at least 0.996 within the range tested; therefore, no correction for collection efficiency is necessary. Experiments performed in the validation study are described in reference 2.

Advantages and Disadvantages

The furfural Girard T reagent derivative has adequate storage stability. Collected samples are analyzed by a quick instrumental method. A disadvantage of the method is the awkwardness in using midget bubblers for collecting personal samples. If the worker's job performance requires much body movement, loss of the collected solution during sampling may occur. Use of a spill proof bubbler is recommended. The bubblers are more difficult to ship than adsorption tubes or filters due to possible breakage and leakage of the bubblers during shipping.

Apparatus

Glass midget bubblers with fritted glass stems.

Personal sampling pump. A calibrated personal sampling pump whose flow rate can be determined to an accuracy of 5%. The sampling pump is protected from splashover or solvent condensation by a trap. The trap is a midget bubbler or impinger with the stem broken off which is used to collect spillage. The trap is attached to the pump with a metal holder. The outlet of the trap is connected to the pump by flexible tubing. Each sampling pump must be calibrated with a representative bubbler and trap in the line to minimize errors associated with uncertainties in the volume sampled.

Thermometer.

Manometer.

HPLC. Equipped with a variable wavelenth uv detector set at 312 nm and a sample injection valve with a 50 µL external sample loop. A selectable wavelength detector with appropriate filter to produce wavelength of 312 nm may also be used.

HPLC column. Packed with Zipax SCX (50 cm long × 2 mm I.D. stainless steel). This column can be obtained from Dupont.

Electronic integrator. Or some other suitable method for measuring peak areas.

Syringes. 2 mL Luerlock.

Vacuum fractional distillation apparatus. For preparing high purity furfural.

Volumetric flasks. 500 mL and other convenient sizes for preparing standard solutions.

Pipets. Convenient sizes for preparing stock standard solutions and for measuring the collection medium.

China marker.

Teflon plugs. Or equivalent for sealing the inlet and outlet of the bubbler stem before shipping.

Reagents

All reagents used must be ACS reagent grade or better.

Furfural.

Citric acid.

Disodium hydrogen phosphate.

Distilled, deionized water.

Ethanol, 95%.

Sodium dihydrogen phosphate monohydrate.

Girard T reagent. (Carboxymethyl)trimethylammonium chloride hydrazide, recrystallized from 95% ethanol.

0.2 M Girard T reagent. Dissolve 5.39 g citric acid, 6.63 g disodium hydrogen phosphate, and 16.77 g of recrystallized Girard T reagent in approximately 400 mL of distilled, deionized water. Transfer the solution to a 500 mL volumetric flask and bring to volume with distilled, deionized water. This solution must be used within 2 weeks and should be stored in the dark in glassware which has been heated as described in section 8.1.

HPLC eluant. 0.028 M Na_2HPO_4 and 0.044 M NaH_2PO_4 in 20% ethanol. Prepare a stock eluent of 39.2 g Na_2HPO_4 and 61.0 g $NaH_2PO_4.H_2O$ made up to 1 L in distilled, deionized water. To prepare the HPLC eluant, combine 100 mL of stock and 200 mL of 95% ethanol. Dilute this solution to 1 L with distilled, deionized water. Filter the solution through a 5 micron Teflon filter and degas prior to use. Slowly bubble helium through the eluant reservoir during use to prevent bacterial growth. This solution is also used for sample dilution prior to analysis.

Stock standard furfural Girard T derivative. Vacuum distill furfural at a reduced pressure of 15 mm Hg. Collect distilling fraction between 60-61°C. Prepare a standard containing 0.976 mg/mL furfural in 0.2 M Girard T reagent by weighing 48.8 mg of freshly distilled furfural into a 50 mL volumetric flask containing approximately 49 mL of Girard T reagent solution. Make up to volume with Girard T reagent solution. This solution must be used within 1 day.

Procedure

Cleaning of equipment. All glassware used for the laboratory analysis should be detergent washed and thoroughly rinsed with tap water and distilled water, and dried. All midget bubblers should be heated in an oxidizing atmosphere at a temperature of approximately 580°C. This procedure will remove organic contaminants.

Collection and shipping of samples. Pipet 15 mL of the buffered Girard T reagent into each midget bubbler. Mark the liquid level on the bubbler with a china marker. The outlet of the midget bubbler is attached to a trap which is used to protect the pump during personal sampling. The trap is a midget impinger or bubbler with the stem broken off which is used to collect spillage. The trap is attached to the pump with a metal holder. The outlet of the trap is connected to the pump by flexible tubing.

Air being sampled should not be passed through any hose or tubing before entering the midget bubbler. A sample size of 120 L is recommended. Sample at a flow rate of 0.5 L/min. Sampling at higher flow rates causes frothing of the collection medium. Set the flow rate as accurately as possible using the manufacturer's directions. Turn the pump on and begin sample collection. Terminate sampling at the predetermined time and record sample flow rate, collection time, and ambient temperature and pressure. If pressure reading is not available, record the elevation. Also record the type of sampling pump used.

The inlet and outlet of the bubbler stem should be sealed by inserting Teflon plugs, or equivalent, in the inlet and outlet. Do not seal with rubber. The standard taper joint of the bubble should be taped securely to prevent leakage during shipping. Care should be taken to minimize spillage or loss of sample by evaporation at all times. The bubbler should be shipped in a suitable container designed to prevent damage in transit. The samples should be shipped to the laboratory as soon as possible.

With each batch or partial batch of ten samples, submit one bubbler containing 15 mL of the collection medium prepared from the same stock as that used for sample collection. This bubbler must be subjected to exactly the same handling as the samples except that no air is drawn through it. Label this bubbler as the blank. Do not ship the material in the trap. If more than 1 mL of material is collected in the trap after sampling, the sample should be considered invalid. A sample of the bulk material should be submitted in a glass container with a Teflon lined cap or equivalent. This sample must not be transported in the same container as the bubblers.

Analysis of samples. Sample analysis should be done in a room with a fairly stable temperature to prevent retention time fluctuations on the ion exchange column. Remove the bubbler stem and tap the stem lightly against the flask to drain the contents into the bubbler flask. If necessary bring the liquid volume to the 15 mL mark with distilled water. Transfer a 5 mL aliquot to a 100 mL flask and bring to volume with the HPLC eluant.

HPLC conditions. The typical operating conditions for the HPLC are:

Column temperature: ambient
Flow rate: 1.0 mL/min
Mobile phase: 0.028 M Na_2HPO_4 and 0.044 M NaH_2PO_4 in 20% ethanol
Detector: uv at 312 nm

Under the above conditions the furfural Girard T derivative will elute in approximately 6.5 min.

Injection. The first step in the analysis is to inject the sample into the high pressure liquid chromatograph. The chromatograph is fitted with a sample injection valve and a 50 μL sample loop. Flush this loop thoroughly with the sample (500 μL), and inject the sample. Duplicate injections should compare within 3%. The area of the sample peak is measured by an electronic integrator or some other suitable form of area measurement, and results are read from a standard curve prepared as discussed below.

Calibration and Standards

A series of standards, varying in concentration over the range corresponding to approximately 0.1 to 3 times the OSHA standard for the substance under study, is prepared and analyzed under the same HPLC conditions and during the same time period as the unknown samples. Curves are established by plotting concentration in mg/15 mL versus peak area. Note: since no internal standard is used in this method, standard solutions must be analyzed at the same time as the samples. This will minimize the effect of known day-to-day variations and variations during the same day of the uv detector response.

Prepare working standards by diluting appropriate aliquots with the buffered Girard T solution to the desired concentration. Prepare at least 5 working standards to cover the range of 0.24-7.32 mg/sample (16-488 μg/mL). This range is based on a 120 L sample. Transfer a 5 mL aliquot of each standard to a 100 mL volumetric flask and bring to volume with HPLC eluant. Dilution and preparation of standards in this way ensures that the derivative/Girard T reagent ratio will be the same as the collected samples. Analyze the samples as described above.

Prepare at least one blank and analyze it at the same time as the standards. Prepare a standard calibration curve by plotting concentration of furfural in mg/15 mL versus peak area.

Calculations

Read the weight, in mg, corresponding to each peak area from the standard curve. No volume correction is needed, because the standard curve is based on mg/15 mL Girard T reagent and the volume of sample injected is identical to the volume of the standards injected, and the final dilution with HPLC eluant is the same for both standards and samples. A correction for the blank must be made for each sample:

mg = mg sample − mg blank

where:

mg sample = mg found in sample bubbler
mg blank = mg found in blank bubbler

A similar procedure is followed for the backup sections. For personal sampling pumps with rotameters only, the following volume correction should be made:

Corrected volume = $ft(P_1T_2/P_2T_1)^{1/2}$

where:

f = flow rate sampled
t = sampling time
P_1 = pressure during calibration of sampling pump (mm Hg)
P_2 = pressure of air sampled (mm Hg)
T_1 = temperature during calibration of sampling pump (°K)
T_2 = temperature of air sampled (°K)

The concentration of furfural in the air sample can be expressed in mg/m^3:

mg/m^3 = (mg)(1000 L/m^3) ÷ corrected air volume sampled in L

Another method of expressing concentration is ppm:

ppm = (mg/m^3)(24.45)(760)(T + 273) ÷ (M.W.)(P)(298)

where:

P = pressure (mm Hg) of air sampled
T = temperature (°C.) of air sampled
24.45 = molar volume (L/mol) at 25° C. and 760 mm Hg
M.W. = molecular weight of furfural
760 = standard pressure (mm Hg)
298 = standard temperature (°K)

References

1. Documentation of NIOSH Validation Tests. National Institute for Occupational Safety and Health, Cincinnati, Ohio (DHEW-NIOSH Publication #77-185), 1977. Available from Superintendent of Documents, U.S. Government Printing Office, Washington, D.C., Order No. 017-033-00231-2.
2. Backup Data Report No. S17 for Furfural, prepared under NIOSH Contract No. 210-76-0123.

FURFURYL ALCOHOL

Synonym: 2-furancarbinol
ACGIH TLV: 10 ppm (40 mg/m³) - Skin
OSHA Standard: 50 ppm (200 mg/m³)

$C_4H_3OCH_2OH$

Method	Sampling Duration	Sampling Location	Useful Range (ppm)	System Cost ($)	Test Cost ($)	Manufacturer
IR	Cont	Portable	1-100	4374	0	Foxboro

NIOSH METHOD NO. S350

(Use as described for n-butyl mercaptan, with 200° C. column temperature)

GLYCIDOL

Synonyms: 2,3-epoxy-1-propanol; epihydric alcohol
ACGIH TLV: 25 ppm (75 mg/m³)
OSHA Standard: 50 ppm (150 mg/m³)

$\overline{OCH_2CHCH_2}OH$

Method	Sampling Duration	Sampling Location	Useful Range (ppm)	System Cost ($)	Test Cost ($)	Manufacturer
IR	Cont	Portable	3-100	4374	0	Foxboro

NIOSH METHOD NO. S70

Principle

A known volume of air is drawn through a charcoal tube to trap the organic vapors present, and the charcoal tube is immediately refrigerated at 4° C. and held at that temperature until the sample is desorbed with tetrahydrofuran. The charcoal in the tube is transferred to a small, stoppered sample container, and the analyte is desorbed with tetrahydrofuran. An aliquot of the desorbed sample is injected into a gas chromatograph. The area of the resulting peak is determined and compared with areas obtained from the injection of standards.

Range and Sensitivity

This method was validated over the range of 73-310 mg/m³ at an atmospheric temperature and pressure of 21° C. and 768 mm Hg, using a 50 L sample. Under the conditions of sample size (50 L) the probable useful range of this method is 15-450 mg/m³ at a detector sensitivity that gives nearly full deflection on the strip chart recorder for a 22 mg sample. The method is capable of measuring much smaller amounts if the desorption efficiency is adequate. Desorption efficiency must be determined over the range used.

The upper limit of the range of the method is dependent on the adsorptive capacity of the charcoal tube. This capacity varies with the concentrations of glycidol and other substances in the air. The first section of the charcoal tube was found to hold 45 mg of glycidol when a test atmosphere containing 300 mg/m^3 of glycidol in air was sampled at 0.93 L/min for 163 min; breakthrough was observed at this time, i.e., the concentration of analyte in the effluent was 5% of that in the influent. (The charcoal tube consists of two sections of activated charcoal separated by a section of urethane foam.) If a particular atmosphere is suspected of containing a large amount of contaminant, a smaller sampling volume should be taken.

Interferences

When the amount of water in the air is so great that condensation actually occurs in the tube, organic vapors will not be trapped efficiently. Preliminary experiments using toluene indicate that high humidity severely decreases the breakthrough volume.

When two or more compounds are known or suspected to be present in the air, such information, including their suspected identities, should be transmitted with the sample. Since tetrahydrofuran is used rather than carbon disulfide to desorb the glycidol from the charcoal, it would not be possible to measure tetrahydrofuran in the sample with this desorbing solvent. If it is suspected that tetrahydrofuran is present, a separate sample should be collected for tetrahydrofuran analysis.

It must be emphasized that any compound which has the same retention time as the analyte at the operating conditions described in this method is an interference. Retention time data on a single column cannot be considered proof of chemical identity. If the possibility of interference exists, separation conditions (column packing, temperature, etc.) must be changed to circumvent the problem.

Precision and Accuracy

The coefficient of variation for the total analytical and sampling method in the range of 73-310 mg/m^3 was 0.080. This value corresponds to a 12 mg/m^3 standard deviation at the OSHA standard level. Statistical information and details of the validation and experimental test procedures can be found in reference 2. On the average the values obtained using the overall sampling and analytical method were 3.1% lower than the "true" values at the OSHA standard level.

Advantages and Disadvantages

The sampling device is small, portable, and involves no liquids. Interferences are minimal, and most of those which do occur can be eliminated by altering chromatographic conditions. The tubes are analyzed by means of a quick, instrumental method. The method can also be used for the simultaneous analysis of two or more substances suspected to be present in the same sample by simply changing gas chromatographic conditions from isothermal to a temperature-programmed mode of operation.

One disadvantage of the method is that the amount of sample which can be taken is limited by the number of mg that the tube will hold before overloading. When the sample value obtained for the backup section of the charcoal tube exceeds 25% of that found on the front section, the possibility of sample loss exists. Furthermore, the precision of the method is limited by the reproducibility of the pressure drop across the tubes. This drop will affect the flow rate and cause the volume to be imprecise, because the pump is usually calibrated for one tube only. A disadvantage of the method is that the sample must be kept refrigerated, or recoveries are poor.

Apparatus

Calibrated personal sampling pump. Whose flow can be determined within 5% at the recommended flow rate.

Charcoal tubes. Glass tube with both ends flame sealed, 7 cm long with 6 mm O.D. and 4 mm I.D., containing 2 sections of 20/40 mesh activated charcoal separated by a 2 mm portion of urethane foam. The activated charcoal is prepared from coconut shells and is fired at 600° C. prior to packing. The adsorbing section contains 100 mg of charcoal, the backup section 50 mg. A 3 mm portion of urethane foam is placed between the outlet end of the tube and the backup section. A plug of silylated glass wool is placed in front of the adsorbing section. The pressure drop across the tube must be less than one inch of mercury at a flow rate of 1 L/min.

Gas chromatograph. Equipped with a flame-ionization detector.

Column. 10 ft × 1/8 in stainless steel, packed with 10% FFAP on 80/100 mesh, acid washed DMCS Chromosorb W.

Electronic integrator. Or some other suitable method for measuring peak area.

Sample containers. 1 mL, with glass stoppers or Teflon-lined caps.

Syringes. 10 µL, and other convenient sizes for making standards.

Pipets. 0.5 mL delivery pipets or 1.0 mL type graduated in 0.1 mL increments.

Volumetric flasks. 10 mL or convenient sizes for making standard solutions.

Reagents

Tetrahydrofuran, chromatographic quality.
Glycidol, reagent grade.
Purified nitrogen.
Prepurified hydrogen.
Filtered compressed air.

Procedure

Cleaning of equipment. All glassware used for the laboratory analysis should be detergent washed and thoroughly rinsed with tap water and distilled water.

Calibration of personal pumps. Each personal pump must be calibrated with a representative charcoal tube in the line. This will minimize errors associated with uncertainties in the sample volume collected.

Collection and shipping of samples. Immediately before sampling, break the ends of the tube to provide an opening at least one-half the internal diameter of the tube (2 mm). The smaller section of charcoal is used as a backup and should be positioned nearest the sampling pump. The charcoal tube should be placed in a vertical direction during sampling to minimize channeling through the charcoal. Air being sampled should not be passed through any hose or tubing before entering the charcoal tube.

A maximum sample size of 50 L is recommended. Sample at a flow of 1.0 L/min or less. The flow rate should be known with an accuracy of at least 5%. The temperature and pressure of the atmosphere being sampled should be recorded. If pressure reading is not available, record the elevation. The charcoal tubes should be capped with the supplied plastic caps immediately after sampling. Under no circumstances should rubber caps be used. One tube should be handled in the same manner as the sample tubes (break, seal, and transport), except that no air is sampled through this tube. This tube should be labeled as a blank.

Capped tubes should be packed tightly and padded before they are shipped to minimize tube breakage during shipping. Capped tubes should be refrigerated during storage and shipping at 4° C. A sample of the bulk material should be submitted to the laboratory in a glass container with a Teflon-lined cap. This sample should not be transported in the same container as the charcoal tubes.

Preparation of samples. In preparation for analysis, each charcoal tube is scored with a file in front of the first section of charcoal and broken open. The glass wool is removed and discarded. The charcoal in the first (larger) section is transferred to a 1 mL stoppered sample container. The separating section of foam is removed and discarded; the second section is transferred to another sample container. These two sections are analyzed separately.

Desorption of samples. Prior to analysis, 0.5 mL of tetrahydrofuran is pipetted into each sample container. Desorption should be done for 30 minutes. Tests indicate that this is adequate if the sample is agitated occasionally during this period.

GC conditions. The typical operating conditions for the gas chromatograph are:

50 mL/min (60 psig) nitrogen carrier gas flow
65 mL/min (24 psig) hydrogen gas flow to detector
500 mL/min (50 psig) air flow to detector
225° C. injector temperature
260° C. detector temperature
155° C. column temperature

Injection. The first step in the analysis is the injection of the sample into the gas chromatograph. To eliminate difficulties arising from blow back or distillation within the syringe needle, one should employ the solvent flush injection technique. The 10 μL syringe is first flushed with solvent several times to wet the barrel and plunger. Three μL of solvent are drawn into the syringe to increase the accuracy and reproducibility of the injected sample volume. The needle is removed from the solvent, and the plunger is pulled back about 0.2 μL to separate the solvent flush from the sample with a pocket of air to be used as a marker. The needle is then immersed in the sample, and a 5 μL aliquot is withdrawn, taking into consideration the volume of the needle, since the sample in the needle will be completely injected. After the needle is removed from the sample and prior to injection, the plunger is pulled back 1.2 μL to minimize evaporation of the sample from the tip of the needle. Observe that the sample occupies 4.9-5.0 μL in the barrel of the syringe. Duplicate injections of each sample and standard should be made. No more than a 3% difference in area is to be expected.

Measurement of area. The area of the sample peak is measured by an electronic integrator or some other suitable form of area measurement, and preliminary results are read from a standard curve prepared as discussed below.

Determination of desorption efficiency. The desorption efficiency of a particular compound can vary from one laboratory to another and also from one batch of charcoal to another. Thus, it is necessary to determine at least once the percentage of the specific compound that is removed in the desorption process, provided the same batch of charcoal is used.

Procedure for determining desorption efficiency. Activated charcoal equivalent to the amount in the first section of the sampling tube (100 mg) is measured into a 2.5 in, 4 mm I.D. glass tube, flame sealed at one end. This charcoal must be from the same batch as that used in obtaining the samples and can be obtained from unused charcoal tubes. The open end is capped with Parafilm. A known amount of the analyte is injected directly into the activated charcoal with a microliter syringe, and the tube is capped with more Parafilm. These tubes must be refrigerated after injection of the analyte onto the charcoal. Previous studies have shown that the desorption efficiency decreases with time if the samples are stored at room temperature. The amount injected is equivalent to that present in a 50 L sample at the selected level.

Six tubes at each of three levels (0.5X, 1X, and 2X the standard) are prepared in this manner and allowed to stand for at least overnight to assure complete adsorption of the analyte onto the charcoal. The tubes should be kept refrigerated, and they are referred to as the samples. A parallel blank tube should be treated in the same manner except that no sample is added to it. The sample and blank tubes are desorbed and analyzed in exactly the same manner as the sampling tube described earlier.

Two or three standards are prepared by injecting the same volume of compound into 0.5 mL of the eluent with the same syringe used in the preparation of the samples. These are analyzed with the sampled. The desorption efficiency (D.E.) equals the average weight in mg recovered from the tube divided by the weight in mg added to the tube, or:

D.E. = average wt (mg) recovered ÷ wt (mg) added

The desorption efficiency is dependent on the amount of analyte collected on the charcoal. Plot the desorption efficiency versus the weight of analyte found. This curve is used to correct for adsorption losses.

Calibration and Standards

It is convenient to express concentration of standards in terms of mg/0.5 mL of tetrahydrofuran, because samples are desorbed in this amount of tetrahydrofuran. The density of the analyte is used to convert mg into μL for easy measurement with a microliter syringe. A series of standards, varying in concentration over the range of interest, is prepared and analyzed under the same GC conditions and during the same time period as the unknown sample. Curves are established by plotting concentration in mg/0.5 mL versus peak area. Note: Since no internal standard is used in the method, standard solutions must be analyzed at the same time that the sample analysis is done. This will minimize the effect of known day-to-day variations and variations during the same day of the FID response.

Calculations

Read the weight, in mg, corresponding to each peak area from the standard curve. No volume corrections are needed, because the standard curve is based on mg/0.5 mL tetrahydrofuran and the volume of sample injected is identical to the volume of the standards injected. Corrections for the blank must be made for each sample:

mg = mg sample − mg blank

where:

mg sample = mg found in front section of sample tube
mg blank = mg found in front section of blank tube

A similar procedure is followed for the backup sections. Add the weights found in the front and backup sections to get the total weight in the sample. Read the desorption efficiency from the curve for the amount found in the front section. Divide the total weight by this desorption efficiency to obtain the corrected mg/sample:

Corrected mg/sample = total weight ÷ D.E.

The concentration of the analyte in the air sampled can be expressed in mg/m^3:

mg/m^3 = (corrected mg)(1000 L/m^3) ÷ air volume sampled in L

Another method of expressing concentration is ppm:

$$\text{ppm} = (\text{mg/m}^3)(24.45)(760)(T + 273) \div (M.W.)(P)(298)$$

where:

P = pressure (mm Hg) of air sampled
T = temperature (°C.) of air sampled
24.45 = molar volume (L/mol) at 25° C. and 760 mm Hg
M.W. = molecular weight (g/mol) of analyte
760 = standard pressure (mm Hg)
298 = standard temperature (°K)

References

1. White, L.D. et al. A convenient optimized method for the analysis of selected solvent vapors in the industrial atmosphere. Amer. Ind. Hyg. Assoc. J. 31: 225, 1970.
2. Documentation of NIOSH Validation Tests, NIOSH Contract No. CDC-99-74-45.

n-HEPTANE

C_7H_{16}

ACGIH TLV: 400 ppm (1600 mg/m³)
OSHA Standard: 500 ppm (2000 mg/m³)

Method	Sampling Duration	Sampling Location	Useful Range (ppm)	System Cost ($)	Test Cost ($)	Manufacturer
DT	Grab	Portable	150-12,000	180	2.25	Bendix
IR	Cont	Portable	10-1000	4374	0	Foxboro
PI	Cont	Fixed	20-2000	4950	0	HNU
GC	1 min	Portable	$1-10^6$	10,000	0	Microsensor

NIOSH METHOD NO. P&CAM 127

(Use as described for acetone, with 80° C. column temperature)

HEXACHLOROETHANE

Synonyms: perchloroethane; carbon hexachloride $\qquad$ CCl_3CCl_3
ACGIH TLV: 10 ppm (100 mg/m³)
OSHA Standard: 1 ppm (10 mg/m³)

Method	Sampling Duration	Sampling Location	Useful Range (ppm)	System Cost ($)	Test Cost ($)	Manufacturer
IR	Cont	Portable	0.06-2	4374	0	Foxboro

NIOSH METHOD NO. P&CAM 127

(Use as described for acetone, with 110° C. column temperature)

n-HEXANE

ACGIH TLV: 50 ppm (180 mg/m³) $\qquad$ C_6H_{14}
OSHA Standard: 500 ppm (1800 mg/m³)

Method	Sampling Duration	Sampling Location	Useful Range (ppm)	System Cost ($)	Test Cost ($)	Manufacturer
DT	Grab	Portable	100-3000	150	2.80	Nat'l Draeger
DT	Grab	Portable	50-6000	165	1.70	Matheson
DT	Grab	Portable	10-12,000	180	2.25	Bendix
IR	Cont	Portable	10-1000	4374	0	Foxboro
PI	Cont	Fixed	0.2-20	4950	0	HNU
GC	1 min	Portable	$1-10^6$	10,000	0	Microsensor

NIOSH METHOD NO. P&CAM 127

(Use as described for acetone, with 52° C. column temperature)

2-HEXANONE

Synonym: methyl n-butyl ketone $CH_3COC_4H_9$
ACGIH TLV: 5 ppm (20 mg/m^3)
OSHA Standard: 100 ppm (410 mg/m^3)

Method	Sampling Duration	Sampling Location	Useful Range (ppm)	System Cost ($)	Test Cost ($)	Manufacturer
IR	Cont	Portable	2-200	4374	0	Foxboro

NIOSH METHOD NO. P&CAM 127

(Use as described for acetone, with 80° C. column temperature)

HEXONE

Synonyms: 4-methyl-2-pentanone; methyl isobutyl ketone; MIBK $CH_3COCH_2CH(CH_3)_2$
ACGIH TLV: 50 ppm (240 mg/m^3)
OSHA Standard: 100 ppm (490 mg/m^3)

Method	Sampling Duration	Sampling Location	Useful Range (ppm)	System Cost ($)	Test Cost ($)	Manufacturer
DT	Grab	Portable	500-10,000	165	1.70	Matheson
DT	Grab	Portable	100-6000	180	2.25	Bendix
IR	Cont	Portable	2-200	4374	0	Foxboro
PI	Cont	Fixed	0.2-20	4950	0	HNU

NIOSH METHOD NO. P&CAM 127

(Use as described for acetone, with 65° C. column temperature)

HYDRAZINE

ACGIH TLV: 0.1 ppm (0.13 mg/m³) - Skin, suspected carcinogen NH_2NH_2
OSHA Standard: 1 ppm (1.3 mg/m³)
NIOSH Recommendation: 0.03 ppm (0.04 mg/m³) ceiling (2 hr)

Method	Sampling Duration	Sampling Location	Useful Range (ppm)	System Cost ($)	Test Cost ($)	Manufacturer
DT	Grab	Portable	0.25-3	150	2.70	Nat'l Draeger
DT	Grab	Portable	0.05-2	180	2.25	Bendix
DT	4 hr	Personal	0.05-3	850	3.10	Nat'l Draeger
CS	Grab	Portable	0.02-1	2465	0.80	MDA
IR	Cont	Portable	0.05-5	4374	0	Foxboro
CS	Cont	Fixed	0.05-1	4950	Variable	MDA
CM	Cont	Port/Fixed	0.005-20	4995	Variable	CEA

NIOSH METHOD NO. S237

Principle

Hydrazine is collected in a midget bubbler containing 0.1 M hydrochloric acid. The resulting solution is made alkaline with 1.2 M sodium hydroxide, and the solution is then reacted with p-dimethylaminobenzaldehyde to form an orange colored acid. The solution is allowed to stand for a half hour before addition of glacial acetic acid. The strong absorption maximum at 480 nm is used as a quantitative measure of hydrazine.

Range and Sensitivity

This method was validated over the range of 0.589-3.44 mg/m³, at an atmospheric temperature and pressure of 22° C. and 763 mm Hg. The probable range of the method is 0.02-4.0 mg/m³ based on the range of standards used to prepare the standard curve. For samples of high concentration where the absorbance is greater than the limits of the standard curve, the samples can be diluted with glacial acetic acid prior to color development to extent the upper limit of the range. A concentration of 0.013 mg/m³ of hydrazine in a 100 L air sample gives a 0.015 absorbance using a 1 cm cell.

Interferences

Hydrazine derivatives may interfere with the method.

Precision and Accuracy

The coefficient of variation for the total analytical and sampling method in the range of 0.589-3.44 mg/m³ was 0.094. This value corresponds to a 0.12 mg/m³ standard deviation at the OSHA standard level. Statistical information and details of the validation and experimental test procedures can be found in reference 2.

A collection efficiency of 0.997, standard deviation of 0.002 (essentially 1.00), was determined for the collection medium. Thus, no correction for collection efficiency is necessary, and it is assumed that no bias is introduced in the sample collection step. There is also no apparent bias in the sampling and analytical method. Thus, coefficient of variation is a satisfactory measure of both accuracy and precision of the sampling an analytical method.

Advantages and Disadvantages

The samples, collected in bubblers, are analyzed by means of a quick, instrumental method. A disadvantage of the method is the awkwardness in using midget bubblers for collecting samples. If the worker's job performance requires much body movement, loss of the collection solution during sampling may occur. Glass bubblers are more difficult to ship than absorption tubes or filters due to possible breakage and leakage of the bubblers during shipping.

Apparatus

Glass midget bubbler. With a stem which has a fritted glass end and contains 10 mL of 0.1 M hydrochloric acid. The fritted end should have porosity approximately equal to that of Corning EC (170-220 micron maximum pore diameter).

Sampling pump. Suitable for sampling at least 1.0 L/min. The pump is protected from splashover or water condensation by a 5 cm (6 mm I.D. and 8 mm O.D.) glass tube loosely packed with a plug of glass wool and inserted between the exit arm of the bubble and the pump.

Integrating volume meter. Such as dry gas or wet test meter.

Thermometer.

Manometer.

Stopwatch.

Spectrophotometer. This instrument must be capable of measuring the absorbance of the developed color at 480 nm.

Matched glass cells or cuvettes, 1 cm path length.

Assorted laboratory glassware. Pipets, volumetric flasks, large test tubes, and graduated cylinders of appropriate capacities.

Reagents

0.1 M Hydrochloric acid. Fill a 1000 mL volumetric flask with approximately 300 mL distilled water, add 8.6 mL concentrated hydrochloric acid, mix, and bring volume to the 1000 mL mark.

Hydrazine, reagent grade.

Methanol, reagent grade.

p-Dimethylaminobenzaldehyde. Prepare a 2.5% solution of p-dimethylaminobenzaldehyde in methanol.

Glacial acetic acid.

1.2 M Sodium hydroxide.

Hydrazine standard solution. Pipet 3 mL hydrazine into a 100 mL volumetric flask. Bring to volume with methanol.

Procedure

Cleaning of equipment. All glassware should be washed in concentrated nitric acid, rinsed with triple distilled water, and rinsed with methanol. A final rinse with triple distilled water is made.

Calibration of personal pump. Personal sampling pump should be calibrated using integrating volume meter or other means.

Collection and shipping of samples. Pipet 10 mL of the 0.1 M hydrochloric acid into each bubbler, and mark the liquid level. Connect a bubbler with a 5 cm (6 mm I.D. and 8 mm O.D.) glass tube containing the glass wool plug, then to the personal sampling pump using short pieces of flexible tubing. The air being sampled should not pass through any tubing or other equipment before entering the bubbler.

Turn the pump on to begin sample collection. Care should be taken to measure the flow rate, time and/or volume as accurately as possible. Record atmospheric pressure and temperature. If pressure reading is not available, record the elevation. The sample should be taken at a flow rate of 1.0 L/min or less. A sample size of 100 L is recommended.

After sampling, the bubbler stem may be removed and cleaned. If necessary, bring the volume of each sample to the 10 mL mark with 0.1 M hydrochloric acid. Wash the stem with approximately 1 mL of 0.1 M hydrochloric acid, adding the wash to the bubbler. The bubblers are sealed with a hard, non-reactive stopper (preferably Teflon or glass). Do not seal with rubber. The stoppers on the bubblers should be tightly sealed to prevent leakage during shipping.

Care should be taken to minimize spillage or loss by evaporation. Whenever possible, hand delivery of the samples is recommended. Otherwise, special bubbler shipping cases designed by NIOSH should be used to ship the samples.

With each batch of ten samples, submit one bubbler from the same lot of bubbler which was used for sample collection and which is subjected to exactly the same handling as the samples except that no air is drawn through it. Label this as a blank. A sample of the bulk material should be submitted to the laboratory in a glass container with a Teflon-lined cap. This sample should not be transported in the same container as the air samples.

Analysis of samples. The sample in each bubbler is analyzed separately. The total volume of the solution should be 11 mL. If the final volume of the solution is greater than 11 mL, an appropriate calculation adjustment must be made. If the volume is less than 11 mL, make it up to 11 mL with 0.1 M hydrochloric acid. Add 1 mL 1.2 M sodium hydroxide to the solution in the bubbler.

Immediately add 10 mL of 2.5% p-dimethylaminobenzaldehyde. Mix the solution and allow to stand for 30 min. Transfer the resulting solution to a 25 mL volumetric flask. Rinse the bubbler with approximately 1 mL of glacial acetic acid, adding rinse to the volumetric flask. Bring the solution to the 25 mL mark with the glacial acetic acid.

Mix the resulting orange solution, and pipet out a 1 mL aliquot of the solution into another clean 25 mL volumetric flask. Bring to volume with glacial acetic acid. Adjust the baseline of the spectrophotometer to zero with distilled water in both cells. Read at 480 nm in the spectrophotometer against a blank prepared in the same fashion as the samples.

Calibration and Standards

Pipet 11 mL of 0.1 M hydrochloric acid into each of six 25 mL volumetric flasks. Carefully pipet 2, 4, 6, 8, and 10 μL of the standard solution into the flasks. Process one flask as a blank. Add 1 mL 1.2 M sodium hydroxide to the flask. Immediately add 10 mL of 2.5% p-dimethylaminobenzaldehyde and allow to stand for 30 min.

After 30 min, bring the volume in the flask to the 25 mL mark with glacial acetic acid. Mix the resulting orange solution, and pipet out a 1 mL aliquot of the solution into another clean 25 mL volumetric flask. Bring to volume with glacial acetic acid.

Adjust the baseline of the spectrophotometer to zero by reading distilled water in both cells. With the wavelength set at 480 nm, read the blank in the sample cell and then read the samples in the same cell or similar cells. Construct a calibration curve by plotting absorbance against μg of hydrazine in the solution.

Calculations

Subtract the absorbance of the blank from the absorbance of each sample. Determine from the calibration curve the μg of hydrazine present in each sample. The concentration of the analyte in the air sampled can be expressed in mg/m^3 (mg/m^3 = μg/L):

mg/m^3 = μg $\div$ air volume sampled in L

Another method of expressing concentration is ppm:

ppm = (mg/m³)(24.45)(760)(T + 273) ÷ (M.W.)(P)(298)

where:

P = pressure (mm Hg) of air sampled
T = temperature (°C.) of air sampled
24.45 = molar volume (L/mol) at 25° C. and 760 mm Hg
M.W. = molecular weight (g/mol) of analyte
760 = standard pressure (mm Hg)
298 = standard temperature (°K)

References

1. Dambrauska, T. and Cornish, H. A modified spectrophotometric method for the determination of hydrazine. Ind. Hyg. J. 23: 151, 1962.
2. Documentation of NIOSH Validation Tests. NIOSH Contract No. CDC-99-74-45.

HYDROGEN CHLORIDE

Synonyms: hydrochloric acid; muriatic acid HCl
ACGIH TLV: 5 ppm (7 mg/m³) ceiling
OSHA Standard: 5 ppm ceiling

Method	Sampling Duration	Sampling Location	Useful Range (ppm)	System Cost ($)	Test Cost ($)	Manufacturer
DT	Grab	Portable	0.5-5000	150	2.70	Nat'l Draeger
DT	Grab	Portable	0.4-20	165	4.60	Matheson
DT	Grab	Portable	0.2-1000	180	2.25	Bendix
DT	8 hr	Personal	2-50	850	3.10	Nat'l Draeger
IR	Cont	Portable	2-10	4374	0	Foxboro
CS	Cont	Fixed	1.5-20	4950	Variable	MDA
CM	Cont	Port/Fixed	0.02-100	4995	Variable	CEA

NIOSH METHOD NO. S246

Principle

A known volume of air is drawn through a midget bubbler containing 10 mL of sodium acetate solution. The resulting solution is diluted to 25 mL with distilled water. The diluted samples are analyzed using a chloride ion specific electrode.

Range and Sensitivity

This method was validated over the range of 3.5-14 mg/m³ at an atmospheric temperature and pressure of 22° C. and 764 mm Hg, using a 15 L sample. The probable useful range of this method is 1-20 mg/m³ for 15 L samples. The upper limit of the range of the method is dependent on the capacity of the midget bubbler. If higher concentrations than those tested are to be sampled, smaller sample volumes should be used. The collection efficiency for hydrogen chloride was determined to be 0.981, standard deviation of 0.005 (essentially 1.00), when sampled for 15 min at 0.94 L/min from a test atmosphere containing 70 mg/m³. Therefore, no correction for collection efficiency is necessary.

Interferences

This method is not specific for hydrogen chloride since any chloride ion which is trapped in the bubbler will be measured and give a positive interference. Sulfide ion must be absent because it poisons the chloride ion electrode. Touch a drop of the sample to a piece of lead acetate paper to check for sulfide ion. If sulfide is present, it is removed by addition of a small amount of powdered cadmium carbonate to the sample. Swirl to disperse the solid and recheck a drop of the sample with lead acetate paper. Avoid a large excess of cadmium carbonate and long contact time with the solution. Filter the sample through a small plug of glass wool and proceed with the analysis.

Other common interfering ions are bromide ion, iodide, and cyanide ion. For interference free operation, the chloride ion level must be at least 300 times the bromide ion level, 2×10^6 the iodide ion level, and 5×10^6 the cyanide ion level. Sufficiently high concentrations of species which form extremely stable complexes with silver ion (such as ammonia and thiosulfate) will also interfere and will result in reading of higher chloride ion activity than actually exists. For less than 1% error, the maximum ratio of ammonia to chloride concentration should be 0.12 and the maximum ratio of thiosulfate ion to chloride ion concentration should be 0.01.

Precision and Accuracy

The coefficient of variation for the total analytical and sampling method in the range of 3.5-14 mg/m³ was 0.064. This value corresponds to a standard deviation of 0.45 mg/m³ at the OSHA standard level. Statistical information and details of the validation and experimental test procedures can be found in reference 2.

A collection efficiency of 0.981 was determined for the collecting medium. On the average the concentrations obtained at the OSHA standard level using the overall sampling and analytical method were 2.7% higher than the "true" concentrations for a limited number of laboratory experiments. Any difference between the "found" and "true" concentrations may not represent a bias in the sampling and analytical method, but rather a random variation from the experimentally determined "true" concentration. Therefore, no recovery correction should be applied to the final result.

Advantages and Disadvantages

Collected samples are analyzed by means of a quick, instrumental method. A disadvantage of the method is the awkwardness in using midget bubblers for collecting personal samples. If the worker's job performance requires much body movement, loss of the collection solution during sampling may occur which would invalidate the sample. If more than 5% of the sample volume is lost, the sample should be discarded.

Apparatus

Glass midget bubbler. Containing the collection medium.

Sampling pump. Suitable for delivering at least 1.0 L/min for 15 min. The pump is protected from splashover or solvent condensation by a 5 cm long by 6 mm I.D. glass tube loosely packed with a plug of glass wool and inserted between the exit arm of the bubbler and the pump.

Integrating volume meter. Such as a dry gas or wet test meter.
Thermometer.
Manometer.
Stopwatch.
Pipets. 1, 2, 3, 5, and 10 mL.
Chloride specific ion electrode. Orion Model 94-17A or equivalent.
Reference electrode. Orion 90-02 double junction, or equivalent.
Expanded scale millivolt-pH meter. Capable of measuring to within 0.5 millivolt.
Polyethylene beakers, 50 mL capacity. Premark the beakers by pipetting 25 mL of distilled water into each beaker and mark the liquid level. Pregraduated polyethylene beakers may be used. However, they should be checked as described above. Discard the water and dry the beakers.
Magnetic stirrer and stirring bars for 50 mL beakers.
Polyethylene containers. These containers should be used to store diluted sodium chloride standards and also for shipping of air samples.
100 mL volumetric flasks.

Reagents

All chemicals must be ACS reagent grade or equivalent.

Doubly distilled water.

Collection medium, 0.5 M sodium acetate solution. Dissolve 82 g of sodium acetate in doubly distilled water and dilute to 2 L.

Sodium chloride, for preparation of standards.

Standard chloride solution. Dissolve 0.584 g of sodium chloride in double distilled water and dilute to 1 L for 0.01 M chloride or 354 μg chloride/mL. Adjust the pH to 5 with glacial acetic acid. This solution is stable for about 2 months. The following more dilute standards should be prepared fresh weekly and kept in polyethylene containers.

Dilute 10 mL 0.01 M chloride to 100 mL with 0.5 M sodium acetate for 0.001 M chloride or 35.4 μg Cl/mL.

Dilute 5 mL 0.01 M chloride to 100 mL with 0.5 M sodium acetate for 0.0005 M chloride or 17.7 μg chloride/mL.

Dilute 3 mL 0.01 M chloride to 100 mL with 0.5 M sodium acetate for 0.0003 M chloride or 10.6 μg chloride/mL.

Dilute 2 mL 0.01 M chloride to 100 mL with 0.5 M sodium acetate for 0.0002 M chloride or 7.1 μg chloride/mL.

Dilute 1 mL 0.01 M chloride to 100 mL with 0.5 M sodium acetate for 0.0001 M chloride or 3.5 μg chloride/mL.

Procedure

Cleaning of equipment. All glassware and plastic ware are washed in detergent solution, rinsed in tap water, and then rinsed with doubly distilled water.

Calibration of personal sampling pumps. Each pump should be calibrated by using an integrating volume meter or other means.

Collection and shipping of samples. Pour 10 mL of the collection medium into the midget bubbler, using a graduated cylinder to measure the volume. Connect the bubbler (via the splashover tube) to the vacuum pump with a short piece of flexible tubing. The air being sampled should not pass through any other tubing or other equipment before entering the bubbler.

Turn the pump on to begin sample collection. Care should be taken to measure the flow rate, time and/or the volume as accurately as possible. Record the atmospheric pressure and the temperature. If the pressure reading is not available, record the elevation. The sample should be taken at a flow rate of 1.0 L/min for 15 min. The flow rate should be known with an accuracy of 5%. The pump rotameter should be observed frequently, and the sampling should be terminated at any evidence of a problem.

Terminate sampling at the predetermined time and note sample flow rate and collection time. After sampling remove the bubbler stem and transfer the contents of the bubbler to a polyethylene container. Rinse the bubbler and bubbler stem with 3-5 mL of collection medium, adding the rinse to the polyethylene container. Seal the polyethylene container with the associated caps just prior to shipment. Care should be taken to minimize spillage or loss by evaporation at all times. Refrigerate samples if analysis cannot be done within a day.

Whenever possible, hand delivery of the samples is recommended. Otherwise, special shipping cases designed by NIOSH should be used to ship the samples. A blank bubbler should be handled in the same manner as the bubblers containing samples (fill, seal, and transport) except that no air is sampled through this bubbler.

Analysis of samples. The sample in each polyethylene container is analyzed separately. Quantitatively transfer the contents of each polyethylene container to a 50 mL polyethylene beaker which has been pre-marked at 25 mL. Rinse the polyethylene container with 2-3 mL of distilled water and add rinse to the beaker. Adjust the pH to 5 with acetic acid and check the pH with pH paper.

Dilute each sample to 25 mL with distilled water and stir the samples with a magnetic stirrer. Lower the chloride ion specific electrode and reference electrode into the stirred solution and record the resulting mv reading (to the nearest 0.5 mv) after it has stabilized (drift less than 0.5 mv/min).

Calibration and Standards

Prepare a series of chloride standard solutions in the pre-marked 50 mL beakers by diluting 10 mL of each of the chloride standards prepared earlier to a volume ot 25 mL with double distilled water, starting with the most dilute standard. Place the chloride ion electrode and the double reference electrode in the stirred solution. Record the resultant mv reading to the nearest 0.5 mv. Plot the mv readings vs. the chloride ion concentrations of the standards on semi-log paper. The chloride ion concentration in $\mu g/25$ mL is plotted on the logarithmic axis.

Calculations

Read the weight in μg corresponding to each mv reading from the standard curve. No volume corrections are needed, because the standard curve is based on $\mu g/25$ mL volume, and the volume of the samples is identical to the volume of the standards. Corrections for the blank must be made for each sample:

$\mu g = \mu g$ sample $- \mu g$ blank

where:

μg sample $= \mu g$ found in sample container
μg blank $= \mu g$ found in blank container

Calculate the μg of hydrogen chloride by multiplying the μg chloride ion found by 1.028, which is the conversion factor to convert μg chloride ion to μg hydrogen chloride. The concentration of the analyte in air sampled can be expressed in mg/m^3 ($mg/m^3 = \mu g/L$):

$mg/m^3 = \mu g \div$ air volume sampled in L

Another method of expressing concentration is ppm:

ppm = (mg/m³)(24.45)(760)(T + 273) ÷ (M.W.)(P)(298)

where:

P = pressure (mm Hg) of air sampled
T = temperature (°C.) of air sampled
24.45 = molar volume (L/mol) at 25° C. and 760 mm Hg
M.W. = molecular weight (g/mol) of analyte
760 = standard pressure (mm Hg)
298 = standard temperature (°K)

References
1. Analytical method for chloride in air. Health Lab. Sci. 12: 253-258, 1975.
2. Documentation of NIOSH Validation Tests. NIOSH Contract No. CDC-99-74-45.

HYDROGEN CYANIDE

Synonyms: hydrocyanic acid; prussic acid   HCN
ACGIH TLV: 10 ppm (10 mg/m³) ceiling - Skin
OSHA Standard: 10 ppm

Method	Sampling Duration	Sampling Location	Useful Range (ppm)	System Cost ($)	Test Cost ($)	Manufacturer
DT	Grab	Portable	2-150	150	2.80	Nat'l Draeger
DT	Grab	Portable	0.5-30,000	165	1.70	Matheson
DT	Grab	Portable	2.5-20,000	180	2.25	Bendix
DT	8 hr	Personal	1-120	850	3.10	Nat'l Draeger
CS	Grab	Portable	0.1-20	2465	0.80	MDA
IR	Cont	Portable	0.8-20	4374	0	Foxboro
CM	Cont	Port/Fixed	0.002-1	4995	Variable	CEA

NIOSH METHOD NO. S250

(Use as described for cyanide; analyze impinger contents but discard filter)

HYDROGEN FLUORIDE

Synonym: hydrofluoric acid HF
ACGIH TLV: 3 ppm (2.5 mg/m^3)
OSHA Standard: 3 ppm
NIOSH Recommendation: 5 mg/m^3 ceiling (15 min)

Method	Sampling Duration	Sampling Location	Useful Range (ppm)	System Cost ($)	Test Cost ($)	Manufacturer
DT	Grab	Portable	1.5-15	150	2.80	Nat'l Draeger
DT	Grab	Portable	1.5-20	180	2.25	Bendix
DT	8 hr	Personal	0.25-30	850	3.10	Nat'l Draeger
CM	Cont	Port/Fixed	0.01-10	4995	Variable	CEA

NIOSH METHOD NO. S176

Principle

A known volume of air is drawn through a midget bubbler containing 10 mL of 0.1 N sodium hydroxide to trap hydrogen fluoride. The resulting solution is made up to 25 mL with 0.1 N sodium hydroxide and transferred to a 50 mL polyethylene beaker. The sample is diluted with 25 mL total ionic strength activity buffer (TISAB). The diluted samples are analyzed using a fluoride ion specific electrode.

Range and Sensitivity

This method was validated over the range of 1.33-4.50 mg/m^3 at an atmospheric temperature and pressure of 22° C. and 761 mm Hg, using a 45 L sample. The probable useful range of this method is 0.245-7.35 mg/m^3 for 45 L samples. The upper limit of the range of the method is dependent on the collection efficiency of the midget bubbler. If higher concentrations than those tested are to be sampled, smaller sample volumes should be used. The collection efficiency for hydrogen fluoride was determined to be 0.991 ± 0.004 when sampled for 30 min at 1.5 L/min from a test atmosphere containing 4.50 mg/m^3.

Interferences

When interfering compounds are known or suspected to be present in the air, such information, including their suspected identities, should be transmitted with the sample. Hydroxide ion is the only significant electrode interference; however, addition of the TISAB eliminates this problem. Very large amounts of complexing metals such as aluminum may result in low readings even in the presence of TISAB.

Precision and Accuracy

The coefficient of variation for the total analytical and sampling method in the range of 1.33-4.50 mg/m^3 was 0.057. This value corresponds to a standard deviation of 0.14 mg/m^3 at the OSHA standard level. Statistical information and details of the validation and experimental test procedures can be found in reference 2.

A collection efficiency of 0.991 ± 0.004 was determined for the collecting medium. On the average the concentrations obtained at the OSHA standard level using the overall sampling and analytical method were 5.5% higher than the "true" concentrations for a limited number of laboratory experiments. Any difference between the "found" and "true" concentrations may not represent a bias in the sampling and analytical method, but rather a random variation from the experimentally determined "true"concentration. Therefore, no recovery correction should be applied to the final result.

Advantages and Disadvantages

The samples collected in bubblers are analyzed by means of a quick, instrumental method. There are no known disadvantages to the method.

Apparatus

Glass midget bubbler. Containing the collection medium. Bubbler stem unit consists of a two-hole rubber stopper to fit 1/8 in O.D. Teflon tubing.

Prefilter unit. Consists of the filter media and a polystyrene 37 mm, 2 piece cassette filter holder. The filter is a 37 mm mixed cellulose ester membrane filter with a 0.8 micron pore size. A cellulose backup pad should not be used, but care must be taken to insure that the filter is sealed tightly to avoid air leaks during sampling.

Sampling pump. Suitable for delivering at least 1.5 L/min for 30 min. The pump is protected from splashover or solvent condensation by a 5 cm long by 6 mm I.D. glass tube loosely packed with a plug of glass wool and inserted between the exit arm of the bubbler and the pump.

Integrating volume meter. Such as a dry gas or wet test meter.

Thermometer.

Manometer.

Stopwatch.

Volumetric flasks. 25 mL or convenient sizes.

Pipets. 1, 2, 4, and 5 mL.

Fluoride specific ion electrode. Orion model 94-09 or equivalent.

Reference electrode. Orion 90-01 single junction, or equivalent calomel or silver/silver chloride electrode.

Expanded scale mv-pH meter. Capable of measuring to within 0.5 mv.

Polyethylene beakers. 50 mL capacity.

Magnetic stirrer and stirring bars for 50 mL beakers.

Reagents

All chemicals must be ACS reagent grade or equivalent. Polyethylene beakers and bottles should be used for holding and storing all fluoride containing solution.

Doubly distilled water.

Glacial acetic acid.

Collection medium, 0.1 N NaOH. Dissolve 4 g of sodium hydroxide in 1 L distilled water.

Sodium hydroxide, 5 N solution. Dissolve 20 g of sodium hydroxide in sufficient distilled water to give 100 mL of solution.

Sodium chloride.

Sodium citrate.

Total ionic strength activity buffer (TISAB). Place 500 mL of doubly distilled water in a 1 L beaker. Add 57 mL of glacial acetic acid, 58 g of sodium chloride, and 0.30 g of sodium citrate. Stir to dissolve. Place beaker in water bath (for cooling) and slowly add 5 N sodium hydroxide until the pH is between 5.0 and 5.5. Cool to room temperature and pour into a 1 L volumetric flask and add doubly distilled water to the mark.

Sodium fluoride, for preparation of standards.

Standard fluoride solution. Dissolve 12.06 mg sodium fluoride in 0.1 N sodium hydroxide (prepared from doubly distilled water) and dilute to 100 mL with 0.1 N sodium hydroxide. This solution is equivalent to 0.0546 mg/mL fluoride. One mL of this solution contains the amount collected at the 0.5X OSHA level, when sampling at 1.5 L/min for 30 min. The 0.5X level standard is made by spiking 10 mL of 0.1 N sodium hydroxide with 1 mL of the standard stock solution. Likewise, the 1X level standard is prepared with 2 mL, and the 2X level is prepared with 4 mL of the standard stock solution. The standards are diluted to 25 mL with 0.1 N sodium hydroxide and 25 mL TISAB is added prior to analysis with the fluoride ion specific electrode.

Procedure

Cleaning of equipment. All glassware and plastic ware are washed in detergent solution, rinsed in tap water, and then rinsed with doubly distilled water.

Calibration of personal sampling pumps. Each pump should be calibrated by using an integrating volume meter or other means.

Collection and shipping of samples. Pour 10 mL of the collection medium into the midget bubbler, using a graduated cylinder to measure the volume. Assemble the prefilter unit in the 2 piece filter cassette holder and close firmly to insure that the center ring seals the edge of the filter. A backup pad should not be used.

Connect the bubbler (via the adsorption tube) to the prefilter assembly and the vacuum pump with a short piece of flexible tubing. The minimum amount of tubing necessary to make the joint between the prefilter and bubbler should be used. The air being sampled should not pass through any other tubing or other equipment before entering the bubbler. Bubbler stem unit consists of a two-hole rubber stopper to fit 1/8 in O.D. Teflon tubing. One piece of tubing conducts the analyte vapor below the level of the collection medium where the analyte is trapped. The short outlet tube is connected to the sampling pump.

Turn the pump on to begin sample collection. Care should be taken to measure the flow rate, time and/or the volume as accurately as possible. Record the atmospheric pressure and the temperature. If the pressure reading is not available, record the elevation. The sample should be taken at a flow rate of 1.5 L/min for 30 min. The flow rate should be known with an accuracy of 5%. The pump rotameter should be observed frequently, and sampling should be terminated at any evidence of a problem.

Terminate sampling at the predetermined time and note sample flow rate and collection time. After sampling, the bubbler stem may be removed and cleaned. Tap the stem gently against the inside wall of the bubbler bottle to recover as much of the sampling solution as possible. Wash the stem with 1-2 mL of the collection medium, adding wash to the bubbler. Transfer the contents of the bubbler to a 50 mL polyethylene bottle. Rinse the bubbler with 2-3 mL of the collection medium and seal the bottle tightly for shipment.

Care should be taken to minimize spillage or loss by evaporation at all times. Refrigerate samples if analysis cannot be done within a day. Whenever possible, hand delivery of the samples is recommended. Otherwise, special bubbler shipping cases designed by NIOSH should be used to ship the samples. A blank bubbler should be handled in the same manner as the bubblers containing samples (fill, seal, and transport) except that no air is sampled through this bubbler. The filter may be discarded since the method is designed to measure gaseous hydrogen fluoride only.

Analysis of samples. The sample in each bubler is analyzed separately. Quantitatively transfer the contents of the polyethylene bottle to a 25 mL volumetric flask. Make up to volume with 0.1 N sodium hydroxide. Transfer the sample to a 50 mL polyethylene beaker. Add 25 mL of TISAB and stir with a magnetic stirrer. Lower the fluoride ion specific electrode and reference electrode into the stirred solution and record the resulting mv reading (to the nearest 0.5 mv) after it has stabilized (drift less than 0.5 mv/min).

Calibration and Standards

Prepare three fluoride standard solutions as described earlier at each of three levels (0.5X, 1X, and 2X the standard). Insert the fluoride ion specific electrode and the reference electrode into one of the standards at the 0.5X level, and stir the solution with the magnetic stirrer. Record the resulting mv reading to the nearest 0.5 mv. Repeat the previous two steps for the remaining standard solutions.

Calibration standards at each level should be repeated twice daily. Average the 6 mv readings at each level. Prepare the standard curve by plotting the three averaged mv readings vs. mg fluoride on semilog paper. Plot the mg fluoride on the log axis.

Calculations

Read the weight, in mg, corresponding to each mv reading from the standard curve. No volume corrections are needed, because the standard curve is based on mg/50 mL volume, and the volume of the samples is identical to the volume of the standards. Corrections for the blank must be made for each sample:

$$mg = mg\ sample - mg\ blank$$

where:

mg sample = mg found in sample bubbler
mg blank = mg found in blank bubbler

Calculate the mg of hydrogen fluoride by multiplying the mg fluoride ion found by 1.05, which is a conversion factor to convert mg fluoride to mg HF. The concentration of the analyte in the air sampled can be expressed in mg/m^3:

$$mg/m^3 = (mg)(1000\ L/m^3) \div air\ volume\ sampled\ in\ L$$

Another method of expressing concentration is ppm:

$$ppm = (mg/m^3)(24.45)(760)(T + 273) \div (M.W.)(P)(298)$$

where:

P = pressure (mm Hg) of air sampled
T = temperature (°C.) of air sampled
24.45 = molar volume (L/mol) at 25° C. and 760 mm Hg
M.W. = molecular weight (g/mol) of analyte
760 = standard pressure (mm Hg)
298 = standard temperature (°K)

References

1. Elfers, L.A. and C.E. Decker. Determination of fluoride in air and stack gas samples by use of an ion specific electrode. Anal. Chem. 40: 1658, 1968.
2. Documentation of NIOSH Validation Tests. NIOSH Contract No. CDC-99-74-45.

HYDROGEN SULFIDE

ACGIH TLV: 10 ppm (14 mg/m^3) $\qquad$ H$_2$S
OSHA Standard: 20 ppm (30 mg/m^3) ceiling
NIOSH Recommendation: 15 mg/m^3 ceiling (10 min)

Method	Sampling Duration	Sampling Location	Useful Range (ppm)	System Cost ($)	Test Cost ($)	Manufacturer
DT	Grab	Portable	0.5-70,000	150	2.60	Nat'l Draeger
DT	Grab	Portable	1-40,000	165	1.70	Matheson
DT	Grab	Portable	0.5-400,000	180	2.25	Bendix
ES	Cont	Personal	10-50	450	0	Bendix
CS	Cont	Personal	1-200	595	0	Dynamation
ES	Cont	Personal	1-1000	695	0	Energetics
DT	8 hr	Personal	0.5-60	850	3.10	Nat'l Draeger
CS	Cont	Fixed	1-100	995	0	Dynamation
ES	Cont	Portable	1-100	1350	0	Rexnord
ES	1 min	Personal	0.1-10	1595	0	Interscan
ES	Cont	Personal	3-50	1625	0	Bacharach
ES	Cont	Portable	0.01-1(100)	2064	0	Interscan
ES	Cont	Portable	2.5-250	2100	0	Energetics
ES	Cont	Fixed	0.1-10(100)	2339	0	Interscan
CS	Grab	Portable	0.5-20	2465	0.80	MDA
CS	Cont	Port/Fixed	1-50	2850	0.75	Houston Atlas
ES	Cont	Port/Fixed	0.1-10(100)	3078	0	Interscan
PI	Cont	Fixed	0.5-50	4950	0	HNU
CS	Cont	Fixed	1-20	4950	Variable	MDA
GC	1 min	Portable	1-10^6	10,000	0	Microsensor

NIOSH METHOD NO. S4

Principle

Hydrogen sulfide is collected by aspirating a measured volume of air through an alkaline suspension of cadmium hydroxide (reference 1). The sulfide is precipitated as cadmium sulfide to prevent air oxidation of the sulfide which occurs rapidly in an aqueous alkaline solution. STRaction 10 is added to the cadmium hydroxide slurry to minimize photo decomposition of the precipitated cadmium sulfide (reference 2). The collected sulfide is subsequently produced by the reaction of the sulfide with an acid solution of N,N-dimethyl-p-phenylenediamine and ferric chloride (references 3-5). Collection efficiency is variable below 10 μg/m^3 and is affected by the type of scrubber, the size of the gas bubbles, and the contact time with the absorbing solution and the concentration of hydrogen sulfide (references 6-8).

Range and Sensitivity

This method was validated over the range of 8.5-63 mg/m^3 at an atmospheric temperature and pressure of 25° C. and 760 mm Hg, using a 2 L sample. Under the conditions of sample size (2 L) the probable useful range of the method is 5-100 mg/m^3. For sample concentrations outside this range the sampling volume should be modified.

Interferences

The methylene blue reaction is highly specific for sulfide at the low concentrations usually encountered in ambient air. Strong reducing agents (e.g., sulfur dioxide) inhibit color development. Even sulfide solutions containing several μg sulfite per mL show this effect and must be diluted to eliminate color inhibition. If sulfur dioxide is absorbed to give a sulfite concentration in excess of 10 μg/mL color formation is retarded. The use of 0.5 mL of ferric chloride solution during analysis eliminates the sulfur dioxide interference up to 40 μg/mL.

Nitrogen dioxide gives a pale yellow color with the sulfide reagents at 0.5 μg/mL or more. No interference is encountered when 0.3 ppm nitrogen dioxide is aspirated through a midget impinger containing a slurry of cadmium hydroxide-cadmium sulfide-STRactan 10. If H$_2$S and nitrogen dioxide are simultaneously aspirated through cadmium hydroxide-STRactan 10 slurry, lower H$_2$S results are obtained, probably because of gas phase oxidation of the H$_2$S prior to precipitation of CdS (reference 8).

Ozone at 57 ppb reduced the recovery of sulfide previously precipitated as CdS by 15% (reference 8). Substitution of other cation precipitants for the cadmium in the absorbent (i.e., zinc, mercury, etc.) will shift or eliminate the absorbance maximum of the solution upon addition of the acid amine reagent. Cadmium sulfide decomposes significantly when exposed to light unless protected by the addition of 1% STRactan to the absorbing solution prior to sampling (reference 2).

The choice of impinger used to trap H$_2$S with the cadmium hydroxide slurry is very important when measuring concentration in the range 5-100 mg/m^3. Impingers or bubblers having fritted end gas delivery tubes are a problem source if the sulfide in solution is oxidized to free sulfur oxygen from the atmosphere. The sulfur collects on the fritted glass membrane and may significantly change the flow rate of the air sample through the system. One way of avoiding this problem is to use a midget impinger with standard glass tapered tips.

Precision and Accuracy

The coefficient of variation for the total analytical and sampling method in the range of 8.5-63 mg/m^3 was 0.121. This value corresponds to a 3.6 mg/m^3 standard-deviation at the OSHA standard level. Statistical information and details of the validation and experimental test procedures can be found in reference 9. On the average the values obtained using the overall sampling and analytical method were 10% higher than the "true" values at the OSHA standard level.

Advantages and Disadvantages

Hydrogen sulfide is readily volatilized from aqueous solution when the pH is below 7.0. Alkaline, aqueous sulfide solutions are very unstable, because sulfide ion is rapidly oxidized by exposure to air.

Cadmium sulfide is not appreciably oxidized even when aspirated with pure oxygen in the dark. However, exposure of an impinger containing cadmium sulfide to laboratory or to more intense light sources produces an immediate and variable photodecomposition. Losses of 50-90% of added sulfide have been routinely reported by a number of laboratories. Even though the addition of STRactan 10 to the absorbing solution controls the photodecomposition (reference 2), it is necessary to protect the impinger from light at all times. This is achieved by the use of low actinic glass impingers, paint on the exterior of the impingers, or an aluminum foil wrapping.

Apparatus

Graduated 25 mL midget impinger. With a standard glass tapered gas delivery tube containing the absorbing solution or reagent. The impinger should be wrapped in aluminum foil to protect the sample from exposure to light.

Calibrated personal sampling pump. Whose flow can be determined within 5% at the recommended flow rate. The sampling pump is protected from splashover or water condensation by an adsorption tube loosely packed with a plug of glass wool and inserted between the exit arm of the impinger and the pump.

Integrating volume meter. Such as dry gas or wet test meter or rotameter capable of measuring 2 L of air at 0.2 L/min with an accuracy of 5%. Instead of these, calibrated hypodermic needles may be used as critical orifices if the pump is capable of maintaining greater than 0.7 atmospheric pressure differential across the needle.

Thermometer.

Manometer.

Stopwatch.

Assorted laboratory glassware.

Colorimeter with red filter or spectrophotometer at 670 nm.

Matched cells, 1 cm path length.

Reagents

All reagents must be ACS analytical reagent quality. Distilled water should conform to the ASTM Standards for Referee Reagent Water. All reagents should be refrigerated when not in use.

Amine-sulfuric acid stock solution. Add 50 mL concentrated sulfuric acid to 30 mL water and cool. Dissolve 12 g of N,N-dimethyl-p-phenylene-diamine dihydrochloride (para-aminodimethylaniline, redistilled if necessary) (10.5 g N,N-dimethyl-p-phenylenediamine oxalate may be used) in acid. Do not dilute. The stock solution may be stored indefinitely under refrigeration.

Amine test solution. Dilute 25 mL of the stock solution to 1 L with 1:1 sulfuric acid.

Ferric chloride solution. Dissolve 100 g of ferric chloride, $FeCl_3.6H_2O$ in water and dilute to 100 mL.

Ethanol, 95%.

STRactan 10. Arabinogalactan, available from Chicago Scientific (Bensenville, IL 60106). Arabinogalactan sold under other brand names may be used.

Cadmium sulfate-STRactan solution. Dissolve 8.6 g of $3CdSO_4.8H_2O$ in approximately 600 mL of water. Add 20 g STRactan 10 and dilute to 1 L.

Sodium hydroxide solution. Dissolve 0.6 g sodium hydroxide in approximately 600 mL of water and dilute to 1 L.

Cadmium hydroxide STRactan absorbing solution. This absorbing solution is prepared by pipeting 5 mL of cadmium sulfate-STRactan solution and 5 mL of sodium hydroxide solution directly into the midget impinger and mixing. This solution is stable for 3-5 days.

Stock sodium sulfide standard. Place 35.28 g of $Na_2S.9H_2O$ into a 1 L volumetric flask and add enough oxygen free distilled water to bring the volume to 1 L. Store under nitrogen and refrigerate. Standardize with standard iodine and thiosulfate solution in an iodine flask under a nitrogen atmosphere to minimize air oxidation. The approximate concentration of the sulfide solution will be 4700 μg sulfide/mL of solution. The exact concentration must be determined by iodine thiosulfate standardization immediately prior to dilution.

Working sodium sulfide solution. Dilute 25 mL of stock solution with oxygen free water to 250 mL. This solution contains the sulfide equivalent to approximately 500 μg/mL of H_2S. Make fresh working sulfide solution daily. The actual concentration of this solution can be determined from the titration results on the stock sodium sulfide standard. For the most accurate results in the iodometric determination of sulfide in aqueous solution, the following general procedure is recommended:

1. Replace the oxygen from the flask by flushing with an inert gas such as carbon dioxide or nitrogen.
2. Add an excess of standard iodine, acidify, and back titrate with standard thiosulfate and starch indicator.

Procedure

Cleaning of equipment. All glassware should be thoroughly cleaned by the following procedure. Wash with a detergent and tap water solution followed by tap water and distilled water rinses. Soak in 1:1 or concentrated nitric acid for 30 min and then follow with tap, distilled, and double distilled water rinses.

Collection and shipping of samples. Prepare 10 mL of absorbing solution as described above directly in the midget impinger. The addition of 5 mL of 95% ethanol to the absorbing solution just prior to aspiration controls foaming for 2 hours (induced by the presence of STRactan 10). In addition, 1 or 2 Teflon demister discs may be slipped up over the impinger air inlet tube to a height approximately 1-2 inches from the top of the tube. Wrap the impinger with aluminum foil. Connect the impinger (via the absorption tube) to the sampling pump with a short piece of flexible tubing.

Air being sampled should not be passed through any other tubing or other equipment before entering the impinger. At the ceiling and peak concentrations, a sample size of 2 L is recommended. Sample for 10 min at a flow of 0.20 L/min. The flow rate should be known with an accuracy of at least 5%. Turn on the pump to begin sample collection. Care should be taken to measure the flow rate, time and/or volume as accurately as possible. The temperature and pressure of the atmosphere being sampled should be recorded. If the pressure reading is not available, record the elevation.

After sampling, the impinger stem must not be removed since it contains CdS deposits. It is necessary to ship the impingers with the stems in so the outlets of the stem should be sealed with Parafilm or other non rubber covers, and the ground glass joints should be sealed (i.e., taped) to secure the top tightly. Care should be taken to minimize spillage or loss by evaporation at all times. Refrigerate samples if analysis cannot be done within a day.

Whenever possible, hand delivery of the samples is recommended. Otherwise, special impinger shipping cases designed by NIOSH should be used to ship the samples. A blank impinger should be handled as the other samples (fill, seal, and transport) except that no air is sampled through this impinger.

Analysis. Remove the impinger top and drain it thoroughly into the impinger bottom. Set aside. Transfer the solution and deposit in the impinger bottom to a 250 mL volumetric flask. Using 50 mL of distilled water, rinse the bottom twice with the aid of a clean rubber policeman on a glass stirring rod. Add the rinse solutions to the volumetric flask. With the aid of the rubber policeman wash the outside of the impinger stem with 20 mL of distilled water and add the washings to the flask and drain 20 mL of distilled water through it into the flask. The total wash water volume should be 90 mL.

Add 15 mL of amine test solution through the impinger inlet tube into the volumetric flask. This is necessary to dissolve the CdS deposited inside the inlet tube. Mix gently to avoid loss of H_2S. Add 0.5 mL of ferric chloride solution and mix. Bring to volume with distilled water. Allow to stand 20 min. Prepare a zero reference solution in the same manner using a 10 mL volume absorbing solution, through which no air has been aspirated. Measure the absorbance of the color at 670 nm in a spectrophotometer or colorimeter set at 100% transmission against the zero reference.

Calibration and Standards

Aqueous sulfide. Place 5 mL of each of the absorbing solutions into each of a series of 250 mL volumetric flasks. Add standard sulfide solution equivalent to 0, 20, 40, 80, 120, 160 µg of hydrogen sulfide to the different flasks. Add 90 mL of distilled water. Add 15 mL of amine-acid test solution to each flask and mix gently. Add 0.5 mL of ferric chloride solution to each flask. Mix, make up to volume, and allow to stand for 20 min. Determine the absorbance in a spectrophotometer at 670 nm against the sulfide-free reference solution. Prepare a standard curve of absorbance verus µg H_2S.

Gaseous sulfide. Cylinders of hydrogen sulfide in dry nitrogen in the range desired are available commercially, and may be used to prepare calibration curves for use at the 10-60 mg/m^3 levels. Nitrogen containing hydrogen sulfide in the 450-600 mg/m^3 range can be diluted to the desired concentrations. Analyses of these known concentrations give calibration curves which simulate all of the operational conditions performed during the sampling and chemical procedure. This calibration curve includes the important correction for collection efficiency at various concentrations of hydrogen sulfide.

Prepare or obtain a cylinder of nitrogen containing hydrogen sulfide in the range of 450-600 mg/m^3. To obtain standard concentrations of hydrogen sulfide, assemble the apparatus consisting of appropriate pressure regulators, needle valves and flow meters for the nitrogen and for a dry air diluent stream. All stainless steel, glass or rubber tubing must be used for the hydrogen sulfide mixture. Flow of hydrogen sulfide in nitrogen is controlled by a needle valve operated in conjunction with a previously calibrated flow meter in the range of 0.2-2.0 L/min. Diluent dry air from a cylinder is controlled by a similar needle valve flow meter combination in the range of 1-20 L/min.

The hydrogen sulfide in nitrogen and the diluent air are combined in a mixing chamber at atmospheric pressure, from which they flow through a baffle tube in which mixing takes place into a 1 L sampling flask which is provided with one or more nipples from which samples can be taken. Sampling is done by connecting a midget impinger to the nipple and drawing a known volume of the mixutre through the impinger for a measured length of time, using a critical orifice to control flow at a constant known rate.

Preparation of simulated calibration curves. The following description represents a typical procedure for air sampling of short duration. The system is designed to provide an accurate measure of hydrogen sulfide in the 10-60 mg/m^3 range. It can be easily modified to meet special needs.

The dynamic range of the colorimetric procedure fixes the total volume of the sample at 2 L; then, to obtain linearity between the absorbance of the solution and the concentration of hydrogen sulfide in mg/m^3, select a constant sampling time. This fixing of the sampling time is desirable also from a practical stand point. In this case, select a sampling time of 10 min. To obtain a 2 L sample of air requires a flow rate of 0.2 L/min. The concentration of standard H_2S in air is computed as follows:

$$C = cf \div (F+f)$$

where:

C = concentration of H_2S in mg/m^3
c = concentration of H_2S in nitrogen, before dilution
F = flow of diluent air, as measured by calibrated flow meter
f = flow of H_2S in nitrogen, as measured by calibrated flow meter

Commercially prepared hydrogen sulfide in nitrogen can be obtained with a known concentration, as analyzed by the laboratory preparing the gas. If it is desired to check this concentration, measured volume of the gas can be bubbled through the absorbing solutions, and the collected sulfide titrated against iodine thiosulfate. The volume of gas can be measured using a wet test meter. If hydrogen sulfide is present at much lower concentrations (1.5-140 $\mu g/m^3$), commercially available permeation tubes containing liquified hydrogen sulfide may be used to prepare calibration tubes (references 8, 11-14).

Calculations

Gaseous sulfide. Using the Beer's law standard curve of absorbance versus μg H$_2$S, determine μg H$_2$S in the sampling impinger corresponding to its absorbance reading at 670. The concentration of H$_2$S in the air sampled can be expressed in mg/m^3 which is numerically equal to $\mu g/L$:

$$mg/m^3 = \mu g/L = \mu g \ H_2S \div \text{air volume sampled in L}$$

Another method of expressing concentration is ppm:

$$ppm = (mg/m^3)(24.45)(760)(T + 273) \div (M.W.)(P)(298)$$

where:

P = pressure (mm Hg) of air sampled
T = temperature (°C.) of air sampled
24.45 = molar volume (L/mol) at 25° C. and 760 mm Hg
M.W. = molecular weight (g/mol) of analyte
760 = standard pressure (mm Hg)
298 = standard temperature (°K)

References

1. Jacobs, M.B., Braverman, M.M., and Hocheiser, S. Ultramicro determination of sulfides in air. Anal. Chem. 29: 1349, 1957.
2. Ramesberger, W.L., and Adams, D.F. Improvements in the collection of hydrogen sulfides in cadmium hydroxide suspension. Env. Sci. Tech. 3: 258, 1969.
3. Mecklenburg, W., and Rozenkranzer, R. Colorimetric determination of hydrogen sulfide. A. Anorg. Chem. 86: 143, 1914.
4. Almy, L.H. Estimation of hydrogen sulfide in proteinaceous food products. J. Amer. Chem. Soc. 47: 1381, 1925.
5. Sheppard, S.E., and Hydson, J.H. Determination of labile sulfide in gelatin and proteins. Ind. Eng. Chem., Anal. Ed., 2: 73, 1930.
6. Bostrom, C.E. The absorption of sulfur dioxide at low concentrations (ppnm) studied by an isotopic tracer method. Air Water Pollut. Int. J. 9: 333, 1965.
7. Bostrom, C.E. The adsorption of low concentrations (ppnm) of hydrogen sulfide in a cadmium hydroxide suspension as studied by an isotopic tracer method. Air Water Pollut. Int. J. 10: 435, 1966.
8. Thomas, B.L., and Adams, D.F. Unpublished information.
9. Documentation of NIOSH Validation Tests. Contract No. CDC-99-74-45.
10. Lodge, J.P., Pate, J.B., Ammons, B.E., Swanson, G.A. The use of hypodermic needles as critical orifices. J. Air. Pollut. Cont. Assoc. 16: 197, 1966.
11. O'Keefe, A.E., and Ortman, G.C. Primary standards for trace gas analysis. Anal. Chem. 38: 760, 1966.
12. O'Keefe, A.E., and Ortman, G.C. Precision picogram dispenser for volatile substances. Anal. Chem. 39: 1047, 1967.
13. Scaringelli, F.P., Frey, S.A., and Saltzman, B.E. Evaluation of Teflon permeation tubes for use with sulfur dioxide. Amer. Ind. Hyg. Assoc. J. 28: 260, 1967.

14. Scaringelli, F.P., Rosenberg, E., and Rehme, K. Stoichiometric Comparison Between Permeation Tubes and Nitrite Ion as Primary Standards for the Colorimetric Determination of Nitrogen Dioxide, presented before the Division of Water, Air and Waste Chemistry of the American Chemical Society, 157th National Meeting, Minneapolis, MN, April, 1969.
15. Kolthoff, I.M., and Elving, P.J., eds. *Treatise on Analytical Chemistry, Part II, Analytical Chemistry of the Elements*, V. 7, Interscience Publishers, New York, 1961.
16. Bock, R. and Puff, H.J. Bestimmung von Sulfid mit einer Sulfidionenempfindlichen Elecktrode. Z. Anal. Chem. 240: 381, 1968.

IRON OXIDE FUME

Synonyms: ferric oxide; rust Fe_2O_3
ACGIH TLV: 3 ppm (5 mg/m^3) as Fe
OSHA Standard: 10 mg/m^3

Method	Sampling Duration	Sampling Location	Useful Range (mg/m^3)	System Cost ($)	Test Cost ($)	Manufacturer
LS	Cont	Personal	0.01-100	1695	0	GCA
LS	Cont	Portable	0.001-100	2785	0	MDA
LS	Cont	Fixed	0.001-200	3990	0	GCA
BA	1,4 min	Portable	0.02-150	4890	0	GCA

NIOSH METHOD NO. S366

Principle
A known volume of air is drawn through a cellulose ester membrane filter to collect iron oxide fume. The samples are ashed using hydrochloric and nitric acids to destroy the filter and other materials in the sample, and the iron oxide is then solubilized in nitric acid. The solutions of samples and standards are aspirated into the oxidizing air acetylene flame of an atomic absorption spectrophotometer (AAS). A hollow cathode lamp for iron is used.

Range and Sensitivity
This method was validated over the range of 3.87-18.19 mg/m^3 using a 145 L sample at an atmospheric temperature of 23° C. and an atmospheric pressure of 764 mm Hg. Under the conditions of sample size (145 L), the working range of the method is estimated to be 1.3-35.7 mg/m^3.
The sensitivity of the sampling and analytical method for a 145 L air sample using the 250 mL final solution volume is 143 µg of iron oxide, corresponding to 0.6 µg/mL iron oxide. The method may be extended to higher values by dilution of the sample. Measurement of lower atmospheric concentrations can be made by using smaller final solution volumes, by longer sampling time, or by scale expansion to increase instrumental response.

Interferences
Iron and other iron compounds which are collected on the cellulose ester membrane filter will

cause a positive interference. The suspected identities of these compounds should be recorded and transmitted with the samples.

Precision and Accuracy

The coefficient of variation for the total analytical and sampling method in the range of 3.87-18.19 mg/m^3 is 0.067. This value corresponds to a 0.70 mg/m^3 standard deviation at the OSHA standard level. Statistical information can be found in reference 3. Details of the test procedures can be found in reference 4.

A collection efficiency of 1.00 was determined for the collecting medium, thus no bias was introduced in the sample collection step, and no correction for collection efficiency is necessary. There was also no apparent bias in the sampling and analytical method. Thus, coefficient of variation is a satisfactory measure of both accuracy and precision of the sampling and analytical method.

Advantages and Disadvantages

The sampling device is small, portable, and involves no liquids. Samples collected on filters are analyzed by means of a quick, instrumental method.

Apparatus

Filter unit. The filter unit consists of a 37 mm diameter, 0.8 um pore size mixed cellulose ester membrane filter, and an appropriate 37 mm three piece cassette filter holder.

Personal sampling pump. A calibrated personal sampling pump whose flow can be determined to an accuracy of 5% at the recommended flow rate.

Atomic absorption spectrophotometer. Equipped with an air acetylene burner head.

Iron hollow cathode lamp.

Oxidant, compressed air.

Fuel, acetylene.

Pressure reducing valve. A 2 gauge, 2 stage pressure reducing valve and appropriate hose connections are needed for each compressed gas tank.

Manometer.

Thermometer.

Glassware, borosilicate.

Phillips beakers, 100 mL.

Watchglass covers.

Pipets. Delivery or graduated, 1, 3, 5, 7, and 10 mL

Volumetric flasks. 250 mL.

Polyethylene bottles.

Five 100 mL polyethylene bottles. For working atomic absorption standards.

One 1000 mL polyethylene bottle. For stock atomic absorption standard.

Hot plate. Adjustable thermostatically controlled hot plate capable of reaching 400° C.

Reagents

All reagents used must be ACS reagent grade or better.

Water, distilled or deionized.

Nitric acid, concentrated.

Nitric acid, dilute. 5 mL concentrated nitric acid diluted to 100 mL with distilled or deionized water.

Hydrochloric acid, 1:1 solution in distilled water.

Iron metal for preparation of AA standards.

Iron oxide powder, Fe_2O_3, 325 mesh.

Procedure

Cleaning of equipment. Before use, all glassware should initially be soaked in a mild detergent solution to remove any residual grease or chemicals. After initial cleaning, glassware must be cleaned with hot concentrated nitric acid and then rinsed thoroughly with tap water and distilled water, in that order, and then dried. For glassware which has previously been subjected to the entire cleaning procedure, a nitric acid rinse will be adequate.

Calibration of personal sampling pumps. Each personal sampling pump must be calibrated with a representative filter cassette in the line. This will minimize errors associated with uncertainties in the sample volume collected.

Collection and shipping of samples. Assemble filter in the three piece filter cassette holder and close firmly to insure that the center ring seals the edge of the filter. The cellulose membrane filter is held in place by a cellulose backup pad. Secure the cassette holder together with tape or shrinkable band. Remove the cassette plugs and attach to the personal sampling pump tubing. Clip the cassette to the worker's lapel. The cassette plugs are replaced after sampling. Air being sampled should not pass through any hose or tubing before entering the filter cassette.

A sample size of 150 L is recommended. Sample at a flow rate of 1.5 L/min. The flow rate should be known with an accuracy of 5%. Since the filter may become plugged by oil mist or overloaded as evidenced by caking, the filter and pump's sampling rate should be checked periodically. When the filter becomes overloaded or when the pump's flow rate cannot be adjusted to 1.5 L/min, terminate sampling. Terminate sampling at the predetermined time and record sample flow rate, collection time, and ambient temperature and pressure. If pressure reading is not available record the elevation. Also record the type of sampling pump used.

With each batch of ten samples, submit one filter from the same lot of filters which was used for sample collection and which is subjected to exactly the same handling as the samples, except that no air is drawn through it. Label this as a blank. The cassettes in which the samples are collected should be shipped in a suitable container, designed to prevent damage in transit.

Analysis of samples. Transfer each sample to a clean 100 mL Phillips beaker. Treat the sample with 3 mL of 1:1 hydrochloric acid to solubilize any iron oxide which may be present. Cover each beaker with a watchglass and heat on a hot plate (140° C.) in a fume hood until most of the acid has evaporated. Repeat twice. Repeat this procedure again three times except use concentrated nitric acid instead of hydrochloric acid. After the digestion process is complete, cover each beaker with a watchglass and heat it on a high temperature hot plate (400° C.) in a fume hood until a white ash appears. Using distilled water, carefully rinse the material on the bottom of the watchglass into the beaker, rinse the sides of the beaker, and allow to evaporate to dryness.

Cool each beaker and dissolve residues in 12.5 mL concentrated nitric acid. Quantitatively transfer the clear solutions to a 250 mL volumetric flask. Rinse each beaker at least three times with 2-3 mL portions of distilled or deionized water and quantitatively transfer each rinsing to the solution in the volumetric flask. Dilute all samples to 250 mL with distilled or deionized water.

Aspirate the solutions into an oxidizing air acetylene flame and record the absorbance at 248.3 nm. The absorbance is proportional to the sample concentration and can be determined from the appropriate calibration curve. When very low concentrations are found in the sample, scale expansion can be used to increase instrument response or the sample can be dried and rediluted to some smaller volume before aspiration. In such a case, use no more acid solution than is necessary to effect a quantitative transfer. Appropriate filter blanks must be analyzed in accordance with the total procedure.

Determination of analytical method recovery. To eliminate any bias in the analytical method, it is necessary to determine the recovery of iron oxide. The analytical method recovery should be determined over the concentration range of interest.

Place one 0.8 micron cellulose ester membrane filter into each of 24 clean 100 mL beakers. Amounts of iron oxide powder (Fe_2O_3) (to the nearest 0.01 mg) equal to the amount collected in a 150 L sample at 2X, 1X and 0.5X the OSHA standard level are added to the beakers. Six at each level and six blanks. Three mg is the amount of Fe_2O_3 used at the 2X level, 1.5 mg at the 1X level; and 0.75 mg at the 0.5X level. The filters are then analyzed according to the method above. Analytical method recovery (A.M.R.) equals the weight in mg found divided by the weight in mg added to the filter, or:

A.M.R. = mg found ÷ mg added

Calibration and Standards

Prepare a 100 µg/mL iron stock standard by dissolving 0.100 g iron metal in 50 mL hot concentrated nitric acid and diluting to 1 L. Transfer to a 1 L polyethylene bottle. From the 100 ppm stock standard, prepare at least five working standards to cover the range from 250-2500 µg/250 mL. Prepare all standard solutions in dilute nitric acid and remake each day. Store the standard solutions in 100 mL polyethylene bottles. Aspirate each of the standard samples and record the absorptions. Prepare a calibration curve by plotting on linear graph paper the absorbance versus the concentration of each standard both before and after the analysis of a series of samples to insure that conditions have not changed.

Calculations

Read the weight, in µg, corresponding to the total absorbance from the standard curve. No volume corrections are needed, because the standard curve is based on µg/250 mL. Corrections for the blank must be made for each sample:

µg = µg sample − µg blank

where:

µg sample = µg found in sample filter
µg blank = µg found in blank filter

Corrections for analytical method recovery (A.M.R.) must be made:

Corrected µg/sample = µg ÷ A.M.R.

Calculate the µg of iron oxide by multiplying the µg iron found by 1.43, which is a conversion factor to convert µg iron to µg iron oxide. For personal sampling pumps with rotameters only, the following air volume correction should be made:

Corrected volume = $ft(P_1T_2/P_2T_1)^{1/2}$

where:

f = flow rate sampled
t = sampling time
P_1 = pressure during calibration of sampling pump (mm Hg)
P_2 = pressure of air sampled (mm Hg)

T_1 = temperature during calibration of sampling pump (°K)
T_2 = temperature of air sampled (°K)

The concentration of iron oxide in the air sample can be expressed in mg/m³ (µg/L = mg/m³):

mg/m³ = µg ÷ air volume sampled in L

References
1. Analytical Methods for Flame Spectrophotometry. Varian Associates, 1972.
2. Methods for Emission Spectrochemical Analysis. ASTM Committee E-2, Philadelphia, 1971.
3. Documentation of NIOSH Validation Tests. Contract No. CDC-99-74-45.
4. Backup Data Report for Iron Oxide Fume. Prepared under NIOSH Contract No. 210-76-0123.

ISOAMYL ACETATE

Synonym: 3-methyl-1-butanol acetate $\qquad$ $CH_3COOCH_2CH_2CH(CH_3)_2$
ACGIH TLV: 100 ppm (525 mg/m³)
OSHA Standard: 100 ppm

Method	Sampling Duration	Sampling Location	Useful Range (ppm)	System Cost ($)	Test Cost ($)	Manufacturer
IR	Cont	Portable	2-200	4374	0	Foxboro
GC	1 min	Portable	$1\text{-}10^6$	10,000	0	Microsensor

NIOSH METHOD NO. P&CAM 127

(Use as described for acetone, with 90° C. column temperature)

ISOAMYL ALCOHOL

Synonym: 3-methyl-1-butanol $(CH_3)_2CHCH_2CH_2OH$
ACGIH TLV: 100 ppm (360 mg/m^3)
OSHA Standard: 100 ppm

Method	Sampling Duration	Sampling Location	Useful Range (ppm)	System Cost ($)	Test Cost ($)	Manufacturer
IR	Cont	Portable	2-200	4374	0	Foxboro

NIOSH METHOD NO. S52

(Use as described for allyl alcohol, with 110° C. column temperature)

ISOBUTYL ACETATE

ACGIH TLV: 150 ppm (700 mg/m^3) $CH_3COOCH_2CH(CH_3)_2$
OSHA Standard: 150 ppm

Method	Sampling Duration	Sampling Location	Useful Range (ppm)	System Cost ($)	Test Cost ($)	Manufacturer
DT	Grab	Portable	100-14,000	165	1.70	Matheson
DT	Grab	Portable	50-6600	180	2.25	Bendix
IR	Cont	Portable	3-300	4374	0	Foxboro
PI	Cont	Fixed	2-200	4950	0	HNU

NIOSH METHOD NO. P&CAM 127

(Use as described for acetone, with 70° C. column temperature)

ISOBUTYL ALCOHOL

Synonym: 2-methyl-1-propanol
ACGIH TLV: 50 ppm (150 mg/m^3)
OSHA Standard: 100 ppm (305 mg/m^3)

$(CH_3)_2CHCH_2OH$

Method	Sampling Duration	Sampling Location	Useful Range (ppm)	System Cost ($)	Test Cost ($)	Manufacturer
DT	Grab	Portable	100-3000	180	2.25	Bendix
IR	Cont	Portable	2-200	4374	0	Foxboro
PI	Cont	Fixed	2-200	4950	0	HNU

NIOSH METHOD NO. S66

(Use as described for n-butyl alcohol)

ISOPHORONE

Synonyms: diisopropylideneacetone; phorone
ACGIH TLV: 5 ppm (25 mg/m^3) ceiling
OSHA Standard: 25 ppm (140 mg/m^3)

$CO(CH=C(CH_3)_2)_2$

Method	Sampling Duration	Sampling Location	Useful Range (ppm)	System Cost ($)	Test Cost ($)	Manufacturer
IR	Cont	Portable	0.8-50	4374	0	Foxboro

NIOSH METHOD NO. P&CAM 127

(Use as described for acetone, with 167° C. column temperature)

ISOPROPYL ACETATE

ACGIH TLV: 250 ppm (950 mg/m³)
OSHA Standard: 250 ppm

$CH_3COOCH(CH_3)_2$

Method	Sampling Duration	Sampling Location	Useful Range (ppm)	System Cost ($)	Test Cost ($)	Manufacturer
DT	Grab	Portable	1000-12,000	165	1.70	Matheson
DT	Grab	Portable	500-7500	180	2.25	Bendix
IR	Cont	Portable	5-500	4374	0	Foxboro
PI	Cont	Fixed	2-200	4950	0	HNU

NIOSH METHOD NO. P&CAM 127

(Use as described for acetone)

ISOPROPYL ALCOHOL

Synonyms: isopropanol; 2-propanol
ACGIH TLV: 400 ppm (980 mg/m³)
OSHA Standard: 400 ppm
NIOSH Recommendation: 800 ppm ceiling (15 min)

$(CH_3)_2CHOH$

Method	Sampling Duration	Sampling Location	Useful Range (ppm)	System Cost ($)	Test Cost ($)	Manufacturer
DT	Grab	Portable	1000-20,000	165	1.70	Matheson
DT	Grab	Portable	25-75,000	180	2.25	Bendix
IR	Cont	Portable	8-800	4374	0	Foxboro
PI	Cont	Fixed	2-200	4950	0	HNU
GC	1 min	Portable	$1-10^6$	10,000	0	Microsensor

NIOSH METHOD NO. S66

(Use as described for n-butyl alcohol, with a 70° C. column temperature. Modify the elution solvent to contain 1% 2-butanol rather than 2-propanol.)

ISOPROPYLAMINE

ACGIH TLV: 5 ppm (12 mg/m^3)
OSHA Standard: 5 ppm

$(CH_3)_2CHNH_2$

Method	Sampling Duration	Sampling Location	Useful Range (ppm)	System Cost ($)	Test Cost ($)	Manufacturer
DT	Grab	Portable	1-60	180	2.25	Bendix
CS	Grab	Portable	0.5-10	2465	0.80	MDA
IR	Cont	Portable	0.2-10	4374	0	Foxboro
PI	Cont	Fixed	0.2-20	4950	0	HNU

NIOSH METHOD NO. P&CAM 221

(See n-butylamine)

ISOPROPYL ETHER

Synonym: diisopropyl ether
ACGIH TLV: 250 ppm (1050 mg/m^3)
OSHA Standard: 500 ppm (2100 mg/m^3)

$(CH_3)_2CHOCH(CH_3)_2$

Method	Sampling Duration	Sampling Location	Useful Range (ppm)	System Cost ($)	Test Cost ($)	Manufacturer
IR	Cont	Portable	10-1000	4374	0	Foxboro

NIOSH METHOD NO. P&CAM 127

(Use as described for acetone with 80° C. column temperature)

LIQUIFIED PETROLEUM GAS

Synonym: LPG
ACGIH TLV: 1000 ppm (1800 mg/m^3)
OSHA Standard: 1000 ppm

Method	Sampling Duration	Sampling Location	Useful Range (ppm)	System Cost ($)	Test Cost ($)	Manufacturer
DT	Grab	Portable	200-8000	180	2.25	Bendix
IR	Cont	Portable	20-2000	4374	0	Foxboro
PI	Cont	Fixed	20-2000	4950	0	HNU

NIOSH METHOD NO. S93

Analytical Method

The LPG vapor present in the atmosphere is measured directly by drawing the air sample into a combustible gas meter properly calibrated with LPG. The meter reading is recorded and the equivalent concentration in ppm is read off the calibration curve. The method has been validated over the range of 478-2006 ppm at 23°C. and 750 mm Hg atmospheric temperature and pressure.

Sampling Equipment

A dual range explosive gas meter of sufficient sensitivity, such as the Davis Vapotester Model M 6, must be used. No sample is actually collected since the atmosphere to be analyzed is usually drawn into the meter using a rubber aspirator bulb. The air can also be drawn into the meter with a personal sampling pump.

Sampling Procedure

The instructions given in the manual for the appropriate combustible gas meter used must be followed carefully for proper results. Typically, the procedure will require the following steps:

The combustion chamber must be swept free of combustible gases and be filled with fresh air. Batteries must be turned on, and the proper voltage applied to the bridge.

The bridge must be balanced to zero deflection on the meter with fresh air in the open chamber. The sample is drawn into the meter, and the meter reading is recorded. Repeat at least three times and report average.

The concentration of LPG in the air samples tested is determined from the calibration curve provided with the properly calibrated meter.

Special Considerations

Any hydrocarbon present in the atmosphere will interfere with the analysis. Where other hydrocarbons are known or suspected to be present in the air, such information, including their suspected identities, should be noted.

Water vapor will impair the proper functioning of the sensing filaments. When testing atmospheres of high humidity, use available cotton filters or special humidity conditioning sample probes to remove excessive dust and moisture from the sample under test.

MAGNESIUM OXIDE FUME

ACGIH TLV: 10 mg/m³
OSHA Standard: 15 mg/m³

MgO

Method	Sampling Duration	Sampling Location	Useful Range (mg/m³)	System Cost ($)	Test Cost ($)	Manufacturer
LS	Cont	Personal	0.01-100	1695	0	GCA
LS	Cont	Portable	0.001-100	2785	0	MDA
LS	Cont	Fixed	0.001-200	3990	0	GCA
BA	1,4 min	Portable	0.02-150	4890	0	GCA

NIOSH METHOD NO. S369

Principle
A known volume of air is drawn through a cellulose membrane filter to collect the analyte. Samples are ashed using nitric acid to destroy the filter and other organic materials in the sample, and the magnesium is then solubilized in nitric acid. The solutions of samples and standards are aspirated into the oxidizing air acetylene flame of an atomic absorption spectrophotometer (AAS). A hollow cathode lamp for magnesium is used.

Range and Sensitivity
This method was validated over the range of 7.48-28.6 mg/m³ using a 150 L sample at atmospheric temperature and pressure of 23° C. and 764 mm Hg. Under the conditions of sample size (150 L), the working range of the method is estimated to be 1-55 mg/m³.

The sensitivity of this method for a 150 L air sample using the 100 mL final solution volume is 0.33 mg/m³ magnesium. The method may be extended to higher values by dilution of the sample. Measurement of lower atmospheric concentrations can be made by using smaller final solution volumes, by longer sampling times, or by scale expansion to increase instrumental response. The sensitivity may be extended by using the resonance line at 285.2 nm.

Interferences
Any magnesium compound which is collected on the cellulose ester membrane filter will cause a positive interference. Si, Al, and Cu (at high concentrations) may interfere with the method. These interferences can be eliminated by adding 0.1-1.0% of La to the sample.

Precision and Accuracy
The coefficient of variation for the total analytical and sampling method in the range of 7.48-28.6 mg/m³ is 0.062. This value corresponds to a 0.93 mg/m³ standard deviation at the

OSHA standard level. Statistical information and details of the validation and experimental test procedures can be found in reference 3.

A collection efficiency of 1.00 was determined for the collecting medium, thus, no bias was introduced in the sample collection step. There was also no apparent bias in the sampling and analytical method. Thus, coefficient of variation is a satisfactory measure of both accuracy and precision of the sampling and analytical method.

Advantages and Disadvantages

The sampling device is small, portable, and involves no liquids. Samples collected on filters are analyzed by means of a quick, instrumental method.

Apparatus

Filter unit. Consists of the filter media and appropriate 37 mm, three piece cassette filter holder.

Personal sampling pump. A calibrated personal sampling pump whose flow can be determined to an accuracy of 5% at the recommended flow rate.

Mixed cellulose ester membrane filter. 37 mm diameter, 0.8 micron pore size.

Atomic absorption spectrophotometer. The instrument must be equipped with an air acetylene burner head.

Magnesium hollow cathode lamp.

Oxidant, compressed air.

Fuel, acetylene.

Pressure reducing valves. A 2 gauge, 2 stage pressure reducing valve and appropriate hose connections are needed for each compressed gas tank used.

Glassware, borosilicate.

100 mL Phillips beakers with watchglass covers.

Pipets. Delivery or graduated, 1, 3, 5, 7, 10, 15, and 20 mL.

100 mL volumetric flasks.

250 mL volumetric flasks.

Nalgene bottles.

Five 100 mL Nalgene bottles. For working atomic absorption standards.

One 1000 mL Nalgene bottle. For stock atomic absorption standard.

Adjustable thermostatically controlled hot plate. Capable of reaching 400° C.

Reagents

All reagents used must be ACS reagent grade or better.

Double distilled water.

Concentrated nitric acid.

Dilute nitric acid. 10 mL concentrated nitric acid diluted to 100 mL with distilled or deionized water.

Magnesium metal. For preparation of AA standards, or commercially prepared aqueous stock standards (1000 μg/mL).

Procedure

Cleaning of equipment. Before use, all glassware should initially be soaked in a mild detergent solution to remove any residual grease or chemicals. After initial cleaning, glassware must be cleaned with hot concentrated nitric acid and then rinsed thoroughly with tap water and distilled water, in that order, and then dried. For glassware which has previously been subjected to the entire cleaning procedure, a nitric acid rinse will be adequate.

Calibration of personal sampling pumps. Each personal sampling pump must be calibrated with a representative filter cassette in the line. This will minimize errors associated with uncertainties in the sample volume collected.

Collection and shipping of samples. Assemble the filter in the three piece filter cassette holder and close firmly to insure that the center ring seals the edge of the filter. The cellulose membrane filter is held in place by a cellulose backup pad. Remove the cassette plugs and attach to the personal sampling pump tubing. Clip the cassette to the worker's lapel. The cassette plugs are replaced after sampling. Air being sampled should not be passed through any hose or tubing before entering the filter cassette.

A sample size of 150 L is recommended. Sample at a flow rate of 1.5 L/min. The flow rate should be known with an accuracy of 5%. Turn the pump on and begin sample collection. Since it is possible for a filter to become plugged by heavy particulate holding or by the presence of oil mists or other liquids in the air, the pump rotameter should be observed frequently, and the sampling should be terminated at any evidence of a problem.

Terminate sampling at the predetermined time and note sample flow rate, collection time, and ambient temperature and pressure. If pressure reading is not available, record the elevation. Carefully record the sample identity and all relevant sampling data. With each batch of ten samples, submit one filter from the same lot of filters which was used for the sample collection and which is subjected to exactly the same handling as the samples, except that no air is drawn through it. Label this as a blank. The cassettes in which the samples are collected should be shipped in a suitable container, designed to prevent damage in transit.

Analysis of samples. Transfer each sample to a clean 100 mL Phillips beaker. Treat the sample in each beaker with 3 mL of concentrated nitric acid to destroy the organic filter matrix. Cover each beaker with a watchglass and heat on a hot plate (140° C.) in a fume hood until most of the acid has evaporated. Repeat this step twice. Cover each beaker with a watchglass and heat it on a high temperature hot plate (400° C.) in a fume hood until a white ash appears. Using distilled water, carefully rinse the material on the bottom of the watchglass into the beaker, rinse sides of beaker, and allow the solution to evaporate to dryness.

Cool each beaker and dissolve residues in 10 mL concentrated nitric acid. Quantitatively transfer the clear solution to a 100 mL volumetric flask. Rinse each beaker at least three times with 2-3 mL portions of distilled water or deionized water and quantitatively transfer each rinse to the solution in the volumetric flask. Dilute all samples to 100 mL with distilled or deionized water.

Aspirate the solutions into an oxidizing air acetylene flame and record the absorbance at 202.5 nm. The absorbance is proportional to the sample concentration and can be determined from the appropriate calibration curve. When very low concentrations are found in the sample, scale expansion can be used to increase instrument response or the sample can be dried and redituted to some smaller volume such as 5 or 10 mL before aspiration. In such a case, use no more acid solution in the previous step than is necessary to effect a quantitative transfer. If greater sensitivity is needed, the analytical wavelength at 285.2 nm may be used. Note: follow instrument manufacturer's recommendations for specific AAS operating parameters.

For samples with concentration levels in excess of 2000 μg/100 mL (20 ppm Mg is the high limit of the linear working range of the standard curve obtained at 202.5 nm) take an aliquot of the remaining sample and dilute to 100 mL with 1:10 nitric acid. Choose the aliquot such that when it is diluted to 100 mL, the dilution ratio brings the concentration of the sample within the linear working range of the calibration curve. Appropriate filter blanks must be analyzed in accordance with the total procedure.

Calibration and Standards

Prepare a 100 μg/mL magnesium stock standard by dissolving 100 mg magnesium metal in 100 mL concentrated nitric acid and diluting to 1000 mL. Transfer to a 1 L Nalgene bottle. From the 100 μg/mL stock magnesium standard solution, prepare at least 7 working standards to cover the linear range of the standard curve of 100-2000 μg/100 mL. Make all standard solutions in dilute nitric acid and prepare fresh each day and store in 100 mL Nalgene bottles.

Aspirate each of the standard samples and record the absorptions. Prepare a calibration curve by plotting on linear graph paper the absorbance versus the concentration of each standard in μg/100 mL. It is advisable to run standards both before and after the analysis on a series of samples to insure that conditions have not changed.

Calculations

Read the weight, in μg, corresponding to the total absorbance from the standard curve. No volume corrections are needed, because the standard curve is based on μg/100 mL. For concentrated samples which require dilution after initial analysis to bring them within the linear range of the standard curve, apply the appropriate dilution factor to obtain the correct weight of magnesium in the sample. Corrections for the blank must be made for each sample:

$$\mu g = \mu g \text{ sample} - \mu g \text{ blank}$$

where:

μg sample = μg found in sample filter
μg blank = μg found in blank filter

Calculate the μg of magnesium oxide by multiplying the μg magnesium found by 1.66, which is a conversion factor to convert μg magnesium to μg magnesium oxide. The concentration of the analyte in the air sample can be expressed in mg/m^3 (μg/L = mg/m^3):

$$\text{mg/m}^3 = \mu g \div \text{air volume sampled in L}$$

References
1. Analytical Methods for Flame Spectrophotometry. Varian Associates, 1972.
2. Methods for Emission Spectrochemical Analysis. ASTM Committee E-2, Philadelphia, 1971.
3. Documentation of NIOSH Validation Tests. Contract No. CDC-99-74-45.

MERCURY VAPOR

ACGIH TLV: 0.05 mg/m³ - Skin
OSHA Standard: 0.1 mg/m³ ceiling
NIOSH Recommendation: 0.05 mg/m³

Hg

Method	Sampling Duration	Sampling Location	Useful Range (mg/m³)	System Cost ($)	Test Cost ($)	Manufacturer
DT	Grab	Portable	0.1-2	150	2.60	Nat'l Draeger
DT	Grab	Portable	0.1-2	165	1.80	Matheson
DT	Grab	Portable	0.05-13.2	180	2.25	Bendix
UV	Cont	Portable	0.05-1	1850	0	Bacharach
GF	10 sec	Portable	0.001-2	3400	0	Jerome

NIOSH METHOD NO. P&CAM 175

Principle

Particulate mercury is collected on a glass fiber prefilter ahead of a two section, solid phase sampling tube. Organic mercury vapor is collected on the first section of the tube and metallic mercury vapor is amalgamated on the second section. The sampling train is purged with filtered air to insure passage of metallic mercury through the filter and Carbosieve B.

The two section tube is then broken between the sections and each section is analyzed separately. The first section represents mercury collected as organic mercury vapor and the second section represents metallic mercury. Each tube section and the filter is analyzed by thermally desorbing the mercury through the absorption cell of the flameless atomic absorption spectrophotometer (modified Coleman Model MAS-50 or equivalent). Absorption signals at the 253.7 nm line are recorded by strip chart recorders. The recorder signals are compared to standard calibration curves covering the required concentration range and the concentration of the mercury in the two sections of the tube is calculated from the signals.

Range and Sensitivity

The analytical range is determined in part by the path length of the cell used for analysis (i.e., 0.001-0.2 μg for a path length of 15.5 cm, 0.01-1.0 μg for 2.5 cm length, and 0.1-2 μg with a 1 cm length cell). The range and sensitivity of the method (as Hg) is the same for the three forms of mercury since the same analytical technique is used for each form.

The sensitivity of the method using the 0-100% transmittance scale for the optical cells is 0.001 μg Hg/sample/1% Abs. for the 15.5 cm cell, 0.005 μg for the 2.5 cm cell, and 0.010 μg for the 1 cm cell. The detection limit will vary with the cell used.

Interferences

Loading the prefilter with excess particulates will increase air flow resistance through the sampler. This may overload battery operated sampling pumps. Excessive amounts of water vapor may interfere if water is condensed in the sampling tubes; however, small amounts of water are removed by purging the tubes with filtered air after sampling as described. Strong oxidizing vapors and gases, particularly chlorine, which attach silver reduce the efficiency of the sampling tube but do not interfere in the analysis of the tube. These interferences should not be a problem under normal sampling conditions.

Precision and Accuracy

Standards are prepared by injecting known amounts of mercury as mercuric nitrate solution directly into tubes containing the silvered substrate followed by drying the tubes at 50° C. for at least 6 hours. Studies using labelled Hg have shown that there are no losses of Hg during a drying period of 7 days. The overall sampling and analytical precision of this revised method is 7.3% relative standard deviation (RSD) for 13 mercury vapor samples collected at 0.2 L/min in an atmosphere containing 65 ng Hg/L. On the average, 5% of the vapor is retained by the Carbosieve B section. This bias is corrected for the calculation. The overall sampling and analytical precision is 10.7% RSD for 6 particulate samples (phenylmercuric acetate) collected at 0.2 L/min in an atmosphere containing 110 ng Hg/L (as phenylmercuric acetate). The overall sampling and analytical precision is 7.4% RSD for 9 dimethyl mercury samples collected at 0.2 L/min in an atmosphere containing 58 ng Hg/L (as dimethyl mercury). The pooled precision (n=40) for the preparation and analysis of 4 groups of standards (20, 41, 81, and 122 ng Hg/tube) analyzed over a two week period is 5.6% RSD.

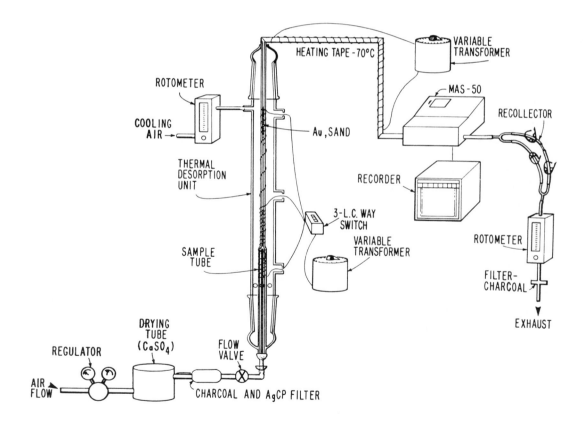

Figure 1. Mercury analysis system

Advantages and Disadvantages

The main advantage is the convenience of the sampling procedure and the ability to analyze particulate, metallic and organic mercury. The efficiency of the silvered substrate for collecting metallic mercury vapor and the efficiency of Carbosieve B for organic mercury vapor allows the use of small sampling tubes with low air flow resistance. The sampling tubes contain no liquids and are easily stored and shipped without mercury losses.

Since each section of the sample is analyzed by the same technique, the need for separate analytical methods is eliminated. Amalgamation of the metallic mercury released from each sample makes the method more selective than other methods. The main disadvantage of the method is the care necessary to avoid contamination of the outer surfaces of the sampling tubes. Although the method is satisfactory for total mercury below 0.001 g, absolute separation of organic and metallic mercury becomes less certain for amounts below this level.

Apparatus

Air sampling tubes. Available from SKC (Eighty Four, PA 15330).
Pipets. 0.005 to 0.05 mL.
Forceps, fine tip.
Glass rod, approximately 3 mm diameter.
Thermal desorption analysis system (Figure 1).
Filter unit for particulate mercury. Consists of the filter media and cassette filter holder.
Two-stage mercury sampling tube (Figure 2).

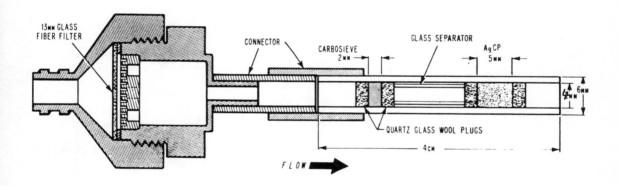

Figure 2. Two-stage mercury sampling tube with glass fiber filter

Calibrated personal sampling pump. Capable of maintaining an accuracy of 5% at the recommended flow rate (0.2 L/min).
Integrating volume meter. Such as a dry gas or wet test meter.
Thermometer.
Manometer.
Stopwatch.
Glass fiber filter, 13 mm.
Flow meters. 0-2 L/min and 0-20 L/min.
Needle valve.
Variable transformers. 22 A and 7.5 A.

Vacuum pump.
Thermal desorption unit (see Figure 3).

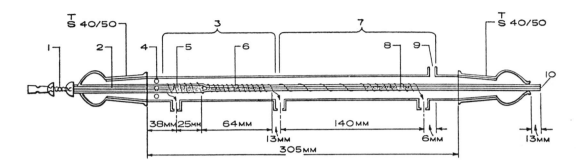

1. LOADING MECHANISM
2. QUARTZ GLASS PLUNGER TUBE
3. SAMPLE DESORPTION SECTION
4. COOLING AIR VENT HOLES
5. SAMPLING TUBE
6. CUPRIC OXIDE (CuO)
7. SECOND DESORPTION SECTION
8. GOLD SECTION
9. COOLING AIR INTAKE
10. OUTLET TO SPECTROPHOTOMETER

Figure 3. Thermal desorption unit

Absorption cells. 15.5 cm, 2.5 cm, and 1.0 cm length, with quartz glass windows.

Flameless atomic absorption mercury analyzer with recorder. Coleman Mercury Analysis System (MAS-50) or equivalent.

Oven capable of maintaining $50 \pm 5°C$.

Reagents

All reagents used should be ACS analytical reagent grade or equivalent.

Silver Chromosorb P (AgCP) tubes.

Stock solution. 100 µg Hg/mL, prepared by dissolving 100 mg of metallic mercury in 50 mL concentrated nitric acid in a 1000 mL Pyrex volumetric flask and diluting to volume with double distilled H_2O. The solution may be stored in the flask for as long as 6 months. Conventional atomic absorption standards or a stable mercury salt may be used.

Standard working solution A. 1 µg/mL, prepared by diluting 1.00 mL of the standard stock solution in 100 mL with 1 mL concentrated HNO_3 and double distilled water. Prepare fresh dilutions every 5 days.

Standard working solution B. 10 µg/mL, prepared as above except with 10 mL of stock solution. Prepare fresh dilutions every 5 days.

Double distilled water.
Nitric acid, HNO_3, concentrated.
Calcium sulfate ($CaSO_4$) anhydrous.
Mersorb charcoal.
Cupric oxide (CuO) rods.
Gold, granular. 35/50 mesh.
Sea sand. 20/40 mesh.

Procedure

Cleaning of equipment. Acid clean all glassware and Teflon before use. Wash in detergent tap water solution and follow with tap water rinse. Soak in concentrated HNO_3 for 30 min and follow with tap, distilled and double distilled water rinses.

Description of the two stage desorption unit. With the exception of the electrical components and the loading spring, the entire thermal desorption unit is made of either quartz or Pyrex glass. A diagram of the unit giving the critical dimensions is shown in Figure 3. Each important part of the thermal desorption unit is numbered in the diagram and the numbers represent the following:

Loading mechanism (Figure 4). Made from 18/7 female glass joint with a steel spring (#1) and a plunger cut to reach the sample desorption section.

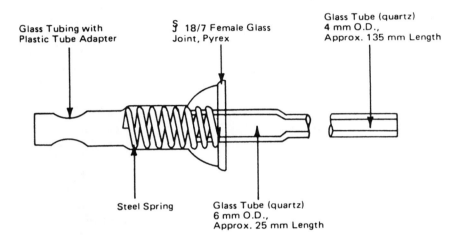

Figure 4. Loading mechanism

Sample desorption section. Made from 8 mm I.D. and 5 mm I.D. quartz tubing. The junction between the two sizes of tubing is tapered on the inside so that sampling tube tips fit snugly against the junction. Twenty-eight coils of 18 gauge Nichrome wire are wrapped around the first desorption section to heat the sampling tubes during the thermal desorption step. The Nichrome wire must be evenly wrapped with approximately 5 mm spacing between coils.

Cooling vent holes. About seven 4 mm holes are placed around the outer jacket of the unit to allow cooling air to flow from the second heating coil to the first coil.

Sampling tube. Sampling tubes are positioned inside the first desorption section during mercury desorption. The spring on the loading mechanism is adjusted to press the sampling tubes lightly in place.

Cupric oxide (CuO). A 40 mm section of rod shaped CuO is placed just downstream from the sampling tube. The CuO is held by a quartz glass wool plug which fits against a crimp in the glass tubing. Although the plugs should be large enough to hold the materials, they should not be packed too tightly.

Second desorption section. This section is made from the 5 mm I.D. quartz glass tubing extending from the first desorption section. The contents are held in place by a crimp in the glass tube and quartz glass wool plugs.

Nichrome wire coils. Thirty coils of size 20 Nichrome wire are wrapped around the second desorption section. The coils are wrapped to allow most of the generated heat to concentrate over the gold granules of the second desorption section (i.e., evenly spaced with 1 mm gap between coils).

Gold section. This section consists of a 25 mm length of 35/50 mesh granular gold mixed one-to-one with 20/40 mesh sea sand. The sand is added to the gold to prevent fusing of the gold granules and to allow better air flow through the section. This section must be at least 15 mm below the entrance of the cooling air.

Cooling air intake. A 6 mm I.D. piece of glass tubing is used for connecting plastic tubing from the cooling air supply to the cooling jacket. The Pyrex glass cooling jacket not only directs the flow of cooling air, but also acts as electrical insulation for the heating coils. Wires to the heating coils enter the cooling jacket through 5 mm holes at the ends of glass nipples on the side of the jacket. The solderless connectors between the wires and the coils are placed inside the jacket to prevent exposing uninsulated wire outside the cooling jacket. The insulated wires to the connectors are sealed in place with a heat resistant sealer.

Outlet. The outlet from the desorption unit is butt-connected with a Tygon overseal to a glass tube which leads to an atomic absorption spectrophotomerer (MAS-50). The end of the quartz tube slides through an opening at the end of the cooling jacket. The opening is kept to a minimum to limit the escape of cooling air.

Installation. A diagram of the installed thermal desorption unit system is shown in Figure 1.

Power connections. The wires to the heating coils should be heavy enough to carry the current necessary to heat the coils (about #12). Power to the heating coils is controlled by a three position switch with an off position, an on position for coil #1 and an on position for coil #2. Power to the switch comes from a 120 VAC input, 22 A variable transformer set at 28 VAC. The voltage should be set so that desorption temperature (in the sample tube) of 500 25° C. is achieved.

Air supply. The air which passes through the heated sections of the thermal desorption unit is supplied at 8 psi. Before entering the desorption unit, the air passes through a drying tube containing anhydrous $CaSO_4$. The dimensions of the drying tube and all subsequent filters may vary but should not interfere with the air flow through the system. The $CaSO_4$ should be changed periodically depending on the humidity of air. The dried air passes through a filter containing activated charcoal and 30/60 mesh silvered Chromosorb P to remove organic vapors and metallic mercury. The dry, filtered air then passes through a needle valve and into the thermal desorption unit, through the flameless atomic absorption optical cell, a rotameter and into a filter containing charcoal which collects any mercury desorbed by the desorption unit.

Cooling air. A flowmeter controls the flow rate at 15 L/min. The air enters the desorption unit from a plastic tube into the air intake near the second desorption section.

Detection system. Mercury desorbed from the gold section of the thermal desorption system enters the optical cell of a flameless atomic absorption spectrophotometer through a glass tube. The glass tube connecting the desorption unit with the optical cell is maintained at 70° C. using a heat tape. Power for the heating tape is supplied by a 120 VAC, 8 A variable transformer.

Collection and shipping of air samples. New sample tubes should be cycled once in the desorption unit. After this treatment, they should be stable indefinitely. White gloves or laboratory wipers should be used when handling the tubes. The sampler should be equipped with a holder for a 13 mm diameter, glass fiber filter preceding the sampling tube to collect particulate mercury. After the filter cassette and sampling tube have been connected, wrap the sampling tube with at least two layers of Teflon tape to minimize contamination.

Immediately after sampling, connect a second filter/tube assembly to original tube with a short piece of clean, uncontaminated tubing. Turn on the pump and using the second tube as a filter, purge the sample tube for 10.0 min at the maximum flow rate of the pump (up to 5 L/min). Turn off the pump and discard the tube that was used as a filter for the original sampling tube.

Remove the sampling tube from the sampler, cap the opening on the cassette and the end of the tube and return the filter/tube assembly to the lab for analysis. Appropriate blank samples should be handled and submitted together with the samples.

Analysis of filters for particulate mercury. Using forceps and wearing surgeon (plastic) gloves, remove the filter from the holder, fold twice and slide it into a sampling tube from which the Carbosieve B (but not the AgCP) has been removed. Place a glass wool plug behind the filter. Analyze the contents of the tube using the procedures for metallic mercury.

Analysis of air sampling tubes for organic and metallic mercury. All operations requiring the handling of the sample tubes should be done using clean white cotton gloves or laboratory wipers. Turn on the MAS-50 analyzer and allow to stabilize. Note: the time required for stabilization varies with individual instruments. Turn on the heating tape. After the MAS-50 has stabilized, open the valve to obtain air flow through the system.

Observing the rotameter, adjust the valve for an air flow of 1 L/min. Turn on the cooling air and adjust cooling air flowmeter to 15 L/min. Turn on the recorder and allow to stabilize. Adjust the MAS-50 and recorder to the desired 0 and 100% T settings. Purge the system before analyzing samples by heating the first desorption section of the desorption unit for 45 sec and then heating the second section for 20 sec. Allow the desorption unit to cool for 7 min before analyzing samples.

Score and break the sampling tube between the second and third glass wool plugs and separate the two tube sections. Do not break the tube containing the filter. Remove the clamp between the loading mechanism and desorption unit and insert the Carbosieve B section of the sampling tube. Push the sampling tube into the first desorption section of the desorption unit using the plunger of the loading mechanism and replace the clamp.

Place an AgCP tube in the recollecting device (Figure 5) and clamp the bypass tube. Turn on the first section of the thermal desorption unit and heat the sampling tube for 45 sec. After the first section is heated, wait 1.0 min before switching the three-way power switch to heat the second desorption section of the thermal desorption unit. This section is usually heated for about 25 sec. Turn the three-way switch to the off position.

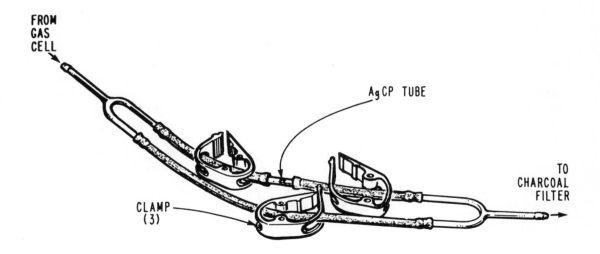

Figure 5. Recollecting device

Allow the thermal desorption unit to cool for about 30 sec and open the clamp on the loading mechanism to unload the used sampling tube. Replace the loading mechanism and clamp. CAUTION: the sampling tube is very hot when it comes out of the desorption unit. Do not touch the tube with bare hands. Drop the hot tube into a dry beaker to cool. Note: if the used sampling tube does not drop when the plunger is removed, a narrow spatula may be used to loosen the tube.

Reload the desorption unit with the AgCP section of the sampling tube after allowing a minimal 7 min cooling period. Replace the recollection tube (immediately downstream of the cell). Repeat the previous steps for the analysis of additional tubes. Recollected samples may be analyzed using a different gas cell, if needed.

Calibration and Standards

Remove the glass wool plug from the open end of a blank AgCP tube using gloved hands and tweezers. Inject the required amount (10-100 L) of mercury solution (A or B) directly on top of the exposed AgCP using a μL pipet while holding the tube in the upright position. Replace the glass wool plug and repack the tube using a glass rod.

Place the wet tubes upright in a small beaker and dry the tubes for at least 6 hours at 50° C. in a drying oven. The absorbances determined using the standard tubes are used to prepare a calibration curve. A minimum of five concentrations replicated three times at each concentration should be used in preparing the calibration curve. Analyze the standards at the same time as samples.

Calculations

To determine total μg of mercury as particulate, organic and metallic mercury, the absorption signals for the filters and sampling tubes are compared with the appropriate calibration curve. Calculate the mercury concentration in μg/m^3 (μg Hg/m^3) for each type of mercury:

$$\mu g\ Hg/m^3 = \mu g\ Hg \div Vs$$

where:

Vs = volume of air sampled in cubic meter at standard conditions of 25° C. and 760 mm Hg (1000 L = 1 m^3; volume = flow rate × time)

Since part of the elemental mercury is retained by Carbosieve B, a correction must be made. This is done for the metallic mercury by multiplying the quantity of mercury found on the AgCP section by 1.05 to obtain the total metallic mercury. The difference between the total metallic mercury and the quantity of mercury found on the AgCP section must be subtracted from the amount found on the Carbosieve B section to obtain the total organic mercury.

References

1. Campbell, E.E., P.E. Trujillo, and G.O. Wood. Development of a Multistage Tandem Air Sampler for Mercury, Quarterly Report No. LA-5340-PR, LA-5467-PR, LA-5188-PR, LA-5105-PR, Los Alamos Scientific Laboratory, Los Alamos, New Mexico, March 31, 1973, September 30, 1973, December 31, 1973 and September 30, 1972.
2. Trujillo, P.E., and E.E. Campbell. Development of a multistage air sampler for mercury. Anal. Chem. 47: 1629, 1975.
3. Hull, R.D., J.C. Haartz, and M. Bolyard. Analytical Evaluation of a multistage sampling device for mercury. To be published.

MESITYL OXIDE

Synonym: 4-methyl-3-penten-2-one
ACGIH TLV: 15 ppm (60 mg/m^3)
OSHA Standard: 25 ppm (100 mg/m^3)

$(CH_3)_2C=CHCOCH_3$

Method	Sampling Duration	Sampling Location	Useful Range (ppm)	System Cost ($)	Test Cost ($)	Manufacturer
IR	Cont	Portable	0.5-50	4374	0	Foxboro

NIOSH METHOD NO. S66

(Use as described for n-butyl alcohol, with 120° C. column temperature)

METHYL ACETATE

ACGIH TLV: 200 ppm (610 mg/m^3)
OSHA Standard: 200 ppm

CH_3COOCH_3

Method	Sampling Duration	Sampling Location	Useful Range (ppm)	System Cost ($)	Test Cost ($)	Manufacturer
DT	Grab	Portable	1000-30,000	165	1.70	Matheson
IR	Cont	Portable	4-400	4374	0	Foxboro
PI	Cont	Fixed	2-200	4950	0	HNU

NIOSH METHOD NO. P&CAM 127

(Use as described for acetone, with 70° C. column temperature)

METHYL ACETYLENE

Synonym: propyne $\qquad$ CH$_3$C≡CH
ACGIH TLV: 1000 ppm (1650 mg/m^3)
OSHA Standard: 1000 ppm

Method	Sampling Duration	Sampling Location	Useful Range (ppm)	System Cost ($)	Test Cost ($)	Manufacturer
IR	Cont	Portable	20-2000	4374	0	Foxboro

NIOSH METHOD NO. S84

Principle
An air sample is pumped into a gas sampling bag with a personal sampling pump. The methyl acetylene content of the samples is determined by gas chromatography.

Range and Sensitivity
This method was validated over the range of 520-1880 ppm at an atmospheric temperature of 21° C. and atmospheric pressure of 763 mm Hg using a 3 L sample volume. The limit of detection is estimated at less than 10 ng methyl acetylene. Using a 2.0 mL gas sampling loop, this corresponds to a level of approximately 2.5 ppm.

Interferences
When two or more compounds are known or suspected to be present in the air, such information, including their suspected identities, should be transmitted with the sample. It must be emphasized that any compound which has the same retention time as the analyte at the operating conditions described in this method is an interference. Retention time data on a single column cannot be considered as proof of chemical identity.

Precision and Accuracy
The coefficient of variation for the total analytical and sampling method in the range of 520-1880 ppm was 0.049. This value corresponds to standard deviation of 49 ppm at the OSHA standard level. Statistical information can be found in reference 1. Details of the test procedure can be found in reference 2.

In validation experiments, this method was found to be capable of coming within 25% of the "true" value on the average 95% of the time over the validation range. The concentrations obtained at 0.5X, 1X, and 2X the OSHA environmental limit averaged 8.3% lower than the dynamically generated test concentrations (n=18). This difference does not represent a bias in the method, but rather a random variation from the experimentally determined "true" concentration. The analytical method recovery was determined to be 0.977 for a collector loading of 500 ppm. In storage stability studies, the mean of samples analyzed after 7 days were within 8.2% of the mean of samples analyzed immediately after collection. Experiments performed in the validation study are described in reference 2.

Advantages and Disadvantages
The sampling device is portable and involves no liquids. Interferences are minimal, and most of

those which do occur can be eliminated by altering chromatographic conditions. The samples in bags are analyzed by means of a quick instrumental method. One disadvantage of the method is that the gas sampling bag is rather bulky and may be punctured during sampling and shipping. It is difficult to ship the samples by air.

Apparatus

Personal sampling pump. A personal sampling pump capable of filling a gas sampling bag at approximately 0.05 L/min is required. Each personal pump should be calibrated to within 5% to minimize errors associated with uncertainties in the sample volume collected. Although sample volume is not used to determine sample concentration, the pump should be calibrated to make certain that the collected sample represents a time weighted average concentration to avoid over filling of the bags; i.e., a maximum sampling time can be determined based on a flow rate and sample volume which is less than 80% of the volume of the bag.

The personal sampling pump must be equipped with an in/out valve. To ensure a leak-free apparatus, adjust the pump so that it delivers at the proper flow rate, and attach the pump outlet to a water manometer with a short piece of flexible tubing. Turn the pump on and observe the water level difference; it should push at least 30 cm of water. If it does not, the pump is incapable of filling the sampling bag and cannot be used.

Gas sampling bag. 5 L capacity, 5-layer sampling bags manufactured by Calibrated Instruments (Ardsley, NY 10502) were found to be satisfactory for sample collection and storage for at least 7 days. The bag is fritted with a metal valve and hose bib. The valves used in validation studies were found to leak when in the open position. It may be necessary to wrap the valve stem connection with Teflon tape to ensure a leak free connection. For the preparation of calibration standards in the laboratory, Saran of Tedlar bags could be used.

Gas chromatograph. Equipped with a flame-ionization detector, 2 mL gas sampling loop, and a temperature programmer.

Column. 4 ft × 1/8 in stainless steel, packed with 50/80 mesh Porapak Q.

Area integrator. An electronic integrator or some other suitable method for measuring peak areas.

Gas tight syringes. Convenient sizes for preparing standards.

Regulators for compressed air. Capable of metering gas at approximately 1 L/min. The regulator should be equipped with a septum tee for standards preparation.

Water manometer.

Thermometer.

Reagents

All reagents used must be ACS reagent grade or better.

Methyl acetylene.

Helium, purified.

Hydrogen, prepurified.

Air, filtered, compressed.

Procedure

Cleaning of sampling bags and checking for leaks. The bags are cleaned by opening the valve and bleeding out the air sample. The use of a vacuum pump is recommended although this procedure can be carried out by manually flattening the bags. The bags are then flushed with air and evacuated. This procedure is repeated at least twice. Bags should be checked for leaks by filling the bag with air until taut, sealing and applying gentle pressure to the bag. Observe for any discernable leaks and any volume changes or slackening of the bag, especially along seams and in the valve stem, for at least a one hour period.

Collection and shipping of samples. Immediately before sampling, attach a small piece of Teflon tubing to the hose bib of the 5 layer gas sampling bag. Rubber tubing should not be used. Unscrew the valve fitting and attach the tubing to the outlet of the sampling pump. Make sure that all connections are tight and leak free. The bag valve must be fully opened during sampling. Air being sampled will pass through the pump and tubing before entering the sampling bag, since a "push" type pump is required. No tubing is attached to the inlet of the pump.

A sample size of 3-4 L is recommended. Sample at a flow rate of 0.05 L/min or less, but not less than 0.01 L/min. The flow rate should be known with an accuracy of at least 5%. Set the flow rate as accurately as possible using the manufacturer's directions. Although the volume of sample collected is not used in determining the concentration, it is necessary to keep the volume to 80% or less of the bag's capacity. Observe the bag frequently to ensure that it is filling properly.

The temperature and pressure of the atmosphere being sampled should be recorded. If pressure reading is not available, record the elevation. Also record sampling time, flow rate, and type of sampling pump used. The gas sampling bag should be labeled appropriately and sealed tightly. Gas sampling bags should be packed loosely and padded before they are shipped to minimize the danger of their being punctunced during shipping. Do not ship the gas by air, unless they are stored in a pressurized cabin.

With each batch or partial batch of ten samples, submit one bag from the same lot of bags used for sample collection. This bag must be subjected to exactly the same handling as the samples except that no air is pumped into it. This bag should be labeled as the blank. Before analysis, 3-4 L of air is metered into the bag using a calibrated source of air equipped with a septum-tee.

GC conditions. The typical operating conditions for the gas chromatograph are:

30 mL/min (25 psig) helium carrier gas flow
ambient injector temperature
150° C. detector temperature
50-150° C. column temperature

Analysis. Attach the gas sampling bag to the sample loop of the gas chromatograph with a short piece of flexible tubing. Open the valve of the bag and fill the loop by using a vaccum pump or manually applying pressure to the sample bag. Allow the loop to attain atmospheric pressure, and inject the sample. Begin temperature programming after the methyl acetylene peak elutes. Temperature program at the maximum rate. The initial temperature should be 50° C. and the final temperature should be 150° C. This is necessary to elute contaminants in the sample. Duplicate injections of each sample and standard should be made. No more than a 3% difference in area is to be expected. A retention time of approximately 3 min is to be expected using the above conditions.

Measurement of area. The area of the sample peak is measured by an electronic integrator or some other suitable form of area measurement, and the results are read from a standard curve as discussed below.

Calibration and Standards

A series of standards, varying in concentration over the range of 100-3000 ppm is prepared and analyzed under the same GC conditions and during the same time period as the unknown samples. Curves are established by plotting concentration in ppm versus peak area.

Completely evacuate and flush several times with air a 5 L gas sampling bag, preferably with the aid of a vacuum pump. Using a calibrated source of air equipped with a septum-tee, meter 3-4 L of air into the bag. Inject appropriate aliquots of methyl acetylene via a gas tight syringe through the septum. Knead the bag to ensure adequate mixing. Prepare at least 5 working standards to cover the range of 100-3000 ppm. The concentration of the bag in ppm equals the volume of methyl acetylene in mL divided by the amount of air in L:

ppm = (mL of methyl acetylene)(1000) ÷ air volume in L

Calculations

Read the concentration in ppm, corresponding to each peak area from the standard curve. A correction for the blank must be made for each sample:

ppm = ppm sample − ppm blank

where:

ppm sample = ppm found in sample bag
ppm blank = ppm found in blank bag

Another method of expressing concentration is mg/m^3:

mg/m^3 = (ppm)(M.W.)(P)(298) ÷ (24.45)(760)(T + 273)

where:

P = pressure (mm Hg) of air sampled
T = temperature (°C.) of air sampled
24.45 = molar volume (L/mol) at 25° C. and 760 mm Hg
M.W. = molecular weight, 40.06 g/mol
760 = standard pressure (mm Hg)
298 = standard temperature (°K)

References

1. Documentation of NIOSH Validation Tests. National Institute for Occupational Safety and Health, Cincinnati, Ohio (DHEW-NIOSH Publication #77-185), 1977. Available from Superintendent of Documents, U.S. Government Printing Office, Washington, D.C., Order No. 017-033-00231-2.
2. Backup Data Report for Methyl Acetylene. Prepared under NIOSH Contract No. 210-76-0123.

METHYL ACRYLATE

ACGIH TLV: 10 ppm (35 mg/m³) - Skin
OSHA Standard: 10 ppm

$CH_2=CHCOOCH_3$

Method	Sampling Duration	Sampling Location	Useful Range (ppm)	System Cost ($)	Test Cost ($)	Manufacturer
DT	Grab	Portable	5-200	150	2.80	Nat'l Draeger
DT	Grab	Portable	100-6500	180	2.25	Bendix
IR	Cont	Portable	0.2-20	4374	0	Foxboro
PI	Cont	Fixed	2-200	4950	0	HNU

NIOSH METHOD NO. P&CAM 127

(Use as described for acetone, with 70° C. column temperature)

METHYLAL

Synonym: dimethoxymethane
ACGIH TLV: 1000 ppm (3100 mg/m³)
OSHA Standard: 1000 ppm

$(CH_3O)_2CH_2$

Method	Sampling Duration	Sampling Location	Useful Range (ppm)	System Cost ($)	Test Cost ($)	Manufacturer
IR	Cont	Portable	20-2000	4374	0	Foxboro

NIOSH METHOD NO. S165

(Use as described for acetonitrile, with 195° C. column temperature)

METHYL ALCOHOL

Synonym: methanol CH₃OH
ACGIH TLV: 200 ppm (260 mg/m³) - Skin
OSHA Standard: 200 ppm
NIOSH Recommendation: 800 ppm ceililng (15 min)

Method	Sampling Duration	Sampling Location	Useful Range (ppm)	System Cost ($)	Test Cost ($)	Manufacturer
DT	Grab	Portable	100-3000	150	2.50	Nat'l Draeger
DT	Grab	Portable	20-60,000	165	2.30	Matheson
DT	Grab	Portable	100-45,000	180	2.25	Bendix
IR	Cont	Portable	4-400	4374	0	Foxboro
PI	Cont	Fixed	2-200	4950	0	HNU
GC	1 min	Portable	1-10⁶	10,000	0	Microsensor

NIOSH METHOD NO. P&CAM 247

Principle

A known volume of air is drawn through a large silica gel tube to trap the organic vapors present. The silica gel in the tube is transferred to a small, stoppered sample container and the analyte is desorbed with water. An aliquot of the desorbed sample is injected into a gas chromatograph. The area of the resulting peak is determined and compared with areas obtained from the injection of standards. This method is similar to NIOSH method S59, but is recommended when sampling is done in high relative humidity or when a large concentration of methanol is expected.

Range and Sensitivity

This method was validated over the range of 135-534 mg/m³ at an atmospheric temperature of 25° C. and a relative humidity of 65-90%. Under the conditions of a 5 L sample size, the probable range of this method is 25-2000 mg/m³. The method is capable of measuring much smaller amounts if the desorption efficiency is adequate. Desorption efficiency must be determined over the range used.

The upper limit of the range of the method is dependent on the adsorptive capacity of the silica gel tube. This capacity varies with the concentration of the analyte and other substances in the air. The first section of the silica gel tube was found to hold 4.6 mg of methanol in a 85-90% relative humidity atmosphere when sampled at a flow rate of 50 mL/min for 175 min.

Breakthrough occurred at this time, i.e., the concentration of the analyte in the effluent was 5% of that in the influent. (The silica gel tube consists of three sections of silica gel held in place by Teflon plugs, each separated by a 12 mm air gap.) If a particular atmosphere is suspected of containing a large amount of contaminant, a smaller sampling volume should be taken.

Interferences

When the amount of water in the air is so great that condensation actually occurs in the tube, organic vapors will not be trapped efficiently. When two or more compounds are known or suspected to be present in the air, such information, including their suspected identities, should be transmitted with the sample. It must be emphasized that any compound which has the same retention time as the specific compound under study at the operating conditions described in this method is an interference. Retention time data on a single column cannot be considered as proof of chemical identity. If the possibility of interference exists, separation conditions (column packing, temperature, etc.) must be changed to circumvent the problem.

Precision and Accuracy

The coefficient of variation for the total analytical and sampling method in the range of 135-534 mg/m^3 was 0.065.

Advantages and Disadvantages

The sampling device is small, portable, and involves no liquids. Interferences are minimal, and most of those which do occur can be eliminated by altering chromatographic conditions. The tubes are analyzed by means of a quick, instrumental method. The method can also be used for the simultaneous analysis of two or more compounds suspected to be present in the same sample by simply changing gas chromatographic conditions from isothermal to a temperature-programmed mode of operation.

One disadvantage of the method is that the amount of sample which can be taken is limited by the number of mg that the tube will hold before overloading. When the sample value obtained for the backup section of the silica gel tube exceeds 25% of that found on the front section, the possibility of sample loss exists. Furthermore, the precision of the method is limited by the reproducibility of the pressure drop across the tubes. This drop will affect the flow rate and cause the volume to be imprecise, because the pump is usually calibrated for one tube only.

The advantage of this method over method S59 is the increased sample weight before overloading and decreased effect of high humidity.

Apparatus

Calibrated personal sampling pump. Whose flow can be determined within 5% at the recommended flow rate.

Silica gel tubes. Glass tube with both ends flame sealed, 15 cm long with 10 mm O.D. and 8 mm I.D., containing 3 sections of 20/40 mesh silica gel held by Teflon plugs and separated by air. The absorbing section contains 700 mg of silica gel held in Teflon plugs and screen. The second section is 12 mm from the absorbing and contains 150 mg of silica gel held by Teflon plugs and screen. The third section is 12 mm from the second and is the same composition as the second. The pressure drop across the tube must be less than one inch of mercury at a flow rate of 1 L/min.

Gas chromatograph. Equipped with a flame-ionization detector.

Column. 10 ft × 1/8 in stainless steel, packed with 10% FFAP on 80/100 Chromosorb W-AW.

Electronic integrator. Or some other suitable method for determining peak areas.

Glass sample containers. 5 mL, with glass stoppers or Teflon-lined caps. If an automatic sample injector is used, the sample injector vials can be used.

Syringes. 10 µL, and other convenient sizes for making standards.

Pipets. 3.0 mL delivery type.

Volumetric flasks. 10 mL or convenient sizes for making standard solutions.

Reagents

Eluent, distilled water.
Methyl alcohol (reagent grade).
Purified nitrogen.

Prepurified hydrogen.
Filtered compressed air.

Procedure

Cleaning of equipment. All glassware used for the laboratory analysis should be detergent washed and thoroughly rinsed with tap water and distilled water.

Calibration of personal pumps. Each personal pump must be calibrated with a representative silica gel tube in the line. This will minimize errors associated with uncertainties in the sample volume collected.

Collection and shipping of samples. Immediately before sampling, break the ends of the tube to provide an opening at least one-half the internal diameter of the tube (4 mm). The smaller sections of silica gel are used as a backup and should be positioned nearest the sampling pump. The silica gel tube should be placed in a vertical direction during sampling to minimize channeling through the silica gel. Air being sampled should not be passed through any hose or tubing before entering the silica gel tube.

A maximum sample size of 5 L is recommended. Sample at a flow rate of 0.20 L/min or less. The flow rate should be known with an accuracy of at least 5%. The temperature and pressure of the atmosphere being sampled should be recorded. If pressure reading is not available, record the elevation. The silica gel tubes should be capped with the supplied plastic caps immediately after sampling. Under no circumstances should rubber caps be used. One tube should be handled in the same manner as the sample tube (break, seal, and transport), except that no air is sampled through this tube. This tube should be labeled as a blank.

Capped tubes should be packed tightly and padded before they are shipped to minimize tube breakage during shipping. A sample of the suspected compound should be submitted to the laboratory in glass containers with Teflon-lined caps. These liquid bulk samples should not be transported in the same container as the silica gel tubes.

Preparation of samples. In preparation for analysis, each silica gel tube is scored with a file in front of the first section of silica gel and broken open. The Teflon plug is removed and discarded. The 700 mg of silica gel in the first section is transferred to a 5 mL stoppered sample container or automatic sample injector vial. The second and third sections are transferred to separate sample containers or vials. These three sections are analyzed separately.

Desorption of samples. Prior to analysis, 3.0 mL of distilled water is pipetted into each sample container. Desorption should be done for 4 hours. Tests indicate that this is adequate if the sample is agitated occasionally during this period. The sample vials should be capped as soon as the water is added to minimize volatilization.

GC conditions. The typical operating conditions for the gas chromatograph are:

30 mL/min (80 psig) nitrogen carrier gas flow
30 mL/min (50 psig) hydrogen gas flow to detector
300 mL/min (50 psig) air flow to detector
200° C. injector temperature
300° C. detector temperature
80° C. column temperature

Injection. The first step in the analysis is the injection of the sample into the gas chromatograph. To eliminate difficulties arising from blow back or distillation within the syringe needle, one should employ the solvent flush injection technique. The 10 μL syringe is first flushed with solvent several times to wet the barrel and plunger. Three μL of solvent are drawn into the syringe to increase the accuracy and reproducibility of the injected sample volume. The needle is

removed from the solvent, and the plunger is pulled back about 0.2 µL to separate the solvent flush from the sample with a pocket of air to be used as a marker. The needle is then immersed in the sample, and a 5 µL aliquot is withdrawn, taking into consideration the volume of the needle, since the sample in the needle will be completely injected. After the needle is removed from the sample and prior to injection, the plunger is pulled back 1.2 µL to minimize evaporation of the sample from the tip of the needle. Observe that the sample occupies 4.9-5.0 µL in the barrel of the syringe. Duplicate injections of each sample and standard should be made. No more than a 3% difference in area is to be expected. An automatic sample injector can be used if it is shown to give reproducibility at least as good as the solvent flush technique.

Measurement of area. The area of the sample peak is measured by an electronic integrator or some other suitable form of area measurement, and preliminary results are read from a standard curve prepared as discussed below.

Determination of desorption efficiency. The desorption efficiency of a particular compound can vary from one laboratory to another and also from one batch of silica gel to another. Thus, it is necessary to determine at least once the percentage of the specific compound that is removed in the desorption process.

Silica gel equivalent to the amount in the first section of the sampling tube (700 mg) is measured into a 5.0 mL sample container. This silica gel must be the same type as that used in obtaining the samples and can be obtained from unused silica gel tubes. A known amount of the analyte is injected directly into the silica gel with a 10 µL syringe, and the container is capped. The amount injected is equivalent to that present in a 5 L sample at the selected level.

At least six tubes at each of three levels (0.5X, 1X, and 2X the standard) are prepared in this manner and allowed to stand for at least overnight to assure complete adsorption of the analyte onto the silica gel. These six tubes are referred to as the samples. A parallel blank tube should be treated in the same manner except that no sample is added to it. The sample and blank tubes are desorbed and analyzed in exactly the same manner as the sampling tube described earlier. The weight of analyte found in each tube is determined from the standard curve. Desorption efficiency is determined by the following equation:

$$D.E. = \text{average wt (mg) recovered} \div \text{wt (mg) added}$$

The desorption efficiency is dependent on the amount of analyte collected on the silica gel. Plot the desorption efficiency versus the weight of analyte found. This curve is used to correct for adsorption losses.

Calibration and Standards

It is convenient to express concentration of standards in terms of mg/3 mL of eluent. A series of standards, varying in concentration over the range of interest, is prepared and analyzed under the same GC conditions and during the same time period as the unknown sample. Curves are established by plotting concentrations in mg/3 mL versus peak area. Note: standard solutions should be analyzed at the same time that the sample analysis is done. This will minimize the effect of variation in FID response.

Calculations

Read the weights, in mg, corresponding to each peak area (area ratio in case of the internal standard method) from the standard curve. Corrections for the blank must be made for each sample:

$$\text{mg} = \text{mg sample} - \text{mg blank}$$

where:

mg sample = mg found in front section of sample tube
mg blank = mg found in front section of blank tube

A similar procedure is followed for the backup sections. Add the weights present in the front and both backup sections of the same sample tube to determine the total weight in the sample. Read the desorption efficiency from the curve for the amount of analyte found in the front section. Divide the total weight by this desorption efficiency to obtain the corrected mg/sample:

Corrected mg/sample = total weight ÷ D.E.

The concentration of analyte in the air sampled can be expressed in mg/m^3, which is numerically equal to μg/L of air:

mg/m^3 = (corrected mg)(1000 L/m^3) ÷ air volume sampled in L

Another method of expressing concentration is ppm:

ppm = (mg/m^3)(24.45)(760)(T + 273) ÷ (M.W.)(P)(298)

where:

P = pressure (mm Hg) of air sampled
T = temperature (°C.) of air sampled
24.45 = molar volume (L/mol) at 25° C. and 760 mm Hg
M.W. = molecular weight (g/mol) of analyte
760 = standard pressure (mm Hg)
298 = standard temperature (°K)

References

1. White, L.D. et al. A convenient optimized method for the analysis of selected solvent vapors in the industrial atmosphere. Amer. Ind. Hyg. Assoc. J. 31: 225, 1970.
2. Documentation of NIOSH Validation Tests. NIOSH Contract No. CDC-99-74-45.
3. Final Report of NIOSH Contract No. HSM-99-71-31. Personal Sampler Pump for Charcoal Tubes, September 15, 1972.
4. Willey, Maurine, DHEW Memorandum. Breakthrough Tests for Methyl Alcohol, January 16, 1976.

METHYLAMINE

CH_3NH_2

Synonym: aminomethane
ACGIH TLV: 10 ppm (12 mg/m³)
OSHA Standard: 10 ppm

Method	Sampling Duration	Sampling Location	Useful Range (ppm)	System Cost ($)	Test Cost ($)	Manufacturer
DT	Grab	Portable	1-60	180	2.25	Bendix
CS	Grab	Portable	0.5-10	2465	0.80	MDA
IR	Cont	Portable	0.2-20	4374	0	Foxboro
PI	Cont	Fixed	0.2-20	4950	0	HNU

NIOSH METHOD NO. P&CAM 221

(See n-butylamine)

METHYL n-AMYL KETONE

$CH_3COC_5H_{11}$

Synonym: 2-heptanone
ACGIH TLV: 50 ppm (235 mg/m³)
OSHA Standard: 100 ppm (465 mg/m³)

Method	Sampling Duration	Sampling Location	Useful Range (ppm)	System Cost ($)	Test Cost ($)	Manufacturer
IR	Cont	Portable	2-200	4374	0	Foxboro

NIOSH METHOD NO. S66

(Use as described for n-butyl alcohol, with 120° C. column temperature. Prepare extraction solvent to contain 1% methanol rather than 2-propanol.)

METHYL BROMIDE

Synonym: bromomethane
ACGIH TLV: 5 ppm (20 mg/m³) - Skin
OSHA Standard: 20 ppm ceiling

CH_3Br

Method	Sampling Duration	Sampling Location	Useful Range (ppm)	System Cost ($)	Test Cost ($)	Manufacturer
DT	Grab	Portable	3-100	150	2.80	Nat'l Draeger
DT	Grab	Portable	2.5-500	165	4.60	Matheson
DT	Grab	Portable	2-200	180	2.25	Bendix
IR	Cont	Portable	0.8-40	4374	0	Foxboro
PI	Cont	Fixed	0.2-20	4950	0	HNU

NIOSH METHOD NO. P&CAM 127

(Use as described for acetone, with 65° C. column temperature)

METHYL CELLOSOLVE

Synonyms: 2-methoxyethanol; glycol monomethyl ether
ACGIH TLV: 5 ppm (16 mg/m³) - Skin
OSHA Standard: 25 ppm (80 mg/m³)

$CH_3OCH_2CH_2OH$

Method	Sampling Duration	Sampling Location	Useful Range (ppm)	System Cost ($)	Test Cost ($)	Manufacturer
IR	Cont	Portable	0.5-50	4374	0	Foxboro

NIOSH METHOD NO. S76

(Use as described for butoxyethanol, with 95° C. column temperature)

METHYL CELLOSOLVE ACETATE

Synonym: 2-methoxyethyl acetate $CH_3COOCH_2CH_2OCH_3$
ACGIH TLV: 5 ppm (24 mg/m^3) - Skin
OSHA Standard: 25 ppm (120 mg/m^3)

Method	Sampling Duration	Sampling Location	Useful Range (ppm)	System Cost ($)	Test Cost ($)	Manufacturer
IR	Cont	Portable	0.5-50	4374	0	Foxboro

NIOSH METHOD NO. P&CAM 127

(Use as described for acetone, with 90° C. column temperature)

METHYL CHLORIDE

Synonym: chloromethane CH_3Cl
ACGIH TLV: 50 ppm (105 mg/m^3)
OSHA Standard: 100 ppm

Method	Sampling Duration	Sampling Location	Useful Range (ppm)	System Cost ($)	Test Cost ($)	Manufacturer
IR	Cont	Portable	3-200	4374	0	Foxboro

NIOSH METHOD NO. S111

(Use as described for dichlorodifluoromethane, with 100° C. column temperature)

METHYL CHLOROFORM

Synonym: 1,1,1-trichloroethane
ACGIH TLV: 350 ppm (1900 mg/m^3)
OSHA Standard: 350 ppm

CCl_3CH_3

Method	Sampling Duration	Sampling Location	Useful Range (ppm)	System Cost ($)	Test Cost ($)	Manufacturer
DT	Grab	Portable	50-600	150	2.00	Nat'l Draeger
DT	Grab	Portable	15-400	165	3.80	Matheson
DT	Grab	Portable	100-500	180	2.25	Bendix
IR	Cont	Portable	7-700	4374	0	Foxboro
PI	Cont	Fixed	20-2000	4950	0	HNU
GC	1 min	Portable	1-10^6	10,000	0	Microsensor

NIOSH METHOD NO. P&CAM 127

(See acetone)

METHYLCYCLOHEXANE

Synonym: cyclohexylmethane
ACGIH TLV: 400 ppm (1600 mg/m^3)
OSHA Standard: 500 ppm

$CH_3C_6H_{11}$

Method	Sampling Duration	Sampling Location	Useful Range (ppm)	System Cost ($)	Test Cost ($)	Manufacturer
IR	Cont	Portable	10-1000	4374	0	Foxboro

NIOSH METHOD NO. P&CAM 127

(Use as described for acetone, with 55° C. column temperature)

METHYLCYCLOHEXANOL

Synonym: hexahydrocresol (50:50 mixture of m- and p-isomers) $CH_3C_6H_{10}OH$
ACGIH TLV: 50 ppm (235 mg/m^3)
OSHA Standard: 100 ppm

Method	Sampling Duration	Sampling Location	Useful Range (ppm)	System Cost ($)	Test Cost ($)	Manufacturer
IR	Cont	Portable	2-200	4374	0	Foxboro

NIOSH METHOD NO. S76

(Use as described for butoxyethanol)

o-METHYLCYCLOHEXANONE

Synonym: 2-methylcyclohexanone $C_6H_9OCH_3$
ACGIH TLV: 50 ppm (230 mg/m^3) - Skin
OSHA Standard: 100 ppm

Method	Sampling Duration	Sampling Location	Useful Range (ppm)	System Cost ($)	Test Cost ($)	Manufacturer
IR	Cont	Portable	2-200	4374	0	Foxboro

NIOSH METHOD NO. S350

(Use as described for n-butyl mercaptan, with 190° C. column temperature)

METHYLENE CHLORIDE

CH_2Cl_2

Synonym: dichloromethane
ACGIH TLV: 100 ppm (360 mg/m³)
OSHA Standard: 500 ppm
NIOSH Recommendation: 75 ppm

Method	Sampling Duration	Sampling Location	Useful Range (ppm)	System Cost ($)	Test Cost ($)	Manufacturer
DT	Grab	Portable	50-3000	150	2.70	Nat'l Draeger
DT	Grab	Portable	10-1000	165	3.80	Matheson
DT	Grab	Portable	50-500	180	2.25	Bendix
DT	4 hr	Personal	12.5-800	850	3.10	Nat'l Draeger
IR	Cont	Personal	10-1000	4374	0	Foxboro
PI	Cont	Fixed	2-200	4950	0	HNU
GC	1 min	Portable	1-10⁶	10,000	0	Microsensor

NIOSH METHOD NO. P&CAM 127

(See acetone)

METHYL FORMATE

$HCOOCH_3$

ACGIH TLV: 100 ppm (250 mg/m³)
OSHA Standard: 100 ppm

Method	Sampling Duration	Sampling Location	Useful Range (ppm)	System Cost ($)	Test Cost ($)	Manufacturer
IR	Cont	Portable	2-200	4374	0	Foxboro

NIOSH METHOD NO. P&CAM 127

(Use as described for acetone)

5-METHYL-3-HEPTANONE

Synonym: ethyl amyl ketone
ACGIH TLV: 25 ppm (130 mg/m³)
OSHA Standard: 25 ppm

$C_2H_5COCH_2CH(CH_3)C_2H_5$

Method	Sampling Duration	Sampling Location	Useful Range (ppm)	System Cost ($)	Test Cost ($)	Manufacturer
IR	Cont	Portable	0.6-50	4374	0	Foxboro

NIOSH METHOD NO. S66

(Use as described for n-butyl alcohol, with 120° C. column temperature)

METHYL IODIDE

Synonym: iodomethane
ACGIH TLV: 2 ppm (10 mg/m³) - Skin, suspected carcinogen
OSHA Standard: 5 ppm (28 mg/m³)

CH_3I

Method	Sampling Duration	Sampling Location	Useful Range (ppm)	System Cost ($)	Test Cost ($)	Manufacturer
DT	Grab	Portable	5-40	165	4.60	Matheson
IR	Cont	Portable	0.8-20	4374	0	Foxboro
PI	Cont	Fixed	0.2-20	4950	0	HNU

NIOSH METHOD NO. S98

Principle

A known volume of air is drawn through a charcoal tube to trap the organic vapors present. The charcoal in the tube is transferred to a small, stoppered sample container and the analyte is desorbed with toluene. An aliquot of the desorbed sample is injected into a gas chromatograph. The area of the resulting peak is determined and compared with areas obtained from the injection of standards.

Range and Sensitivity

This method was validated over the range of 17-52 mg/m³ at an atmospheric temperature and pressure of 25° C. and 735 mm Hg, using a 50 L sample. Under the conditions of sample size (50

L) the probable range of this method is 3-84 mg/m^3 at a detector sensitivity that gives nearly full deflection on the strip chart recorder for a 2 mg sample. The method is capable of measuring much smaller amounts if the desorption efficiency is adequate. Desorption efficiency must be determined over the range used.

The upper limit of the range of the method is dependent on the adsorptive capacity of the charcoal tube. This capacity varies with the concentrations of the analyte and other substances in the air. The first section of the charcoal tube was found to hold 12.5 mg of the analyte when a test atmosphere of 52 mg/m^3 of the analyte in dry air was sampled at 1 L/min for 4 hours. Breakthrough did not occur at this time, i.e., the concentration of the analyte in the effluent was less than 5% of that in the influent. (The charcoal tube consists of two sections of activated charcoal separated by a section of urethane foam.) If a particular atmosphere is suspected of containing a large amount of contaminant, a smaller sampling volume should be taken.

Interferences

When the amount of water in the air is so great the condensation actually occurs in the tube, organic vapors will not be trapped efficiently. Preliminary experiments with toluene indicate that high humidity severely decreases the breakthrough volume. When two or more compounds are known or suspected to be present in the air, such information, including their suspected identities, should be transmitted with the sample.

It must be emphasized that any compound which has the same retention time as the specific compound under study at the operating conditions described in this method is an interference. Retention time data on a single column cannot be considered as proof of chemical identity. If the possibility of interference exists, separation conditions (column packing, temperature, etc.) must be changed to circumvent the problem.

Precision and Accuracy

The coefficient of variation for the total analytical and sampling method in the range of 17-52 mg/m^3 was 0.070. This value corresponds to a standard deviation of 2.0 mg/m^3 at the OSHA standard level. Statistical information and details of the validation and experimental test procedures can be found in reference 2. The average values obtained using the overall sampling and analytical method were 8.6% less than the "true" value at the OSHA standard level.

Advantages and Disadvantages

The sampling device is small, portable, and involves no liquids. Interferences are minimal, and most of those which do occur can be eliminated by altering chromatographic conditions. The tubes are analyzed by means of a quick, instrumental method. The method can also be used for the simultaneous analysis of two or more compounds suspected to be present in the same sample by simply changing gas chromatographic conditions from isothermal to a temperature-programmed mode of operation.

One disadvantage of the method is that the amount of sample which can be taken is limited by the number of mg that the tube will hold before overloading. When the sample value obtained for the backup section of the charcoal tube exceeds 25% of that found in the front section, the possibility of sample loss exists. Furthermore, the precision of the method is limited by the reproducibility of the pressure drop across the tubes. This drop will affect the flow rate and cause the volume to be imprecise, because the pump is usually calibrated for one tube only.

Apparatus

Calibrated personal sampling pump. Whose flow can be determined accurately (5%) at the recommended flow rate.

Charcoal tubes. Glass tube with both ends flame sealed, 7 cm long with 6 mm O.D. and 4 mm I.D., containing 2 sections of 20/40 mesh activated charcoal separated by a 2 mm portion of urethane foam. The activated charcoal is prepared from coconut shells and is fired at 600° C. prior to packing. The absorbing section contains 100 mg of charcoal, the backup section 50 mg. A 3 mm portion of urethane foam is placed between the outlet end of the tube and the backup section. A plug of silylated glass wool is placed in front of the absorbing section. The pressure drop across the tube must be less than one inch of mercury at a flow rate of 1 L/min.

Gas chromatograph. Equipped with a flame-ionization detector.

Column. 10 ft × 1/8 in stainless steel, packed with Chromosorb 101.

Electronic integrator. Or some other suitable method for measuring peak areas.

Glass sample containers. 2 mL, with glass stoppers or Teflon-lined caps. If an automatic sample injector is used, the sample injector vials can be used.

Syringes. 10 µL, and other convenient sizes for making standards.

Pipets. 1.0 mL delivery type.

Volumetric flasks. 10 mL or convenient sizes for making standard solutions.

Reagents

Eluant, toluene (chromatographic grade).
Methyl iodide (99%).
Purified nitrogen.
Prepurified hydrogen.
Filtered compressed air.

Procedure

Cleaning of equipment. All glassware used for the laboratory analysis should be detergent washed and thoroughly rinsed with tap water and distilled water.

Calibration of personal pumps. Each personal pump must be calibrated with a representative charcoal tube in the line. This will minimize errors associated with uncertainties in the sample volume collected.

Collection and shipping of samples. Immediately before sampling, break the ends of the tube to provide an opening at least one-half the internal diameter of the tube (2 mm). The smaller section of charcoal is used as a backup and should be positioned nearest the sampling pump. The charcoal tube should be placed in a vertical direction during sampling to minimize channeling through the charcoal. Air being sampled should not be passed through any hose or tubing before entering the charcoal tube.

A maximum sample size of 50 L is recommended. Sample at a flow of 1 L/min or less. The flow rate should be known with an accuracy of at least 5%. The temperature and pressure of the atmosphere being sampled should be recorded. If pressure reading is not available, record the elevation. The charcoal tubes should be capped with the supplied plastic caps immediately after sampling. Under no circumstances should rubber caps be used. One tube should be handled in the same manner as the sample tubes (break, seal, and transport), except that no air is sampled through this tube. This tube should be labeled as a blank.

Capped tubes should be packed tightly and padded before they are shipped to minimize tube breakage during shipping. A sample of the suspected compound should be submitted to the laboratory in glass containers with Teflon-lined caps. These liquid bulk samples should not be transported in the same container as the charcoal tubes.

Preparation of samples. In preparation for analysis, each charcoal tube is scored with a file in front of the first section of charcoal and broken open. The glass wool is removed and discarded. The charcoal in the first (larger) section is transferred to a 2 mL stoppered sample container or automatic sample injector vial. The separating section of foam is removed and discarded; the second section is transferred to another sample container or vial. These two sections are analyzed separately.

Desorption of samples. Prior to analysis, 1.0 mL of toluene is pipetted into each sample container. Desorption should be done overnight or for an equivalent period of time. The sample vials should be capped as soon as the solvent is added to minimize volatilization.

GC conditions. The typical operating conditions for the gas chromatograph are:

30 mL/min (80 psig) nitrogen carrier gas flow
30 mL/min (50 psig) hydrogen gas flow to detector
300 mL/min (50 psig) air flow to detector
200° C. injector temperature
300° C. detector temperature
190° C. column temperature

Injection. The first step in the analysis is the injection of the sample into the gas chromatograph. To eliminate difficulties arising from blow back or distillation within the syringe needle, one should employ the solvent flush injection technique. The 10 μL syringe is first flushed with solvent several times to wet the barrel and plunger. Three μL of solvent are drawn into the syringe to increase the accuracy and reproducibility of the injected sample volume. The needle is removed from the solvent, and the plunger is pulled back about 0.2 μL to separate the solvent flush from the sample with a pocket of air to be used as a marker. The needle is then immersed in the sample, and a 5 μL aliquot is withdrawn, taking into consideration the volume of the needle, since the sample in the needle will be completely injected. After the needle is removed from the sample and prior to injection, the plunger is pulled back 1.2 μL to minimize evaporation of the sample from the tip of the needle. Observe that the sample occupies 4.9-5.0 μL in the barrel of the syringe. Duplicate injections of each sample and standard should be made. No more than a 3% difference in area is to be expected. An automatic sample injector can be used if it is shown to give reproducibility at least as good as the solvent flush technique. In this case 2 μL injections are satisfactory.

Measurement of area. The area of the sample peak is measured by an electronic integrator or some other suitable form of area measurement, and preliminary results are read from a standard curve prepared as discussed below.

Determination of desorption efficiency. The desorption efficiency of a particular compound can vary from one laboratory to another and also from one batch of charcoal to another. Thus, it is necessary to determine at least once the percentage of the specific compound that is removed in the desorption process, provided the same batch of charcoal is used.

Activated charcoal equivalent to the amount in the first section of the sampling tube (100 mg) is measured into a 2.0 mL sample container. This charcoal must be from the same batch as that used in obtaining the samples and can be obtained from unused charcoal tubes. A 350 mg/mL stock solution of the analyte in toluene is prepared. A known amount of this solution is injected directly into the activated charcoal with a 10 μL syringe, and the container is capped. The amount injected is equivalent to that present in a 50 L sample at the selected level. It is not practical to inject the neat liquid directly because the amounts to be added would be too small to measure accurately.

At least six tubes at each of three levels (0.5X, 1X, and 2X the standard) are prepared in this manner and allowed to stand for at least overnight to assure complete adsorption of the analyte onto the charcoal. These six tubes are referred to as the samples. A parallel blank tube should be treated in the same manner except that no sample is added to it. The sample and blank tubes are desorbed and analyzed in exactly the same manner as the sampling tube described above. The weight of analyte found in each tube is determined from the standard curve. Desorption efficiency is determined by the following equation:

D.E. = average wt (mg) recovered ÷ wt (mg) added

The desorption efficiency is dependent on the amount of analyte collected on the charcoal. Plot the desorption efficiency versus the weight of analyte found. The curve is used to correct for adsorption losses.

Calibration and Standards

It is convenient to express concentration of standards in terms of mg/mL of eluent. A series of standards, varying in concentration over the range of interest, is prepared and analyzed under the same GC conditions and during the same time period as the unknown samples. Curves are established by plotting concentration in mg/mL versus peak area. Note: standard solutions should be analyzed at the same time that the sample analysis is done. This will minimize the effect variations of FID response.

Calculations

Read the weight, in mg, corresponding to each peak area from the standard curve. No volume corrections are needed, because the standard curve is based on mg/mL eluent and the volume of sample injected is identical to the volume of the standards injected. Corrections for the blank must be made for each sample:

mg = mg sample − mg blank

where:

mg sample = mg found in front section of sample tube
mg blank = mg found in front section of blank tube

A similar procedure is followed for the backup sections. Add the weights found in the front and backup sections of the same sample tube to determine the total weight in the sample. Read the desorption efficiency from the curve for the amount of analyte found in the front section. Divide the total weight by this desorption efficiency to obtain the corrected mg/sample:

Corrected mg/sample = total weight ÷ D.E.

The concentration of the analyte in the air sampled can be expressed in mg/m^3, which is numerically equal to μg/L of air:

mg/m^3 = (corrected mg)(1000 L/m^3) ÷ air volume sampled in L

Another method of expressing concentration is ppm:

ppm = (mg/m^3)(24.45)(760)(T + 273) ÷ (M.W.)(P)(298)

where:

P = pressure (mm Hg) of air sampled
T = temperature (°C.) of air sampled
24.45 = molar volume (L/mol) at 25° C. and 760 mm Hg
M.W. = molecular weight (g/mol) of analyte
760 = standard pressure (mm Hg)
298 = standard temperature (°K)

References

1. White, L.D. et al. A convenient optimized method for the analysis of selected solvent vapors in the industrial atmosphere. Amer. Ind. Hyg. Assoc. J. 31: 225, 1970.
2. Documentation of NIOSH Validation Tests. NIOSH Contract No. CDC-99-74-45.

METHYLISOBUTYLCARBINOL

Synonyms: 4-methyl-2-pentanol; isobutylmethylcarbinol $(CH_3)_2CHCH_2CHOHCH_3$
ACGIH TLV: 25 ppm (100 mg/m³) - Skin
OSHA Standard: 25 ppm

Method	Sampling Duration	Sampling Location	Useful Range (ppm)	System Cost ($)	Test Cost ($)	Manufacturer
IR	Cont	Portable	0.5-50	4374	0	Foxboro

NIOSH METHOD NO. S52

(Use as described for allyl alcohol, with 120° C. column temperature)

METHYL METHACRYLATE

ACGIH TLV: 100 ppm (410 mg/m³)
OSHA Standard: 100 ppm

$CH_2=C(CH_3)COOCH_3$

Method	Sampling Duration	Sampling Location	Useful Range (ppm)	System Cost ($)	Test Cost ($)	Manufacturer
DT	Grab	Portable	50-500	150	2.80	Nat'l Draeger
DT	Grab	Portable	50-2750	180	2.25	Bendix
IR	Cont	Portable	2-200	4374	0	Foxboro
PI	Cont	Fixed	2-200	4950	0	HNU

NIOSH METHOD NO. S264

(Use as described for ethyl silicate)

α-METHYLSTYRENE

ACGIH TLV: 50 ppm (240 mg/m³)
OSHA Standard: 100 ppm

$C_6H_5C(CH_3)=CH_2$

Method	Sampling Duration	Sampling Location	Useful Range (ppm)	System Cost ($)	Test Cost ($)	Manufacturer
IR	Cont	Portable	2-200	4374	0	Foxboro

NIOSH METHOD NO. P&CAM 127

(Use as described for acetone, with 115° C. column temperature)

MONOMETHYLANILINE

Synonym: N-methylaniline
ACGIH TLV: 0.5 ppm (2 mg/m³) - Skin
OSHA Standard: 2 ppm

$CH_3C_6H_5NHCH_3$

Method	Sampling Duration	Sampling Location	Useful Range (ppm)	System Cost ($)	Test Cost ($)	Manufacturer
DT	Grab	Portable	2.5-45	180	2.25	Bendix
IR	Cont	Portable	0.4-4	4374	0	Foxboro
PI	Cont	Fixed	0.2-20	4950	0	HNU

NIOSH METHOD NO. P&CAM 168

(See aniline)

MORPHOLINE

Synonym: diethylenimide oxide
ACGIH TLV: 20 ppm (70 mg/m³) - Skin
OSHA Standard: 20 ppm

$OCH_2CH_2NHCH_2CH_2$

Method	Sampling Duration	Sampling Location	Useful Range (ppm)	System Cost ($)	Test Cost ($)	Manufacturer
DT	Grab	Portable	1-60	180	2.25	Bendix
IR	Cont	Portable	0.4-40	4374	0	Foxboro

NIOSH METHOD NO. P&CAM 221

(Use as described for n-butylamine, with 200° C. column temperature)

NICKEL

ACGIH TLV: 1 mg/m³ for metal; 0.1 mg/m³ for soluble compounds
OSHA Standard: 1 mg/m³
NIOSH Recommendation: 0.015 mg/m³

Ni

Method	Sampling Duration	Sampling Location	Useful Range (mg/m³)	System Cost ($)	Test Cost ($)	Manufacturer
DT	Grab	Portable	0.25-1	150	3.50	Nat'l Draeger

NIOSH METHOD NO. P&CAM 298

Principle
A known volume of air is drawn through a cellulose ester membrane filter to collect particulate nickel metal and compounds. The filters are then ashed using nitric and perchloric acids, with hydrofluoric acid added if needed to dissolve silicates, and diluted to a known volume. The solutions are analyzed, along with standard nickel solutions, by atomic absorption spectroscopy using a heated graphite atomizer.

Range and Sensitivity
The analytical working range is 1-6 μg of Ni dissolved in 25 mL. This corresponds to 0.005-0.030 mg Ni/m³ in a 200 L air sample. The range may be extended by diluting the ashed sample to volumes smaller or larger than 25 mL as long as the combination of nickel concentration in solution and aliquot size to the graphite atomizer delivers 0.002-0.012 μg Ni/aliquot, providing the method of standard additions is not used. This working range applies at the 341.5 nm absorption line; the 232 nm line is about three times more sensitive but may be more subject to nonspecific absorbance.

The instrumental sensitivity at 341.5 nm is about 0.00027 μg Ni/aliquot for 1% (e.g., 0.0053 g/mL for a 0.05 mL aliquot). This corresponds to about 0.100 A for a 0.05 mL aliquot of 0.125 g/mL solution. The detection limit is approximately 0.0005 μg Ni/aliquot, corresponding to 0.00125 mg/m³ for a 200 L air sample dissolved in 25 mL of solution with a 0.05 mL aliquot.

Interferences
There are no known interferences

Precision and Accuracy
Instrumental precision, as determined from three standards each analyzed in triplicate (0.001 μg Ni/aliquot), was 10.3% RSD. Interlaboratory precision, determined from two replicate field samples for each of three laboratories, ranged from 6.7% RSD to 26% RSD on three sets of samples. Recovery of standard nickel solutions added to clean filters was 90.4 ± 4.0% and 100.8 ± 3.5% for two laboratories at 16.2 μg Ni/filter.

Advantages and Disadvantages
The major advantage is the sensitivity of the instrumental method. Also the analysis is specific for nickel and easy to perform. The equipment required is relatively expensive.

Apparatus

Sampling equipment. Consists of filter unit (0.8 micron cellulose ester membrane filter, 37 mm diameter, in a plastic cassette filter holder) and personal sampling pump capable of sampling at 1.5-2 L/min. The pump must be calibrated with a representative filter in line, using a soap bubble flowmeter or wet or dry test meter, and its flow rate must be known accurately to within 5%.

Atomic absorption spectrometer. Equipped with a heated graphite atomizing rod, tube or furnace. Reproducible control of times and temperatures during the drying, charring and atomizing cycles is essential and a minimum atomization temperature of 2700° C. is required. Accessories needed for the spectrometer are:

hollow cathode lamp for nickel
readout device (recorder or digital peak height analyzer)
gas control system for argon or nitrogen purge gas
pipetting system (automatic or manual, 5-50 μL, as appropriate to atomizer size)
background correction system (H_2 or D_2 simultaneous correction is preferred but manual correction using a non-absorbing line is sufficient)

Glassware, borosilicate.
Phillips beakers. 125 mL, with watchglass covers.
Volumetric flasks. 25 mL or other appropriate size.
Adjustable, thermostatically controlled hotplate. Capable of reaching a surface temperature of 400° C.
Perchloric acid hood.

Reagents

Wherever possible, reagents used should be ACS reagent grade or better.
Distilled or deionized water.
Nitric acid. 70% (w/w), redistilled in glass.
Perchloric acid. 60% (w/w).
Hydrofluoric acid. 50% (w/w).
Dilute nitric acid. 2% (w/w). Add 2 mL 70% HNO_3 to distilled or deionized water and dilute to 100 mL.
Stock standard nickel solution, 1000 ppm Ni. Dissolve 1.000 g pure Ni metal in a minimum volume of 70% HNO_3 and dilute to 1 L with 2% HNO_3. Alternatively, commercially prepared standard may be used.
Working standards. Prepare a 10 ppm Ni standard solution by dilution of the 1000 ppm stock solution with 2% HNO_3. This solution is stable for several months if stored in a polyethylene bottle and is used to prepare fresh weekly the working standards which should cover the range 0.05-0.3 ppm Ni (i.e., 0.05, 0.10, 0.15, 0.20, 0.25 and 0.30 ppm Ni in 2% HNO_3).
Argon or nitrogen gas in compressed gas cylinder.

Procedure

Cleaning of equipment. New glassware must be cleaned by soaking in hot, concentrated nitric acid followed by thorough rinsing with distilled or deionized water. Then, after each use, the glassware should be washed with, in order, detergent solution, tap water, diluted nitric acid (soak 4 hours or longer) and distilled or deionized water.

Collection and shipping of samples. Assemble the filter in the cassette and close firmly to insure that a seal is made around the edge of the filter (the filter is supported by a cellulose backup pad). Apply a shrinkable cellulose band to the outside of the cassette. Remove the plugs from the cassette and attach to the personal sampling pump by means of flexible tubing. Clip the cassette, face down, to the worker's lapel. Air should not pass through any hose or tubing before entering the cassette.

Take the sample at an accurately known flow rate in the range 1.5-2 L/min. A sample size of 200-400 L is recommended. Check the pump frequently during sampling to insure that the flow rate has not changed. If sampling problems preclude the accurate measurement of air volume, discard the sample. Record the sampling time, flow rate and ambient temperature and pressure. With each batch of ten samples or less, submit one blank filter from the same lot used for sampling which was subjected to the same handling as the sample filters except that no air was pulled through it. Ship the cassettes so as to prevent excessive vibration or jarring.

Analysis of samples. Open the cassette filter holder and carefully remove the cellulose membrane filter with tweezers and transfer it to a Phillips beaker. Discard the cellulose backup pad. To each beaker containing a sample or blank filter, add 5 mL 70% HNO_3, cover with a watchglass and allow to sit on a low temperature (110-140° C.) hotplate for 6-12 hours. Then reduce the volume of solution to a few drops by increasing hotplate temperature or removing the watchglass.

Add 2 mL 70% HNO_3, 1 mL 60% $HClO_4$ and 5 drops 50% HF. Cover with a watchglass and heat on a 400° C. hotplate until dense fumes of perchloric acid persist. Using distilled water, rinse the watchglass and sides of the beaker and evaporate to dryness. Cool the beaker and dissolve the residue in 2% HNO_3. Quantitatively transfer the solution to a 25 mL volumetric flask and dilute to volume with 2% HNO_3. Inject samples into the graphite atomizer and operate in accordance with the manufacturer's instructions within the following limitations:

Aliquot size. 5-50 µL, depending on atomizer dimensions.

Inert gas flow. Increased sensitivity will be obtained if flow is interrupted during atomization step.

Readout mode. Peak height (analog or digital) is more precise on some spectrometers. The peak is fast (ca. 3 sec) and narrow.

Wavelength. Either 232 nm or 341.5 nm may be used; 232 nm is more sensitive but also more prone to background absorbance.

Background correction. The necessity for background correction should be determined on a representative sample. If needed, use D_2 or H_2 lamp (simultaneously or sequentially with Ni lamp) or a nonabsorbing line (231.5 for the 232 nm Ni line).

Dry cycle. At least 20 sec at 100° C. Longer time and ramp program may be needed for larger aliquots.

Ash (char) cycle. 15-30 sec at 700° C. Longer time needed for high solids content in sample. Ramping is desirable in any case to prevent spattering of any liquid remaining after dry cycle.

Atomize cycle. 6-8 sec at 2700-3000° C. Times as short as 4-5 sec may be used at the higher temperatures. Do not atomize below 2600° C. as atomization is too slow for reproducible atomization.

Graphite atomizer. May be either non-pyrolytic or pyrolytic; slightly better sensitivity with pyrolytic.

Analyze a series of standards before and after each set of samples and analyze one mid range standard after each ten samples. Analyze all solutions in triplicate and take the mean absorbance as proportional to concentration. Filter blanks and reagent blanks should also be analyzed in the same manner.

Calibration and Standards

The calibration curve is constructed from the absorbances measured for the standard solutions analyzed under the same instrumental conditions as the samples as described earlier. If the sample matrix has a high dissolved solids content, the absorbance of the nickel may be suppressed. In that case, one of two alternatives must be followed: prepare the standards in a chemically similar matrix or use the method of standard additions to at least two representative samples to establish the slope of the calibration curve in that matrix. Note: standard additions must be restricted to a concentration range producing a linear calibration curve.

Calculations

Read the concentration of the sample or blank solution from the calibration curve. Express concentration in μg Ni/mL. Subtract from the sample concentration the appropriate blank concentration. Multiply the corrected sample concentration by the volumetric dilution factor (e.g., 25 mL) to obtain total μg Ni/sample. For personal sampling pumps with rotameters, the air volume should be corrected:

corrected volume = $ft(P_1T_2/P_2T_1)^{1/2}$

where:

f = sample flow rate in L/min
t = sampling time in min
P_1 = pressure (mm Hg) of atmosphere when pump was calibrated
P_2 = pressure (mm Hg) of atmosphere sampled
T_1 = temperature (K) of atmosphere when pump was calibrated
T_2 = temperature (K) of atmosphere sampled

Divide μg Ni/sample by the air volume, in L, to obtain μg Ni/L air, which is equal to mg Ni/m^3 air.

Reference

1. Begnoche, B.C. and T.H. Risby. Determination of metals in atmospheric particulates using low-volume sampling and flameless atomic absorption spectrometry. Anal. Chem. 47: 1041-1045, 1975.

NITRIC ACID

ACGIH TLV: 2 ppm (5 mg/m^3) HNO_3
OSHA Standard: 2 ppm

Method	Sampling Duration	Sampling Location	Useful Range (ppm)	System Cost ($)	Test Cost ($)	Manufacturer
DT	Grab	Portable	1-50	150	2.80	Nat'l Draeger
DT	Grab	Portable	0.2-40	180	2.25	Bendix

NIOSH METHOD NO. S319

Principle

Samples are collected by drawing a known volume of air through a midget impinger containing distilled water. The contents of the impinger are analyzed by direct potentiometry using an ion specific electrode. The concentration of nitric acid in the sample is determined by a comparison of the potential obtained for the sample solution to a calibration curve obtained by measurement of standard solutions. A calibration standard is run before and after each sample to assure that reliable results are obtained.

Range and Sensitivity

This method was validated over the range 2.60-10.8 mg/m^3 using a 180 L air sample. The lower limit of detection for nitrate using the nitrate ion specific electrode is 6×10^{-6} M or 0.4 µg/mL according to the electrode manufacturer.

Interferences

Constituents of the buffer solution complex, precipitate, decompose or otherwise remove the common interfering anions including bromide, chloride, fluoride, iodide, phosphate, and nitrite (reference 1). However, high concentrations of any of these species should be avoided.

Precision and Accuracy

The coefficient of variation for the total sampling and analytical method in the range of 2.60-10.8 mg/m^3 was 0.0824. This value corresponds to a 0.4 mg/m^3 standard deviation at the OSHA standard level. Statistical information and details of the validation and experimental test procedures can be found in references 2 and 3. A collection efficiency of 97% was determined for HNO_3 in water; thus, no bias was introduced in the sample collection stem from this source.

The average concentration "found" at the OSHA standard levels using the overall sampling and analytical method were 4.3% lower than the "true" concentrations found for a limited number of samples analyzed by an alternate method (reference 3). In addition, samples were found to be stable when stored for 7 days. Thus, coefficient of variation is a satisfactory measure of the precision of the sampling method.

Advantages and Disadvantages

Advantages of the method include simplicity, specificity, and speed. Disadvantages of the method are associated with the collection device, the midget impinger. Not only is it inconvenient, but there is the possible loss of sample if the impinger is not maintained in an upright position during sampling.

Apparatus

Personal sampling pump. A calibrated personal sampling pump, the flow of which can be determined within 5% at the recommended flow rate. The pump is protected from splashover or water condensation by a piece of glass wool loosely packed within the tubing used to connect the pump and midget impinger.

Midget impinger.
Distilled water.
Pipet. 20 mL, or other suitable device for filling the impinger.
Barometer.
Thermometer.
Stopwatch.

Nitrate liquid ion-exchange membrane electrode. Orion 92-07 or equivalent.
Double junction reference electrode. Orion 90-02 or equivalent.
Digital pH/mV meter. Orion model 801 or equivalent.
Volumetric flasks. 25 mL.
Pipets. 20 and 5 mL.
Beakers. 50 mL.
Magnetic stirrer and Teflon-coated stirring bars.

Reagents

All reagents must be ACS reagent grade or better.

Buffer solution. Dissolve 4.3 g aluminum sulfate, 1.6 g boric acid, 3.9 g silver sulfate, and 2.4 g sulfamic acid in distilled water and dilute to 500 mL. The purpose of this buffer is explained in reference 1.

Nitric acid stock solution. Pipet 5 mL of concentrated nitric acid into a 500 mL volumetric flask and dilute to the mark with distilled water. The concentration of the stock solution should be determined accurately by titration of 15 mL aliquots with 0.1 N sodium hydroxide using phenolphthalein as an indicator. Duplicate titrations should agree within 0.1 mL. Calculate the molarity of the stock solution using the equation:

$$\text{Molarity} = (\text{molarity of NaOH})(\text{mL of NaOH}) \div 15 \text{ mL}$$

Nitric acid working solution. Pipet 5 mL of the nitric acid stock solution into a 500 mL volumetric flask and dilute to the mark with distilled water.

Working standards. Working standards corresponding to the amounts of nitric acid collected at the 0.5, 1, and 2X the OSHA standard can be prepared by pipeting 5, 10, and 20 mL respectively of the nitric acid working solution into three 25 mL volumetric flasks, each of which contains 5 mL of the buffer solution, and diluting to the mark with distilled water. The concentrations of the working standards are calculated by multiplying the concentration of the nitric acid stock solution by the appropriate dilution factors.

Saturated sodium sulfate solution. Add sufficient sodium sulfate to distilled water to form a saturated solution. The supernatant is used for the outer chamber filling solution in the double junction reference electrode.

Procedure

Cleaning of equipment. All glassware should be washed with detergent and rinsed thoroughly with distilled water.

Calibration of personal pump. Each personal pump must be calibrated with a representative midget impinger in the line.

Collection and shipping of samples. Connect a midget impinger containing 20 mL of distilled water to a sampling pump using a piece of flexible tubing. The tubing should be loosely plugged with a piece of glass wool to protect the sampling pump from splashover and condensation. The impinger must be maintained in a vertical position during sampling. Air being sampled should not be passed through any hose or tubing before entering the impinger.

A sample size of 180 L is recommended. Sample at a flow rate of 1.0 L/min for 180 min. The flow rate should be known with an accuracy of at least 5%. Record the ambient temperature and barometric pressure. Remove the impinger stem and tap the stem gently against the inside wall of the impinger bottle to recover as much of the sampling solution as possible. Rinse the impinger stem with 1 mL of distilled water and add the wash to the impinger bottle.

Seal the impinger with a hard, non-reactive stopper (preferably Teflon). Do not seal with rubber. The stoppers on the impingers should be tightly sealed to prevent leakage during shipping. With each batch of ten samples submit one impinger containing 20 mL of distilled water. This impinger should be handled in the same manner as the samples except that no air is drawn through it. Label this impinger as the blank.

Analysis of samples. Pipet 5 mL of the buffer solution into a 25 mL volumetric flask. Transfer the contents of the impinger to the volumetric flask and dilute to the mark with distilled water. Empty the contents of the volumetric flask into a 50 mL beaker. Add a magnetic stirring bar.

Immerse the nitrate ion electrode and reference electrode in the sample solution and record the mv reading. Both samples and standards should be stirred while readings are taken. The reading should be taken after the meter has stabilized to 0.1 mv/10 sec. Follow the instrument manufacturer's instruction manual for proper operation and measurement procedures. Samples should be interspersed between standards as described below. Appropriate blanks must be analyzed by the same procedure used for the samples.

Calibration and Standards

Nitric acid standards should be prepared as described earlier. A series of at least three standards should be analyzed prior to the measurement of samples. Plot the potentials in mv developed in the standards (on the linear axis) against the molarity of the standards (on the log axis) on semilogarithmic graph paper. The slope of the best straight line fitted into the standards measurements should be equal to 59.2 ± 1 mv/10-fold change in concentration. If the slope of the line does not fall within these limits, consult the troubleshooting section of the electrode manufacturer's instruction manual.

Samples should be interspersed between standards having concentrations which bracket the concentrations of the samples. At the conclusion of the analysis of a series of samples, calculate the average response in mv for each standard. Plot the data as described above. The resulting calibration line is used to calibrate the concentrations of the samples.

Calculations

Determine the molarity of nitric acid in the sample from the calibration line. Calculate the weight of nitric acid in μg corresponding to the molarity of the sample using the equation:

$$\mu g = (\text{molarity})(1.575 \times 10^6)$$

Corrections for the blank must be made for each sample impinger:

$$\mu g = \mu g \text{ sample} - \mu g \text{ blank}$$

where:

μg sample = μg found in sample impinger
μg blank = μg found in blank impinger

Determine the volume of air sampled at ambient conditions in L based on the appropriate information, such as flow rate in L/min multiplied by sampling time. If a pump using a rotameter for flow rate control was used for sample collection, a pressure and temperature correction must be made for the indicated flow rate. The expression for this correction is:

Corrected volume = $ft(P_1T_2/P_2T_1)^{1/2}$

where:

    f = sample flow rate
    t = sampling time
    P_1 = atmospheric pressure during calibration of sampling pump (mm Hg)
    P_2 = atmospheric pressure of air sampled (mm Hg)
    T_1 = ambient temperature during calibration of sampling pump (°K)
    T_2 = ambient temperature of air sampled (°K)

The concentration of nitric acid in the sampled air can be expressed in mg/m³ (= μg/L) as follows:

mg/m³ = μg ÷ air volume sampled in L

Another method of expressing concentration is ppm (corrected to standard conditions of 25° C. and 760 mm Hg):

ppm = (mg/m³)(24.45)(760)(T + 273) ÷ (M.W.)(P)(298)

where:

    P = barometric pressure (mm Hg) of air sampled
    T = temperature (°C.) of air sampled
    24.45 = molar volume (L/mol) at 25° C. and 760 mm Hg
    M.W. = molecular weight
    760 = standard pressure (mm Hg)
    298 = standard temperature (°K)

References

1. P.J. Milham, A.S. Awad, R.E. Paull, and J.H. Bull. Analyst 95: 751, 1970.
2. Documentation of NIOSH Validation Tests, National Institute for Occupational Safety and Health, Cincinnati, Ohio (DHEW-NIOSH-Publication No. 77-185), 1977. Available from Superintendent of Documents, U.S. Government Printing Office, Washington, D.C., Order No. 017-033-00231-2.
3. Backup Data Report for Nitric Acid. No. S319, prepared under NIOSH Contract No. 210-76-0123.
4. Final Report. NIOSH Contract HSM-99-71-31, Personal Sampler Pump for Charcoal Tubes, September 15, 1972.

NITRIC OXIDE

ACGIH TLV: 25 ppm (30 mg/m³) NO
OSHA Standard: 25 ppm

Method	Sampling Duration	Sampling Location	Useful Range (ppm)	System Cost ($)	Test Cost ($)	Manufacturer
DT	Grab	Portable	0.5-5000	150	2.60	Nat'l Draeger
DT	Grab	Portable	10-300	165	2.30	Matheson
DT	Grab	Portable	2-200	180	2.25	Bendix
DT	4 hr	Personal	1.25-350	850	3.10	Nat'l Draeger
ES	1 min	Personal	0.25-25	1765	0	Interscan
ES	Cont	Portable	0.1-50	2100	0	Energetics
ES	Cont	Portable	1-100(500)	2145	0	Interscan
ES	Cont	Fixed	0.25-25(250)	2553	0	Interscan
ES	Cont	Port/Fixed	0.25-25(250)	3260	0	Interscan
IR	Cont	Portable	4-50	4374	0	Foxboro
CM	Cont	Port/Fixed	0.002-10	4995	Variable	CEA

NIOSH METHOD NO. P&CAM 231

Principle

Nitrogen dioxide (NO_2) and nitric oxide (NO) are collected from air in a three section sorbent tube. The NO_2 is absorbed in the first section, which contains triethanolamine (TEA) impregnated on molecular sieve. The NO is converted to NO_2 by a proprietary oxidizer in the second section. The NO_2 thus formed from the NO is absorbed in the third section by another bed of TEA-impregnated molecular sieve. The first and third sections are desorbed with solutions of TEA in water and the nitrite in these solutions is determined spectrophotometrically by the Griess-Saltzman reaction (reference 1). The nitrite found in the first section is reported as NO_2 and the nitrite in the third section is reported as NO.

Range and Sensitivity

The linear range of the standard curve is from 0.5-18 µg of nitrite in 10 mL of desorbing solution, which corresponds in this method to a range of 0.8-30 ppm of NO_2 or NO in a 1 L sample of air. The sensitivity is 0.4 µg/10 mL for an absorbance of 0.04. The upper limit of the range can be extended by taking smaller aliquots for analysis, or by diluting intensely colored solutions with water.

Interferences

Inorganic nitrites cause positive interference. Nitric acid and nitrates do not interfere. Ammonia does not interfere.

Precision and Accuracy

The average recovery for 22 samples in the range of 0.5-5 ppm of NO_2 was greater than 96% and the coefficient of variation was 0.07. For 18 samples the average recovery of NO varied with the amount of NO collected. The recovery was 100% at 12.5 ppm. At 25 ppm only 84% recovery was achieved, and at 50 ppm only 67%. However, the coefficient of variation over the range was only 0.06. The recovery may vary depending upon the sample flow rate and the properties of the particular log of oxidizer used. Each laboratory should determine the efficiency of the sampling tubes employed. The accuracy of the overall sampling and analytical method has not been determined.

Advantages and Disadvantages

Both nitrogen dioxide and nitric oxide are collected simultaneously. This method is simple and convenient for field sampling. Samples can be stored at ambient temperature for at least 10 days without any effect on the results. At 50 ppm of NO the collection efficiency is poor (about 67%) because the oxidizer is consumed. If high humidity or water mist is present, the breakthrough volume can be severely reduced. If water condenses in the tube, NO_2 and NO may not be collected quantitatively.

Apparatus

Solid sorbent tubes. Sorbent tubes are made in the following manner. Using a gas-oxygen torch, heat a section of 5 mm I.D., 7 mm O.D. Pyrex glass tubing and pull it apart to form a tube approximately 15 cm long with a taper 2 cm long. Seal the tapered end of the tube in the flame. Allow it to cool, then insert a small plug of glass wool through the open end of the tube; push the glass wool through the open end of the tube with a thin wooden stick and pack gently. Weigh 400 mg of TEA sorbent and pour the material into the tube. Gently tap the tube on the table top several times to ensure uniform packing. Insert another small plug of glass wool to keep the TEA sorbent in place. For the next section, pour 800 mg of oxidizer into the tube. Again tap the tube and insert a plug of glass wool; pack lightly. Insert another plug of glass wool, maintaining an air gap of 12 mm between these two plugs. Weigh 400 mg of TEA sorbent and pour the material into the tube. Carefully tap the tube and gently pack another glass wool plug without closing the 12 mm air gap. Seal the open end of the tube with the torch (Figure 1).

Personal sampling pump. A pump that can provide a flow rate of 50 mL/min within 5% accuracy is required. The pump should be calibrated with a representative sorbent tube in the sampling line. A dry or wet test meter or glass rotameter that will determine the flow rate to within 5% may be used for the calibration.

Spectrophotometer. Capable of measurements at 540 nm.

Matched glass cells or cuvettes. 1 cm path length.

Assorted laboratory glassware. Pipets, glass stoppered graduated cylinders, and volumetric flasks of appropriate sizes.

Reagents

Oxidizer. Proprietary material No. 1900277 from the Draegerwerk Company of West Germany, supplied through its U.S. distributor, National Mine Safety Company, or the equivalent.

TEA Sorbent. Place 25 g of triethanolamine in a 250 mL beaker; add 4 g of glycerol, 50 mL of acetone and sufficient distilled water to bring the volume up to 100 mL. To the mixture add about 50 mL of type 13X, 30/40 mesh molecular sieve. Stir and let stand in a covered beaker for about 30 min. Decant the excess liquid, and transfer the molecular sieve to a porcelain pan. Place the pan under a heating lamp until most of the moisture has evaporated. Complete the drying in an oven at 110° C. for 1 hr. The sorbent should be free flowing. Store it in a closed glass container.

Desorbing solution. Dissolve 15.0 g of triethanolamine in approximately 500 mL of distilled water, add 0.5 mL of n-butanol, and dilute to 1 L.

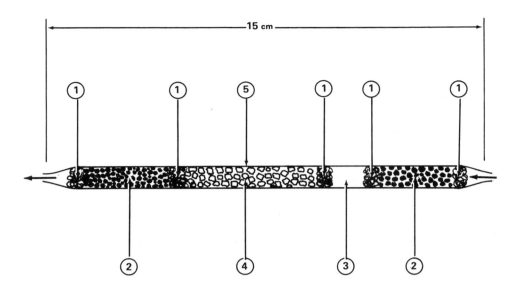

1. GLASS WOOL PLUGS
2. TEA SORBENT, 400 mg
3. AIR GAP, 12 mm
4. OXIDIZER, 800 mg
5. GLASS TUBE, 5 mm i.d.

SORBENT TUBE FOR NO_2 and NO

Hydrogen peroxide, 0.02%. Dilute 0.2 mL of 30% hydrogen peroxide to 250 mL with distilled water.

Sulfanilamide solution. Dissolve 10 g of sulfanilamide in 400 mL of distilled water. Add 25 mL of concentrated phosphoric acid, mix well, and dilute to 500 mL.

NEDA solution. Dissolve 0.5 g of N-(1-naphthyl)ethylenediamine dihydrochloride in 500 mL of distilled water.

Nitrite stock standard solution (100 µg/mL). Dissolve 0.1500 g of reagent grade sodium nitrite in distilled water an dilute to 1 L.

Procedure

Cleaning of equipment. Wash all glassware with detergent solution, soak in nitric acid, rinse in tap water and distilled water, and then rinse thoroughly with double distilled water.

Collection and shipping of samples. Before sampling, break open the ends of the sorbent tube to provide an opening that is approximately one-half the internal diameter of the tube. The air must flow through the 12 mm air space before it flows through the oxidizer. Therefore attach the end of the tube without the air gap between the oxidizer section and TEA sorbent section to the pump with a length of small diameter Tygon tubing. Mount the tube in a vertical position to avoid channeling. The air being sampled should not pass through any hose or tubing before it enters the sorbent tube.

Turn on the pump to begin sample collection. Sample at a flow rate of 50 mL/min or less to obtain a maximum sample volume of 1 L. Measure the flow rate and time, or volume, as accurately as possible. If a low flow rate pump is used, set the rate to an approximate value and record the initial final stroke counter readings. Obtain the sample volume by multiplying the number of strokes by the stroke volume. Measure and record the temperature and pressure of the atmosphere being sampled.

Cap the sorbent tubes with 7 mm I.D. plastic caps immediately after sampling. (Masking tape can be substituted for the plastic caps.) With each batch of samples, submit one blank sorbent tube. This tube is handled in the same manner as the other tubes (break, seal, and transport) except that no air is drawn through it. When more than ten samples are submitted, include an additional blank for every ten samples. Pack the capped sorbent tubes tightly and pad them to minimize breakage during shipping.

Analysis of samples. With tweezers remove and discard the glass wool plugs from an exposed sorbent tube and transfer each TEA sorbent bed to separate, 25 mL glass stoppered graduated cylinders. Label the graduated cylinder as to the location of the TEA sorbent with respect to the oxidizer section. To each graduated cylinder add enough of the desorbing solution to make the volume up to 20 mL, and shake the mixture vigorously for about 30 sec.

Allow a few minutes for the solids to settle, and then transfer 10 mL to another 25 mL glass-stoppered graduated cylinder. Develop the color of the solution for 10 min in the same manner as described below for the preparation of the standard curve. From the standard curve determine the amount of nitrite in the 10 mL aliquot.

Determination of collection and desorption efficiencies. The collection and desorption efficiencies of a given compound can vary from one laboratory to another and also from one batch of sorbent tubes to another. Thus, it is necessary to determine at least once the percentage of sample collected and then removed in the desorption process. Results indicate that the recovery of NO varies with the amount of NO collected, particularly at higher concentrations (for example, at 50 ppm).

Sorbent tubes from the same batch as that used in obtaining samples are used in this determination. Known volumes of NO_2 and NO are injected into a bag containing a known volume of air. The bag is made of Tedlar (or another material that will not absorb NO_2 or NO) and should have a gas sampling valve and a septum injection port. The concentrations of NO_2 and NO in the bag may be calculated at room temperature and pressure. A measured volume is then sampled through a sorbent tube with a calibrated sampling pump. At least 5 tubes are prepared in this manner. These tubes are desorbed and analyzed in the same manner as the samples.

Calibration and Standards

Dilute 2 mL of the nitrite stock standard (100 μg/mL) with the desorbing solution to prepare a solution with a nitrite concentration of 2 μg/mL. To a series of 25 mL glass stoppered graduated cylinders add 1, 3, 5, 7, and 9 mL of the dilute standard solution. Add enough of the absorbing solution to bring the volume in each cylinder up to 10 mL to prepare working standards with nitrite concentrations of 2, 6, 10, 14, and 18 μg/10 mL.

To each graduated cylinder, add 1 mL of the 0.02% hydrogen peroxide solution, 10 mL of the sulfanilamide solution, and 1.4 mL of the NEDA solution, with thorough mixing after the addition of each reagent. Allow 10 min for complete color development. Measure the absorbance of the solutions at 540 nm, using a reagent blank in the reference cell. Prepare a standard curve by plotting absorbance versus weight of nitrite (in μg) in 10 mL of the desorbing solution.

Calculations

From the standard curve, read the weight of nitrite (in μg) in 10 mL of the desorbing solution corresponding to the absorbance of the sample solution. Multiply this weight by 2 to determine the total amount (in μg) of nitrite extracted with 20 mL of desorbing solution from the sorbent section being analyzed. The calibration procedure is based upon the empirical observation that

0.63 mole of sodium nitrite produces the same absorbance in the color development solution as 1 mole of NO$_2$ (reference 2). Divide the amount of nitrite desorbed from the sorbent material by 0.63 to determine the apparent amount of NO$_2$ collected in the sorbent section. These calculations are summarized in the following equation:

$$W = (\mu g\ NO_2)(2) \div 0.63$$

where:

W = weight (in μg) of NO$_2$ found

Correct the amount of NO$_2$ calculated above for the amount of NO$_2$, if any, found on the corresponding sorbent section of a blank tube to obtain the amount of NO$_2$ in the sample, as follows:

$$Ws = W - Wb$$

where:

Ws = corrected weight (in μg) of NO$_2$ in sample
Wb = weight (in μg) of NO$_2$ in the corresponding section of a blank tube

The concentration of NO$_2$ in ppm by volume in the air sample is calculated as follows:

$$ppm = (Ws)(24.45)(760)(T + 273) \div (V)(M.W.)(P)(298)$$

where:

V = volume (L) of air sampled
P = pressure (mm Hg) of air sampled
T = temperature (°C.) of air sampled
24.45 = molar volume (L/mol) at 25° C. and 760 mm Hg
M.W. = molecular weight

The ppm of NO$_2$ found in the third section (downstream from the oxidizer) is reported as ppm of NO.

References

1. Saltzman, B.E. Colorimetric microdetermination of nitrogen dioxide in the atmosphere. Anal. Chem. 26: 1949, 1954.
2. Blacker, J.H. Triethanolamine for collecting nitrogen dioxide in the TLV range. Amer. Ind. Hyg. Assoc. J. 34: 390, 1973.
3. NIOSH Sampling Data Sheet No. 32.01. NIOSH Manual of Sampling Data Sheets, Measurements Research Branch, Division of Physical Sciences and Engineering, National Institute for Occupational Safety and Health, December 22, 1975.
4. Wiley, M.A., C.S. McCammon, Jr., and L.J. Doemeny. A Solid Sorbent Personal Sampling Method for the Simultaneous Collection of Nitrogen Dioxide and Nitric Oxide in Air, presented at the American Industrial Hygiene Association Conference, Atlanta, Georgia, May, 1976.

NITROBENZENE

ACGIH TLV: 1 ppm (5 mg/m³) - Skin
OSHA Standard: 1 ppm

$C_6H_5NO_2$

Method	Sampling Duration	Sampling Location	Useful Range (ppm)	System Cost ($)	Test Cost ($)	Manufacturer
IR	Cont	Portable	0.4-4	4374	0	Foxboro

NIOSH METHOD NO. S217

Principle

A known volume of air is drawn through a silica gel tube to trap the organic vapors present. The silica gel in the tube is transferred to a small, stoppered sample container, and the analyte is desorbed with methanol. An aliquot of the desorbed sample is injected into a gas chromatograph. The area of the resulting peak is determined and compared with areas obtained for standards.

Range and Sensitivity

This method was validated over the range of 3.11-12.45 mg/m³ at an atmospheric temperature and pressure of 23° C.. and 765 mm Hg, using a 55 L sample. Under the conditions of sample size (55 L) the probable useful range of this method is 0.5-15 mg/m³. The method is capable of measuring much smaller amounts if the desorption efficiency is adequate. Desorption efficiency must be determined over the range used.

The upper limit of the range of the method is dependent on the adsorptive capacity of the silica gel tube. This capacity varies with the concentrations of the analyte and other substances in the air. The first section of the silica gel tube was found to hold at least 2.78 mg of analyte when a test atmosphere containing 12.45 mg/m³ of analyte in air was sampled at 0.930 L/min for 240 min; at that time the concentration of nitrobenzene in the effluent was less than 5% of that in the influent. (The silica gel tube consists of two sections of silica gel separated by a section of urethane foam.) If a particular atmosphere is suspected of containing a large amount of contaminant, a smaller sampling volume should be taken.

Interferences

Silica gel has a high affinity for water, so organic vapors may not be trapped efficiently in the presence of a high relative humidity. This effect may be important even though there is no visual evidence of condensed water in the silica gel tube. When intefering compounds are known or suspected to be present in the air, such information, including their suspected identities, should be transmitted with the sample.

It must be emphasized that any compound which has the same retention time as the analyte at the operating conditions described in this method is an interference. Retention time data on a single column cannot be considered proof of chemical identity. If the possibility of interference exists, separation conditions (column packing, temperature, etc.) must be changed to circumvent the problem.

Precision and Accuracy

The coefficient of variation for the total analytical and sampling method in the range of

3.11-12.45 mg/m³ was 0.058. This value corresponds to a standard deviation of 0.29 mg/m³ at the OSHA standard level. Statistical information and details of the validation and experimental test procedures can be found in reference 1.

On the average the concentrations obtained at the OSHA standard level using the overall sampling and analytical method were 0.6% lower than the "true" concentrations for a limited number of laboratory experiments. Any difference between the "found" and "true" concentrations may not represent a bias in the sampling and analytical method, but rather a random variation from the experimentally determined "true" concentration. Therefore, no recovery correction should be applied to the final result.

Advantages and Disadvantages

The sampling device is small, portable, and involves no liquids. Interferences are minimal, and most of those which do occur can be eliminated by altering chromatographic conditions. The tubes are analyzed by means of a quick, instrumental method. The method can also be used for the simultaneous analysis of two or more compounds suspected to be present in the same sample by simply changing gas chromatographic conditions.

One disadvantage of the method is that the amount of sample which can be taken is limited by the number of mg that the tube will hold before overloading. When the sample value obtained for the backup section of the silica gel tube exceeds 25% of that found in the front section, the possibility of sample loss exists. Furthermore, the precision of the method is limited by the reproducibility of the pressure drop across the tubes. This drop will affect the flow rate and cause the volume to be imprecise, because the pump is usually calibrated for one tube only.

Apparatus

Calibrated personal sampling pump. Whose flow can be determined within 5% at the recommended flow rate.

Silica gel tubes. Glass tube with both ends flame sealed, 7 cm long with 6 mm O.D. and 4 mm I.D., containing 2 sections of 20/40 mesh silica gel separated by a 2 mm portion of urethane foam. The adsorbing section contains approximately 150 mg of silica gel, the backup section approximately 75 mg. A 3 mm portion of urethane foam is placed between the outlet end of the tube and the backup section. A plug of silylated glass wool is placed in front of the adsorbing section. The pressure drop across the tube must be less than one inch of mercury at a flow rate of 1 L/min.

Gas chromatograph. Equipped with a flame-ionization detector.

Column. 10 ft × 1/8 in stainless steel, packed with 10% FFAP on 80/100 mesh, acid washed DMCS Chromosorb W.

Electronic integrator. Or some other suitable method for measuring peak areas.

Sample containers. 2 mL, with glass stoppers or Teflon-lined caps. If an automatic sample injector is used, the associated vials may be used.

Syringes. 10 μL, and other convenient sizes for making standards.

Pipets. 1.0 mL delivery pipets.

Volumetric flasks. 10 mL or convenient sizes for making standard solutions.

Reagents

Methanol, reagent grade.
Nitrobenzene, reagent grade.
n-Hexane, reagent grade.
Prepurified hydrogen.
Filtered compressed air.
Purified nitrogen.

Procedure

Cleaning of equipment. All glassware used for the laboratory analysis should be detergent washed and thoroughly rinsed with tap water and distilled water.

Calibration of personal pumps. Each personal pump must be calibrated with a representative silica gel tube in the line. This will minimize errors associated with uncertainties in the sample volume collected.

Collection and shipping of samples. Immediately before sampling, break the ends of the tube to provide an opening at least one-half the internal diameter of the tube (2 mm). The smaller section of silica gel is used as a backup and should be positioned nearest the sampling pump. The silica gel tube should be placed in a vertical direction during sampling to minimize channeling through the silica gel. Air being sampled should not be passed through any hose or tubing before entering the silica gel tube.

A maximum sample size of 50 L is recommended. Sample at a flow of 1.0 L/min or less. The flow rate should be known with an accuracy of at least 5%. The temperature and pressure of the atmosphere being sampled should be recorded. If pressure reading is not available, record the elevation. The silica gel tubes should be capped with the supplied plastic caps immediately after sampling. Under no circumstances should rubber caps be used.

With each batch of ten samples, submit one tube from the lot of tubes which was used for sample collection and which is subjected to exactly the same handling as the samples except that no air is drawn through it. Label this as a blank. Capped tubes should be packed tightly and padded before they are shipped to minimize tube breakage during shipping. A sample of the bulk material should be submitted to the laboratory in a glass container with a Teflon lined cap. This sample should not be transported in the same container as the silica gel tubes.

Preparation of samples. In preparation for analysis, each silica gel tube is scored with a file in front of the first section of silica gel and broken open. The glass wool is removed and discarded. The silica gel in the first (larger) section is transferred to a 2 mL stoppered sample container. The separating section of foam is removed and discarded; the second section is transferred to another stoppered container. These two sections are analyzed separately.

Desorption of samples. Prior to analysis, 1.0 mL of methanol is pipetted into each sample container. The sample is desorbed for 30 min. Tests indicate that this is adequate if the sample is agitated occasionally during this period. If an automatic sample injector is used, the sample vials should be capped as soon as the solvent is added to minimize volatilization.

GC conditions. The typical operating conditions for the gas chromatograph are:

50 mL/min (60 psig) nitrogen carrier gas flow
65 mL/min (24 psig) hydrogen gas flow to detector
500 mL/min (50 psig) air flow to detector
245° C.. injector temperature
275° C.. detector temperature
160° C.. column temperature

Injection. The first step in the analysis is the injection of the sample into the gas chromatograph. To eliminate difficulties arising from blow back or distillation within the syringe needle, one should employ the solvent flush injection technique. The 10 μL syringe is first flushed with solvent several times to wet the barrel and plunger. Three μL of solvent are drawn into the syringe to increase the accuracy and reproducibility of the injected sample volume. The needle is removed from the solvent, and the plunger is pulled back about 0.2 μL to separate the solvent flush from the sample with a pocket of air to be used as a marker. The needle is then immersed in the sample, and a 5 μL aliquot is withdrawn, taking into consideration the volume of the needle, since the sample in the needle will be completely injected. After the needle is removed from the sample and prior to injection, the plunger is pulled back 1.2 μL to minimize evaporation of the sample from the tip of the needle. Observe that the sample occupies 4.9-5.0 μL in the barrel of the syringe. Duplicate injections of each sample and standard should be made. No more than a 3% difference in area is to be expected. An automatic sample injector can be used if it is shown to give reproducibility at least as good as the solvent flush technique.

Measurement of area. The area of the sample peak is measured by an electronic integrator or some other suitable form of area measurement, and preliminary results are read from a standard curve prepared as discussed below.

Determination of desorption efficiency. The desorption efficiency of a particular compound can vary from one laboratory to another and also from one batch of silica gel to another. Thus, it is necessary to determine at least once the percentage of the specific compound that is removed in the desorption process, provided the same batch of silica gel is used.

Silica gel equivalent to the amount in the first section of the sampling tube (approximately 150 mg) is measured into a 2.5 in, 4 mm I.D. glass tube, flame sealed at one end. This silica gel must be from the same batch as that used in obtaining the samples and can be obtained from unused silica gel tubes. The open end is capped with Parafilm. A known amount of hexane solution of nitrobenzene containing 60 mg/mL is injected directly into the silica gel with a microliter syringe. Cap the silica gel tube with more Parafilm. When using an automatic sample injector, the sample injector vials, capped with Teflon faced septa, may be used in place of the glass tubes.

The amount injected is equivalent to that present in a 60 L air sample at the selected level. Six tubes at each of three levels (0.5X, 1X, and 2X OHSA standard levels) are prepared in this manner and allowed to stand for at least overnight to assure complete adsorption of the analyte onto the silica gel. These tubes are referred to as the samples. A parallel blank tube should be treated in the same manner except that no sample is added to it. The sample and blank tubes are desorbed and analyzed in exactly the same manner as the sampling tube described earlier.

Two or three standards are prepared by injecting the same volume of compound into 1.0 mL of methanol with the same syringe used in the preparation of the samples. These are analyzed with the samples. The desorption efficiency equals the average weight in mg recovered from the tube divided by the weight in mg added to the tube, or:

D.E. = average wt (mg) recovered $\div$ wt (mg) added

The desorption efficiency is dependent on the amount of analyte collected on the silica gel. Plot the desorption efficiency versus the weight of analyte found. The curve is used to correct for adsorption losses.

Calibration and Standards

It is convenient to express concentration of standards in terms of mg/1.0 mL methanol, because samples are desorbed in this amount of methanol. The density of the analyte is used to calculate mg from μL for easy measurement with a microliter syringe. A series of standards, varying in concentration over the range of interest, are prepared and analyzed under the same GC conditions and during the same time period as the unknown sample. Curves are established by plotting concentration in mg/1.0 mL versus peak area. Note: since no internal standard is used in the method, standard solutions must be analyzed at the same time that the sample analysis is done. This will minimize the effect of known day-to-day variations and variations during the same day of the FID response.

Calculations

Read the weight, in mg, corresponding to each peak area from the standard curve. No volume corrections are needed, because the standard curve is based on mg/1.0 mL methanol and the volume of sample injected is identical to the volume of the standards injected. Corrections for the blank must be made for each sample:

mg = mg sample − mg blank

where:

mg sample = mg found in front section of sample tube
mg blank = mg found in front section of blank tube

A similar procedure is followed for the backup sections. Add the weights found in the front and backup sections to get the total weight in the sample. Read the desorption efficiency from the curve for the amount found in the front section. Divide the total weight by this desorption efficiency to obtain the corrected mg/sample:

Corrected mg/sample = total weight ÷ D.E.

The concentration of the analyte in the air sampled can be expressed in mg/m^3:

mg/m^3 = (corrected mg)(1000 L/m^3) ÷ air volume sampled in L

Another method of expressing concentration is ppm:

ppm = (mg/m^3)(24.45)(760)(T + 273) ÷ (M.W.)(P)(298)

where:

P = pressure (mm Hg) of air sampled
T = temperature (°C.) of air sampled
24.45 = molar volume (L/mol) at 25° C.. and 760 mm Hg
M.W. = molecular weight (g/mol) of analyte
760 = standard pressure (mm Hg)
298 = standard temperature (°K)

Reference
1. Documentation of NIOSH Validation Tests. NIOSH Contract No. CDC-99-74-45.

NITROETHANE

ACGIH TLV: 100 ppm (310 mg/m³)
OSHA Standard: 100 ppm

$C_2H_5NO_2$

Method	Sampling Duration	Sampling Location	Useful Range (ppm)	System Cost ($)	Test Cost ($)	Manufacturer
IR	Cont	Portable	2-200	4374	0	Foxboro
GC	1 min	Portable	$1-10^6$	10,000	0	Microsensor

NIOSH METHOD NO. S211

(Use as described for 1-chloro-1-nitropropane, with 120° C. column temperature. Substitute XAD-2 resin for the Chromosorb 108 in the sampling tubes.)

NITROGEN DIOXIDE

ACGIH TLV: 3 ppm (6 mg/m³)
OSHA Standard: 5 ppm (9 mg/m³)
NIOSH Recommendation: 1 ppm (1.8 mg/m³) ceiling (15 min)

NO_2

Method	Sampling Duration	Sampling Location	Useful Range (ppm)	System Cost ($)	Test Cost ($)	Manufacturer
DT	Grab	Portable	0.5-100	150	2.60	Nat'l Draeger
DT	Grab	Portable	0.5-1000	165	2.30	Matheson
DT	Grab	Portable	0.2-200	180	2.25	Bendix
DT	8 hr	Personal	1-100	850	3.10	Nat'l Draeger
ES	2 min	Personal	0.05-5	1765	0	Interscan
ES	Cont	Portable	0.02-10	2100	0	Energetics
ES	Cont	Portable	0.02-2(50)	2295	0	Interscan
CS	Grab	Portable	0.5-20	2465	0.80	MDA
ES	Cont	Fixed	0.05-5(50)	2533	0	Interscan
ES	Cont	Port/Fixed	0.05-5(50)	3260	0	Interscan
IR	Cont	Portable	0.2-20	4374	0	Foxboro
CS	Cont	Fixed	1-20	4950	Variable	MDA
CM	Cont	Port/Fixed	0.002-10	4995	Variable	CEA

NIOSH METHOD NO. P&CAM 231

(See nitric oxide)

NITROGLYCOL

Synonyms: ethylene glycol dinitrate; EGDN $O_2NOCH_2CH_2ONO_2$
ACGIH TLV: 0.05 ppm (0.3 mg/m³) - Skin
OSHA Standard: 1 mg/m³ ceiling
NIOSH Recommendation: 0.1 mg/m³ ceiling (20 min)

Method	Sampling Duration	Sampling Location	Useful Range (ppm)	System Cost ($)	Test Cost ($)	Manufacturer
DT	Grab	Portable	0.25	150	2.40	Nat'l Draeger

NIOSH METHOD NO. P&CAM 203

Principle

The nitrate esters, nitroglycol and nitroglycerin, are sorbed on Tenax-GC, a porous organic polymer, as a known volume of air containing the analytes is drawn through a small bed of the sorbent in a glass tube. The nitrate esters are extracted from the sorbent with ethanol; the resulting solution is analyzed by gas chromatography with an electron-capture detector. Areas under the nitrate ester peaks from sample unknowns are compared to areas under the peaks from nitrate ester standards.

Range and Sensitivity

The minimum amount of nitroglycol (EGDN) detectable in a single injection by the gas chromatographic method is substantially smaller than 0.25 ng. This sensitivity easily permits the detection of EGDN in a 10 L air sample containing 0.01 mg/m³. The minimum amount of nitroglycerin (NG) detectable in a single injection by the gas chromatographic method is smaller than 2.5 ng. This sensitivity permits the detection of NG in a 100 L air sample containing 0.01 mg/m³.

The capacity of the sorbent tubes is at least 1 mg of either EGDN or NG. Thus, it is unlikely that an excessive amount of sample will ever be collected even when the concentration greatly exceeds the OSHA standard of 1 mg/m³. This retention of EGDN or NG by the sorbent tubes is not significantly different when samples are taken from atmospheres of low and high relative humidities.

Interferences

Ethylene glycol mononitrate, if present in high concentration relative to that of EGDN, produces a gas chromatographic peak that tails into the EGDN peak. This interference, although troublesome, does not prohibit a reasonably satisfactory analysis for EGDN. The combined selectivity of Tenax-GC and electron capture detection practically eliminated other interferences. Water vapor does not interfere.

Precision and Accuracy

The coefficient of variation of the analytical method for the determination of EGDN is estimated to be about 0.03 at an EGDN concentration of 2 mg/m³ in 15 L samples. The precision of the measurement of NG concentration is approximately the same at an NG concentration of 0.27 mg/m³.

The precision of the overall sampling and analytical method has not been adequately determined. It has been estimated that the coefficient of variation for the total method is about 0.10 at an EGDN concentration of 2 mg/m^3 and an NG concentration of 0.3 mg/m^3. Since both EGDN and NG are sorbed on Tenax-GC at ambient temperature with 100% efficiency, it can be expected that the overall precision will be determined by variations in the measurement of sample volume.

In laboratory measurements not involving a personal sampling pump, the average recovery of EGDN from vapor samples at concentrations of 0.27-3.93 mg/m^3 in 30-90 L samples was 104% and that of NG was 102% at a concentration of 0.28 mg/m^3 in 60 L samples.

Advantages and Disadvantages

The sampling device is small and portable and involves no liquids. The tubes are analyzed by means of gas chromatography, a rapid instrumental method. Interferences in the chromatographic method are minimal.

The principal difficulties are those associated with the use of electron-capture detection. They include the limited linear range of electron-capture detectors and operating variables such as the effects of column bleed, moisture, and oxygen. The effects of these variables may be minimized by analyzing standards at the same time as samples. The electron-capture detector is about 500 times as sensitive to the nitrate esters as a hydrogen flame-ionization detector, and it provides a substantial advantage in specificity.

The precision of the method is limited by the reproducibility of the pressure drop across the tubes and, therefore, the flow rate through the tubes. Because the pump is usually calibrated for one particular tube, differences in flow rates can occur when sampling through other tubes and can cause sample volumes to vary. The method can be applied for a wide range of sample sizes. However, repeated dilutions of the ethanol extract may be necessary to bring the amount of sample injected into the proper range for the electron-capture detector.

Apparatus

Properly calibrated personal sampling pump. Calibrate the pump with a representative sorbent tube in the sampling line using a wet or dry test meter or a glass rotameter capable of measuring the appropriate flow rates (less than 1 L/min) within an accuracy of 5%.

Sorbent tubes. The sorbent tubes are 70 mm long and 5 mm I.D. They contain two sections of 35/60 mesh Tenax-GC separated and held in place by glass wool plugs. The front section contains 100 mg of sorbent and the backup section contains 50 mg. Since the pressure drop must be limited to 1 in of Hg at 1 Hg at 1.0 L/min, it is necessary to avoid overpacking with glass wool. (Tenax-GC is distributed by Applied Science Laboratories, Inc. State College, PA.)

Gas chromatograph. Equipped with an electron-capture detector, a 2.5 ft by 0.25 in glass column packed with 10% OV-17 on 60/80 mesh Gas Chrom Q, and an integrator, or an equivalent means of measuring areas under peaks.

Assorted supplies. A volumetric flask and pipettes, a 10 μL syringe, and other normally available laboratory supplies and equipment are needed.

Reagents

The only reagent required is absolute ethanol for extraction of the sorbent. The gas chromatograph requires a supply of helium and an argon-methane mixture, or other carrier or purge gases as required for the particular instrument used.

Procedure

Cleaning of equipment. All glassware used for the laboratory analysis should be detergent washed and thoroughly rinsed with tap water and distilled water. Particular attention should be paid to the cleaning of the μL syringe with ethanol.

Collection and shipping of samples. Immediately before sampling, break the ends of the tube to provide an opening at least 2 mm. The smaller section of sorbent is used as a backup and should be positioned nearest the sampling pump. The sorbent tube should be placed in a vertical position with the larger section of sorbent pointing up during sampling to minimize channeling of EGDN and NG through the sorbent tube. Air being sampled should not be passed through any hose or tubing before entering the sorbent tube. This is particularly important with the nitrate esters since they are strongly sorbed on most surfaces, including glass.

The flow rate and time (or volume) must be measured as accurately as possible. The sample should be taken at a flow rate of 1 L/min or less to attain the total sample volume required. The minimum volume that must be collected to permit detection of NG at the OSHA standard (or EGDN at a still lower concentration) is only 0.1 L. At the OSHA standard, sample volumes as large as 1000 L can be collected without causing trouble in the analysis. At the time personal samples are taken, relatively large volumes of air also should be sampled through other sorbent tubes. These bulk air samples will be used by the analyst to identify possible interferences before the personal samples are analyzed.

The temperature and pressure of the atmosphere being sampled should be measured and recorded if either should differ greatly from standard values (25° C. and 760 mm Hg). The Tenax-GC tubes should be capped with the supplied plastic caps immediately after sampling. Under no circumstances should rubber caps be used. One tube should be handled in the same manner as the sample tubes (break, seal, and transport), except that no air is sampled through this tube. This tube should be labeled as a blank.Capped tubes should be packed tightly and padded before they are shipped to minimize tube breakage during shipping.

Preparation of samples. The sorbent tube is scored with a file near the front end and is broken open. The first glass wool plug and the front (100 mg) section of the sorbent are transferred to a small vial or test tube. The remaining glass wool plug and the backup section of sorbent are transferred to another small vial. Two mL of ethanol is added to each vial, the vials are stoppered, and the contents are shaken for about 1 min. It may be necessary to dilute the ethanol solution before taking a sample aliquot for injection if the concentrations of the nitrate esters are high.

Column. 2.5 ft × 0.25 in glass column, 10% OV-17 on 60/80 mesh Gas Chrom Q.

GC conditions. Typical operating conditions for the gas chromatograhic analysis are:

100 mL/min helium carrier gas
125 mL/min argon/methane purge gas
130° C. column temperature
160° C. injection port temperature
280° C. detector temperature

Avoid an injector temperature above 160° C. or a column temperature above 150° C. to prevent NG decomposition (reference 1).

Injection. The first step in the analysis is the injection of the sample into the gas chromatograph. To eliminate difficulties arising from blow back or distillation within the syringe needle, one should employ the solvent flush injection technique. The 10 μL syringe is first flushed with solvent several times to wet the barrel and plunger. Two μL of solvent are drawn into the syringe to increase the accuracy and reproducibility of the injected sample volume. The needle is removed from the solvent, and the plunger is pulled back about 0.4 μL to separate the solvent flush from the sample with a pocket of air to be used as a marker. The needle is then immersed in the sample, and a 5 μL aliquot is withdrawn to the 7.4 μL mark (2 μL solvent + 0.4 μL air + 5 μL sample = 7.4 μL). After the needle is removed from the sample and prior to injection, the plunger is pulled back a short distance to minimize evaporation of the sample from the tip of the needle. Duplicate injections of each sample and standard should be made. No more than a 3% difference in area is to be expected. Automatic sampling device may also be used.

Measurement of area. The area under the sample peak is measured by an electronic integrator or some other suitable form of area measurement, and preliminary results are read from a standard curve prepared as discussed below.

Determination of desorption efficiency. Only small differences in desorption efficiency in the described procedure are to be expected from differences between lots of Tenax-GC. If the efficiency of desorption should be questioned for some other reason, it can be checked following the procedure detailed in NIOSH method no. P&CAM 127 (see acetone). However, since it is not possible to measure accurately the small quantities of the nitrate esters that are to be added to sorbent tubes, 15 μL aliquots of appropriate ethanol solutions are added. Ethanol is also the solvent used to desorb the nitrate esters from the sorbent.

Calibration and Standards

If pure NG and EGDN are available, standard solutions in ethanol may be prepared and analyzed under the same gas chromatographic conditions and during the same time period as the unknown samples. In the course of the development of this method, several reference materials were obtained from manufacturers of nitrate ester products. This experience indicated that the following materials can be obtained for the preparation of standards:

Solutions of known concentrations of NG and EGDN in ethanol were provided by a dynamite manufacturer. A sample of analyzed dynamite containing EGDN but no NG was also provided by a dynamite manufacturer. Since dynamite is a heterogenous material, care must be taken to obtain a representative sample.

Pharmaceutical grade sublingual NG tablets may be obtained from wholesaler drug dealers. These tablets are manufactured to meet U.S. Pharmacopeia specifications, which require that the actual NG content be between 80 and 112% of the stated value. The actual NG content of a specific lot of tablets may be obtained from the manufacturer. For use, the tablets are weighed, crushed, and extracted with ethanol, the solution is filtered and diluted to a fixed volume, and aliquots are injected. An alternate source was a material of a known NG content called nitroglycerin lactose trituration, provided by a pharmaceutical manufacturer. This material is used for quality control in the manufacture of NG tablets.

An experimental comparison of the NG and EGDN contents of these standard materials gave a coefficient of variation of 0.04 for the NG contents and 0.02 for the EGDN contents. It is necessary that standards be analyzed concurrently with samples to minimize the effects of variations in detector response and other factors. Calibration curves are established by plotting concentration in mg/2.0 mL versus peak area.

Calculations

Read the weight, in mg, corresponding to each peak area from the standard curve. No volume corrections are needed, because the standard curve is based on mg/2.0 mL and the volume of sample injected is identical to the volume of the standards injected. Corrections for the blank must be made for each sample:

mg = mg sample − mg blank

where:

mg sample = mg found in front section of sample tube
mg blank = mg found in front section of blank tube

A similar procedure is followed for the backup sections. Add the corrected amounts present in the front and backup sections of the same sample tube to determine the total measured amount of EGDN or NG in the sample. The concentration of each compound in air may be expressed in mg/m^3:

mg/m^3 = (corrected mg)(1000 L/m^3) ÷ air volume sampled in L

Another method of expressing concentration is ppm:

ppm = (mg/m^3)(24.45)(760)(T + 273) ÷ (M.W.)(P)(298)

where:

P = pressure (mm Hg) of air sampled
T = temperature (°C.) of air sampled
24.45 = molar volume (L/mol) at 25° C. and 760 mm Hg
M.W. = molecular weight
760 = standard pressure (mm Hg)
298 = standard temperature (°K)

References

1. Barrett, W.J., H.K. Dillon, and R.H. James. Sampling and Analysis of Four Organic Compounds Using Solid Sorbents, Southern Research Institute, Birmingham, Alabama, Final Report for Contract No. HSM-99-73-63 to National Institute for Occupational Safety and Health, Division of Laboratories and Criteria Development, Physical and Chemical Analysis Branch, Cincinnati, Ohio, 1974, pp. 28-52, 97-104.
2. White, L.D. et al. A convenient optimized method for the analysis of selected solvent vapors in the industrial atmosphere. Amer. Ind. Hyg. Assoc. J. 31: 225, 1970.

NITROMETHANE

ACGIH TLV: 100 ppm (250 mg/m³)
OSHA Standard: 100 ppm

CH_3NO_2

Method	Sampling Duration	Sampling Location	Useful Range (ppm)	System Cost ($)	Test Cost ($)	Manufacturer
IR	Cont	Portable	2-200	4374	0	Foxboro
GC	1 min	Portable	$1-10^6$	10,000	0	Microsensor

NIOSH METHOD NO. S211

(Use as described for 1-chloro-1-nitropropane, with 100° C. column temperature. Substitute Chromosorb 106 for the Chromosorb 108 in the sampling tubes.)

NITROTOLUENE

Synonym: 2-, 3-, or 4-nitrotoluene
ACGIH TLV: 2 ppm (11 mg/m³) - Skin
OSHA Standard: 5 ppm (30 mg/m³)

$CH_3C_6H_4NO_2$

Method	Sampling Duration	Sampling Location	Useful Range (ppm)	System Cost ($)	Test Cost ($)	Manufacturer
IR	Cont	Portable	1.4-20	4374	0	Foxboro

NIOSH METHOD NO. S217

(Use as described for nitrobenzene)

OCTANE

Synonym: n-octane $\quad C_8H_{18}$
ACGIH TLV: 300 ppm (1450 mg/m³)
OSHA Standard: 500 ppm (2350 mg/m³)

Method	Sampling Duration	Sampling Location	Useful Range (ppm)	System Cost ($)	Test Cost ($)	Manufacturer
DT	Grab	Portable	100-2500	150	2.60	Nat'l Draeger
DT	Grab	Portable	150-12,000	180	2.25	Bendix
IR	Cont	Portable	10-1000	4374	0	Foxboro
PI	Cont	Fixed	2-200	4950	0	HNU
GC	1 min	Portable	$1-10^6$	10,000	0	Microsensor

NIOSH METHOD NO. P&CAM 127

(Use as described for acetone, with 52° C. column temperature)

OZONE

ACGIH TLV: 0.1 ppm (0.2 mg/m³) $\quad O_3$
OSHA Standard: 0.1 ppm

Method	Sampling Duration	Sampling Location	Useful Range (ppm)	System Cost ($)	Test Cost ($)	Manufacturer
DT	Grab	Portable	0.05-300	150	2.70	Nat'l Draeger
DT	Grab	Portable	0.05-3	165	2.30	Matheson
DT	Grab	Portable	0.05-3	180	2.25	Bendix
CS	Grab	Portable	0.025-0.5	2465	0.80	MDA

NIOSH METHOD NO. S8

Principle

Air containing ozone is drawn through a midget impinger containing 10 mL of 1% potassium iodide in 1 N sodium hydroxide. A stable product is formed that can be stored with little loss for several days. The analysis is completed in the laboratory by the addition of phosphoric sulfamic acid reagent, which liberates the iodine. The yellow iodine color is read in a spectrophotometer at 352 nm

Range and Sensitivity

This method has been validated over the range 0.1-0.4 mg/m^3. This is the limit of the useful range for the 45 L sample size. The method is capable of measuring smaller or larger concentrations if the sampling size is adjusted, but it has not been validated for other sample sizes.

Interferences

Chlorine, hydrogen peroxide, organic peroxides, and various other oxidants will liberate iodine by this method. The response to nitrogen dioxide is limited to 10% by the use of sulfamic acid in the procedure to destroy nitrite, thus minimizing any error due to the collection of NO$_2$. The negative interferences from reducing gases such as sulfur dioxide and hydrogen sulfide are very serious (probably on a mole-to-mole equivalency). The procedure is very sensitive to reducing dusts that may be present in the air or on the glassware. Losses of iodine also occur even on clean glass surfaces and thus the manipulations should minimize this exposure.

Precision and Accuracy

The coefficient of variation for the total analytical and sampling method in the range of 0.1-0.4 mg/m^3 is 0.0806. The standard deviation at the OSHA standard level is 0.0077 mg/m^3. Statistical information and details of the validation and experimental test procedures can be found in reference 2. On the average, the values obtained using the overall sampling and analytical method were the same as the true value at the OSHA standard level. These data are based on validation experiments in which neutral buffered KI was used to determine the actual ozone concentration.

Advantages and Disadvantages

The method is simple, accurate and precise. Those compounds listed under interferences will liberate iodine by this method. A delay of several days is permissible between sampling and completion of analysis. However, the relationship between delay and analytical results has not been established.

Apparatus

Midget impinger. With 1 mm I.D. nozzle, graduation marks and Teflon-coated ground glass stopper.

Personal sampling pump. Capable of drawing the required sample flow for intervals of up to 45 min.

Air metering device. Capable of measuring a flow of 1-2 L/min, including a prefilter to prevent liquid droplets from reaching the measuring device.

Spectrophotometer. Capable of measuring the yellow color at 352 nm with stoppered tubes or cuvettes.

Assorted laboratory glassware.

Reagents

The reagents described must be made up using ACS reagent grade or better grade of chemical.

Double distilled water.

Absorbing reagent. Dissolve 40.0 g of sodium hydroxide in just less than 1 L of water, then dissolve 10.0 g of potassium iodide and make the mixture to 1 L. Store in a clean glass bottle with a screw cap (with inert liner) or rubber stopper (previously boiled for 30 min in alkali and washed). Age for at least 1 day before using. The reagent should be used within 2 weeks.

Acidifying reagent. Dissolve 5.0 g of sulfamic acid in 100 mL of water, then add 84 mL of 85% phosphoric acid and dilute to 200 mL.

Standard potassium iodate solution. Dissolve 0.758 g of potassium iodate in water and dilute to

1 L. One mL of this stock solution is equivalent to 260 µL of ozone. Prepare a dilute standard solution just before it is required by pipetting exactly 5 mL of stock solution into a 50 mL volumetric flask and making to mark with distilled water.

Procedure

Cleaning of equipment. All glassware should be cleaned with dichromate cleaning solution followed by 3 tap and 3 distilled water rinses.

Collection of samples. Assemble a train composed of midget impinger, a rotameter, and pump. Use ground glass connections upstream from the impinger. Insure that all fittings are vacuum tight. Pipette exactly 10 mL of the absorbing solution into the midget impinger and sample at a flow rate of 1 L/min. note the volume of air sampled. Record atmospheric temperature and pressure. Cap the impinger with the stopper supplied for shipment to the laboratory.

Analysis. If the liquid level is below the 10 mL mark in the impinger, and distilled water to bring the level up to the mark. Add 2 mL of the acidifying reagent with a volumetric pipette. Swirl gently to expel the carbon dioxide released immediately. Then cap and shake vigorously until bubbling ceases. It is imperative that all of the carbon dioxide is expelled, otherwise more may be released while the UV measurement is being made.

Place the stoppered impinger in a water bath at room temperature for 5-10 min to dissipate the heat of neutralization. Transfer a portion of the sample to a cuvette and determine the absorbance at 352 nm. A cuvette containing distilled water is used as the reference. Do not delay the reading since reducing impurities sometimes causes rapid fading of the color. Prepare a reagent blank by adding 2 mL of the acidifying reagent to mL of unexposed absorbing reagent. Cool and determine the blank absorbance. The blank absorbance should be determined each day and should be subtracted from the absorbance of the samples.

Samples may be aliquoted before or after acidification if very large concentrations of oxidant are expected. In the former case dilute the aliquot to 10 mL with unexposed absorbing reagent and proceed in the usual manner. In the latter case dilute the aliquot to 10 mL with reagent blank mixture. Aliquoting after acidification is not as reliable as before acidification and should be used only to save a sample when unexpectedly large concentrations of oxidant are encountered. The calculations should include the aliquoting factor.

Calibration and Standards

Add the freshly prepared, dilute standard iodate solution in graduated amounts of 0.1-0.5 mL (measured accurately in a graduated pipette or small burette) to a series of 25 mL glass-stoppered volumetric flasks. Make at least 4 standards. Add alkaline potassium iodide solution (measured accurately in a graduated pipette) to make the total volume of each exactly 10 mL. Acidify and determine the absorbance of each standard as with the samples.

Calculations

Plot the absorbance of the standards (corrected for the blank) against the concentration of dilute standard iodate solution. One mL of the diluted standard iodate solution (in a total of 10 mL absorbing solution) is equivalent to 51 µg of ozone. To readily determine ambient ozone concentrations in units of mg/m^3, it is convenient to plot absorbance versus mg equivalents of ozone/10 mL of absorbing solution. Upon determining the absorbance of the sample solutions, use the standard absorbance curve to find the ozone equivalence. The concentration of ozone in the air sample equals the ozone equivalence divided by the air volume in L:

ozone (mg/m^3) = ozone equivalence ÷ air volume in L

The reaction of ozone with alkaline KI is not quantitative and the yield of iodine is concentration dependent. The following correction equation was developed during validation of the method:

ozone (mg/m^3) = $0.038 + 1.038(A) - 0.010505(A^2)$

where:

A = concentration of ozone (also in mg/m^3) read from the absorbance curve.

Another method of expressing concentration is ppm defined as μL of ozone/L of air:

ppm ozone = (mg/m^3)(24.45)(760)(T + 273) ÷ (48)(P)(298)

where:

P = pressure (mm Hg) of air sampled
T = temperature (°C.) of air sampled
24.45 = molar volume (L/mol) at 25° C. and 760 mm Hg
48 = molecular weight (g/mol) of ozone
760 = standard pressure (mm Hg)
298 = standard temperature (°K)

References

1. Selected Methods for the Measurement of Air Pollutants. USDHEW Public Health Service Publication No. 999-AP-11, 1965.
2. Documentation of NIOSH Validation Tests. Contract No. CDC-99-74-45.

n-PENTANE

ACGIH TLV: 600 ppm (1800 mg/m^3) C_5H_{12}
OSHA Standard: 1000 ppm (2295 mg/m^3)

Method	Sampling Duration	Sampling Location	Useful Range (ppm)	System Cost ($)	Test Cost ($)	Manufacturer
DT	Grab	Portable	100-1500	150	2.60	Nat'l Draeger
DT	Grab	Portable	100-5000	165	3.80	Matheson
DT	Grab	Portable	110-1600	180	2.25	Bendix
IR	Cont	Portable	20-2000	4374	0	Foxboro
PI	Cont	Fixed	20-2000	4950	0	HNU
GC	1 min	Portable	1-10^6	10,000	0	Microsensor

NIOSH METHOD NO. P&CAM 127

(Use as described for acetone, with 52° C. column temperature)

2-PENTANONE

Synonym: methyl propyl ketone $CH_3COC_3H_7$
ACGIH TLV: 200 ppm (700 mg/m^3)
OSHA Standard: 200 ppm

Method	Sampling Duration	Sampling Location	Useful Range (ppm)	System Cost ($)	Test Cost ($)	Manufacturer
IR	Cont	Portable	4-400	4374	0	Foxboro

NIOSH METHOD NO. P&CAM 127

(Use as described for acetone)

PETROLEUM DISTILLATE

Synonyms: petroleum naphtha; hydrocarbons (120-147 °C. b.p.)
OSHA Standard: 500 ppm (2000 mg/m^3)

Method	Sampling Duration	Sampling Location	Useful Range (ppm)	System Cost ($)	Test Cost ($)	Manufacturer
DT	Grab	Portable	2-23	150	2.70	Nat'l Draeger
IR	Cont	Portable	10-1000	4374	0	Foxboro
PI	Cont	Fixed	20-2000	4950	0	HNU

NIOSH METHOD NO. P&CAM 127

(Use as described for acetone, with 85° C. column temperature and n-decane as internal standard)

PHENOL

Synonyms: carbolic acid; hydroxybenzene $\qquad$ C_6H_5OH
ACGIH TLV: 5 ppm (19 mg/m³) - Skin
OSHA Standard: 5 ppm
NIOSH Recommendation: 60 mg/m³ ceiling (15 min)

Method	Sampling Duration	Sampling Location	Useful Range (ppm)	System Cost ($)	Test Cost ($)	Manufacturer
DT	Grab	Portable	5	150	2.80	Nat'l Draeger
DT	Grab	Portable	0.5-25	165	2.30	Matheson
DT	Grab	Portable	0.4-62.5	180	2.25	Bendix
IR	Cont	Portable	0.2-10	4374	0	Foxboro
CM	Cont	Port/Fixed	0.001-10	4995	Variable	CEA

NIOSH METHOD NO. S330

Principle

A known volume of air is drawn through a midget bubbler containing 15 mL of 0.1 N sodium hydroxide to trap the phenol vapors present. The resulting solution is acidified with sulfuric acid. An aliquot of the collected sample is injected into a gas chromatograph. The area of the resulting peak is determined and compared with the areas for standards.

Range and Sensitivity

This method was validated over the range of 9.46-37.8 mg/m³ at an atmospheric temperature and pressure of 22°C. and 760 mm Hg, using a 100 L sample. Under the conditions of sample size (100 L) the probable useful range of this method is 5-60 mg/m³ at a detector sensitivity that gives nearly full deflection on the strip chart recorder for a 6 mg sample. The method is capable of measuring much smaller amounts if the analytical method recovery is adequate.

Interferences

It must be emphasized that any compound which has the same retention time as the analyte at the operating conditions described in this method is an interference. Retention time data on a single column cannot be considered proof of chemical identity. If the possibility of interference exists, separation conditions (column packing, temperature, etc.) must be changed to circumvent the problem.

Precision and Accuracy

The coefficient of variation for the total analytical and sampling method in the range of 9.46-37.8 mg/m³ was 0.068. This value corresponds to a 1.3 mg/m³ standard deviation at the OSHA standard level. Statistical information and details of the validation and experimental test procedures can be found in reference 1.

A collection efficiency of 1.00 ± 0.01 was determined for the collection medium. On the average the concentrations obtained at the OSHA standard level using the overall sampling and analytical method were 2.6% lower than the "true" concentrations for a limited number of laboratory experiments. Any difference between the "found" and "true" concentrations may not represent a bias in the sampling and analytical method, but rather a random variation from the experimentally determined "true" concentration. Therefore, no recovery correction should be applied to the final result.

Advantages and Disadvantages

The samples collected in bubblers are analyzed by means of a quick, instrumental method.

Apparatus

Glass midget bubbler. Containing the collection medium.

Sampling pump. Suitable for delivering at least 1 L/min for 100 min. The pump is protected from splashover or solvent condensation by a 5 cm long by 6 mm I.D. glass tube loosely packed with a plug of glass wool and inserted between the exit arm of the bubbler and the pump.

Integrating volume meter. Such as a dry gas or wet test meter.

Thermometer.

Manometer.

Stopwatch.

Gas chromatograph. Equipped with a flame-ionization detector.

Column. 4 ft long × 1/4 in O.D. stainless steel, packed with 35/60 mesh Tenax.

Electronic integrator. Or some other suitable method for mesuring peak areas.

Syringes. 10 µL, and other convenient sizes for making standards and injecting samples into the GC.

Volumetric flasks. Convenient sizes for making solutions.

Pipets. 15 mL or other convenient sizes.

Reagents

Distilled water.

Phenol, reagent grade.

Sulfuric acid, reagent grade.

Collection medium, 0.1 N sodium hydroxide. Dissolve 4.0 g of sodium hydroxide in distilled water and dilute to a final volume of 1 L.

Purified nitrogen.

Purified hydrogen.

Filtered compressed air.

Procedure

Cleaning of equipment. All glassware used for the laboratory analysis should be detergent washed and thoroughly rinsed with tap water and distilled water.

Calibration of personal sampling pumps. Each pump should be calibrated by using an integrating volume meter or other means.

Collection and shipping of samples. Pour 15 mL of the collection medium into each midget bubbler. Connect the midget bubbler with a 5 cm glass adsorption tube (6 mm I.D. and 8 mm O.D.) containing the glass wool plug, then to the personal sampling pump using short pieces of flexible tubing. The air being sampled should not pass through any tubing or other equipment before entering the bubbler.

Turn the pump on to begin sample collection. Care should be taken to measure the flow rate, the time and/or the volume as accurately as possible. Record the atmospheric pressure and the temperature. If the pressure reading is not available, record the elevation. The sample should be taken at a flow rate of 1 L/min. A sample size of 100 L is recommended.

After sampling, the bubbler stem may be removed and cleaned. Tap the stem gently against the inside wall of the bubbler bottle to recover as much of the sampling solution as possible. Wash the stem with 1 mL of distilled water, adding the wash to the bubbler. The bubblers are sealed with a hard, non reactive stopper (preferable Teflon or glass). Do not seal with rubber. The stoppers on the bubblers should be tightly sealed to prevent leakage during shipping.

Care should be taken to minimize spillage or loss by evaporation. Whenever possible, hand delivery of the sample is recommended. Otherwise, special bubbler shipping cases designed by NIOSH should be used to ship the samples. A blank bubbler should be handled in the same manner as the bubblers containing samples (fill, seal, and transport) except that no air is sampled through this bubbler.

Analysis of samples. The sample in each bubbler is analyzed separately. Transfer the solution to a 25 mL volumetric flask. Rinse the bubbler twice with 1 mL of distilled water and add the rinses to the flask. Add 0.1 mL of concentrated sulfuric acid to the flask and mix. Check with pH paper to make sure that the pH is less than 4. Dilute to mark with distilled water and mix.

GC conditions. The typical operating conditions for the gas chromatograph are:

50 mL/min (60 psig) nitrogen carrier gas flow
65 mL/min (24 psig) hydrogen gas flow to detector
500 mL/min (50 psig) air flow to detector
215° C. injector temperature
225° C. detector temperature
200° C. column temperature

The glass inlet on the GC should be cleaned at the end of each day with water and acetone rinses. Reinsert the glass inlet into the injection port and let it bake out overnight.

Injection. The first step in the analysis is the injection of the sample into the gas chromatograph. To eliminate difficulties arising from blow back or distillation within the syringe needle, one should employ the solvent flush injection technique. The 10 μL syringe is first flushed with solvent several times to wet the barrel and plunger. Three μL of solvent are drawn into the syringe to increase the accuracy and reproducibility of the injected sample volume. The needle is removed from the solvent, and the plunger is pulled back about 0.2 μL to separate the solvent flush from the sample with a pocket of air to be used as a marker. The needle is then immersed in the sample, and a 5 μL aliquot is withdrawn, taking into consideration the volume of the needle, since the sample in the needle will be completely injected. After the needle is removed from the sample and prior to injection, the plunger is pulled back 1.2 μL to minimize evaporation of the sample from the tip of the needle. Observe that the sample occupies 4.9-5.0 μL in the barrel of the syringe. Duplicate injections of each sample and standard should be made. No more than a 3% difference in area is to be expected. An automatic sample injector can be used if it is shown to give reproducibility at least as good as the solvent flush method.

Measurement of area. The area of the sample peak is measured by an electronic integrator or some other suitable form of area measurement, and preliminary results are read from a standard curve prepared as discussed below.

Procedure for preparing standard solutions. Six standards at each of three levels (0.5X, 1X, and 2X the OSHA standard) are prepared by introducing a known amount of analyte into 15 mL of 0.1 N sodium hyroxide in a 25 mL volumetric flask. The amount introduced is equivalent to that present in a 1 L air sample. The standards are acidified with 0.1 mL of concentrated sulfuric acid and made up to volume with distilled water. The solution should be checked with pH paper to make sure that its pH is less than 3. A parallel blank is prepared in the same manner, except that no analyte is added. The standards and blank are analyzed in exactly the same manner as the samples as decribed above.

Calibration and Standards

It is convenient to express concentration of standards in terms of mg/sample. A series of standards, varying in concentration over the range of interest, are prepared and analyzed under the same GC conditions and during the same time period as the unknown samples. Curves are established by plotting concentration in mg/sample versus peak area. Note: since no internal standard is used in the method, standard solutions must be analyzed at the same time that the sample analysis is done. This will minimize the effect of known day-to-day variations and variations during the same day of the FID response.

Calculations

Read the weight, in mg, corresponding to each peak area from the standard curve. No volume corrections are needed, because the standard curve is based on mg/sample and the volume of sample injected is identical to the volume of the standards injected. Corrections for the blank must be made for each sample:

mg = mg sample − mg blank

where:

mg sample = mg found in sample bubbler
mg blank = mg found in blank bubbler

The concentration of the analyte in the air sampled can be expressed in mg/m^3:

mg/m^3 = (corrected mg)(1000 L/m^3) ÷ air volume sampled in L

Another method of expressing concentration is ppm:

ppm = (mg/m^3)(24.45)(760)(T + 273) ÷ (M.W.)(P)(298)

where:

P = pressure (mm Hg) of air sampled
T = temperature (°C.) of air sampled
24.45 = molar volume (L/mol) at 25° C. and 760 mm Hg
M.W. = molecular weight (g/mol) of analyte
760 = standard pressure (mm Hg)
298 = standard temperature (°K)

Reference

1. Documentation of NIOSH Validation Tests. NIOSH Contract No. CDC-99-74-45.

PHENYL ETHER

Synonyms: diphenyl ether; phenoxybenzene $C_6H_5OC_6H_5$
ACGIH TLV: 1 ppm (7 mg/m^3)
OSHA Standard: 1 ppm

Method	Sampling Duration	Sampling Location	Useful Range (ppm)	System Cost ($)	Test Cost ($)	Manufacturer
IR	Cont	Portable	0.04-2	4374	0	Foxboro

NIOSH METHOD NO. P&CAM 127

(Use as described for acetone, with 215° C. column temperature)

PHENYLHYDRAZINE

ACGIH TLV: 5 ppm (20 mg/m^3) - Skin, suspected carcinogen $C_6H_5NHNH_2$
OSHA Standard: 5 ppm

Method	Sampling Duration	Sampling Location	Useful Range (ppm)	System Cost ($)	Test Cost ($)	Manufacturer
IR	Cont	Portable	2-20	4374	0	Foxboro

NIOSH METHOD NO. S160

Principle

Phenylhydrazine is collected in a standard midget bubbler containing 0.1 N hydrochloric acid. The resulting solution is reacted with phosphomolybdic acid to form a bluish green colored complex. The strong absorption maximum at 730 nm is used as a quantitative measure of phenylhydrazine.

Range and Sensitivity

This method was validated over the range of 10.37-44.8 mg/m^3, at an atmospheric temperature and pressure of 22° C. and 761 mm Hg. The probable range of the method is 5-45 mg/m^3 based on the range of standards used to prepare the standard curve. For samples of high concentration where the absorbance is greater than the limits of the standard curve, the samples can be diluted with hydrochloric acid prior to color development to extend the upper limit of the range. A concentration of 5 mg/m^3 of phenylhydrazine can be determined in a 100 L air sample based on a difference of 0.05 absorbance unit from the blank using a 1 cm cell. Greater sensitivity could be obtained by use of a longer path length cell.

Inteferences

Any hydrazine derivative will potentially interfere with this method. Other reducing agents such as ferrous salts may also interfere. Aldehydes and ketones in the air may interfere, because of the potential formation of phenylhydrazones in the acid medium in the bubbler.

Precision and Accuracy

The coefficient of variation for the total analytical and sampling method in the range of 10.37-44.8 mg/m^3 was 0.060. This value corresponds to a 1.32 mg/m^3 standard deviation at the OSHA standard level. Statistical information and details of the validation and experimental test procedures can be found in reference 2.

A collection efficiency of 1.00 ± 0.01 was determined for the collection medium; thus, no bias was introduced in the sample collection step. There was no apparent bias in the sampling and analytical method for which an analytical method recovery correction was made. Thus coefficient of variation is a satisfactory measure of both accuracy and precision of the sampling and analytical method.

Advantages and Disadvantages

The samples, collected in bubblers, are analyzed by means of a quick, instrumental method. Phenylhydrazine is slowly air-oxidized and is difficult to maintain known purity. Therefore, all analytical work is done using phenylhydrazine hydrochloride.

Apparatus

Glass, standard midget bubbler. With a stem which has a fritted glass end and containing 0.1 N HCl. The fritted end should have porosity approximately equal to that of Corning EC (170-220 micron maximum pore diameter).

Sampling pump. Suitable for delivering at least 1 L/min for 100 min. The pump is protected from splashover or water condensation by a 5 cm (6 mm I.D. and 8 mm O.D.) glass tube loosely packed with a plug of glass wool and inserted between the exit arm of the bubbler and the pump.

Integrating volume meter. Such as a dry gas or wet test meter.

Thermometer.

Manometer.

Stopwatch.

Spectrophotometer. This instrument should be capable of measuring the developed color at 730 nm.

Matched glass cells or cuvettes. 1 cm path length.

Assorted glassware. Pipets, volumetric flasks, and graduated cylinders of appropriate capacities.

Reagents

All reagents must be ACS reagent grade or better.

Distilled water.

Collection medium, 0.1 N HCl. Fill a 1000 mL volumetric flask with approximately 300 mL distilled water, add 8.6 mL concentrated HCl, mix and bring volume to the 1000 mL mark.

Phosphomolybdic acid solution (PMA). Dissolve 15 g PMA in 500 mL distilled water, allow to stand one day and filter before use through a fluted paper filter.

Phenylhydrazine hydrochloride.

Phenylhydrazine hydrochloride, standard solution. Weigh accurately 0.1000 g phenylhydrazine hydrochloride in a 100 mL volumetric flask and fill to mark with 0.1 N HCl.

Procedure

Cleaning of equipment. No specialized cleaning of glassware is required.

Calibration of personal pump. Personal sampling pump should be calibrated by using an integrating volume meter or other means.

Collection and shipping of samples. Pour 15 mL of the 0.1 N HCl into each bubbler. Connect a bubbler with a 5 cm glass adsorption tube (6 mm I.D. and 8 mm O.D.) containing the glass wool plug, then to the personal sampling pump using short pieces of flexible tubing. The air being sampled should not pass through any tubing or other equipment before entering the bubbler. Turn the pump on to begin sample collection. Care should be taken to measure the flow rate, the time and/or the volume as accurately as possible. Record the atmospheric pressure and the temperature. If the pressure reading is not available, record the elevation. The sample should be taken at a flow rate of 1 L/min. A sample size of 100 L is recommended.

After sampling, the bubbler stems are lifted out of the bubbler bottom and rinsed with approximately 5 mL distilled water. The bubblers are sealed with a hard, non reactive stopper (preferably Teflon or glass). Do not seal with rubber. The stoppers on the bubblers should be tightly sealed to prevent leakage during shipping. Care should be taken to minimize spillage or loss by evaporation at all times. Whenever possible, hand delivery of the samples is recommended. Otherwise, special bubbler shipping cases designed by NIOSH should be used to ship the samples. A blank bubbler should be handled in the same manner as the other samples (fill, seal, and transport) except that no air is sampled through this bubbler.

Analysis of samples. The sample in each bubbler is analyzed separately. Transfer the solution to a 50 mL volumetric flask. Rinse bubbler twice with 5 mL distilled water each and add rinses to volumetric flask. Add 10 mL phosphomolybdic acid solution to volumetric flask. This may be carefully measured in a 10 mL graduated cylinder. Dilute to the 50 mL mark with distilled water. Pipet a 3 mL aliquot into a 10 mL volumetric flask and dilute to 10 mL with distilled water. Read at 730 nm in the spectrophotometer against a blank prepared from the 0.1 N HCl in the same fashion as the samples.

Calibration and Standards

To six 50 mL volumetric flasks, add 15 mL 0.1 N HCl using a 25 mL graduated cylinder. Carefully pipet 1, 2, 3, 6, and 9 mL of the standard solution into the flasks. Process one flask as a blank. Add to this mixture 10 mL of phosphomolybdic acid solution using a 10 mL graduated cylinder. Continue as described under analysis of samples.

Adjust the baseline of the spectrophotometer to zero by reading distilled water in both cells. Read the blank in the sample cell, then read the samples at 730 nm. Construct a calibration curve by plotting absorbance against equivalent mg of phenylhydrazine in the color developed solution. Since the calibration curve is made using aliquots of phenylhydrazine hydrochloride, the equivalent weight of phenylhydrazine must be calculated:

equivalent mg = (mg phenylhydrazine.HCl weighed out in standard)(108.14)(mL standard) ÷ (144.6)(100 mL)

where:

108.14 = molecular weight of phenylhydrazine
144.6 = molecular weight of phenylhydrazine.HCl

Calculations

Subtract the absorbance of the blank from the absorbance of each sample. Determine from the calibration curve the mg of phenylhydrazine present in each sample. The concentration of the analyte in the air sampled can be expressed in mg/m^3:

mg/m^3 = (mg phenylhydrazine)(1000 L/m^3) ÷ air volume sampled in L

Another method of expressing concentration is ppm:

ppm = (mg/m^3)(24.45)(760)(T + 273) ÷ (M.W.)(P)(298)

where:

P = pressure (mm Hg) of air sampled
T = temperature (°C.) of air sampled
24.45 = molar volume (L/mol) at 25° C. and 760 mm Hg
M.W. = molecular weight (g/mol) of analyte
760 = standard pressure (mm Hg)
298 = standard temperature (°K)

References

1. L. Feinsilver, J. Perregrino, and C. Smith, Jr. Estimation of hydrazine and three of its methyl derivatives. Ind. Hyg. J. 20: 26-31, 1959.
2. Documentation of NIOSH Validation Tests. NIOSH Contract No. CDC-99-74-45.

PHOSGENE

Synonyms: carbonyl chloride; chloroformyl chloride $COCl_2$
ACGIH TLV: 0.1 ppm (0.4 mg/m^3)
OSHA Standard: 0.1 ppm
NIOSH Recommendation: 0.2 ppm ceiling (15 min)

Method	Sampling Duration	Sampling Location	Useful Range (ppm)	System Cost ($)	Test Cost ($)	Manufacturer
DT	Grab	Portable	0.04-75	150	2.80	Nat'l Draeger
DT	Grab	Portable	0.1-20	165	2.30	Matheson
DT	Grab	Portable	0.1-9	180	2.25	Bendix
CS	Grab	Portable	0.01-0.4	2465	0.80	MDA
CS	Cont	Portable	0.01-0.4	2520	Variable	MDA
IR	Cont	Portable	0.06-0.4	4374	0	Foxboro
CS	Cont	Fixed	0.02-0.4	4950	Variable	MDA

NIOSH METHOD NO. P&CAM 219

Principle
4,4'-Nitrobenzylpyridine in diethylphthalate reacts with traces of phosgene to produce a brilliant red color. The addition of an acid acceptor such as N-phenylbenzylamine stablizes the color and increases the sensitivity. The absorbance is determined at 475 nm (reference 1).

Range and Sensitivity
Sampling efficiency is 99% or better. Five mg of phosgene can be detected; the minimum sample size is 25 L. Therefore, 0.02 mg/m^3 phosgene should be detectable with a 250 L sample. 20 mg/m^3 upper limit should be within the range of most laboratory photometers for a 25 L air sample.

Interferences
Other acid chlorides, alkyl and aryl derivatives which are substituted by active halogen atoms, and sulfate esters are known to produce color with this reagent. However, most of these interferences can be removed in a pre-scrubber containing an inert solvent such as Freon 113 cooled by an ice bath. This method is not subject to interference from likely concentrations of chloride, hydrogen chloride, chlorine dioxide or simple chlorinated hydrocarbons such as carbon tetrachloride, chloroform, and tetrachloroethylene. A slight depression of color density has been noted under high humidity conditions.

Precision and Accuracy
The accuracy and precision of this method have not been determined. Calibration is a major problem because of the reactivity of phosgene. Flow dilution systems can be used but require anhydrous air for dilution. Permeation type calibration standards may be employed, if available.

Advantages and Disadvantages
Advantages include high sensitivity, standard laboratory equipment, and relative simplicity. Disadvantages include potential interferences, relative changes in color formation with various lots of reagents and the need for frequent calibration checks.

Apparatus
Midget impingers.
Vacuum source, aspirator or pump.
Flow meter (wet test meter).
Polyethylene bottles and beakers.
Bausch and Lomb Spec 20, or equivalent photometer.
Assorted laboratory glassware and tubing. No plasticized tubing (such as Tygon) should be used.

Reagents
4,4'-nitrobenzylpyridine. Aromil Chemical Co. (Baltimore, MD) or Aldrich Chemical Co. (Milwaukee, WI).
N-phenylbenzylamine. ACS reagent grade or better.
Diethylphthalate. Special selection may be required since some lots of the phthalate produce unstable color.
Color reagent. Reagent is made up of a solution of 2.5 g 4,4'-nitrobenzylpyridine, 5 g N-phenylbenzylamine and 992.5 g diethylphthalate. The concentration of color-forming reagents in diethylphthalate is critical as less color is developed with either a decreased or increased concentration.

Procedure

Cleaning of equipment. All glassware and plastic ware are to be doubly rinsed with isopropanol (IPA) and dried. If IPA rinse(s) does not thoroughly clean apparatus, use standard glassware cleaning procedures before IPA rinses.

Collection of samples. Atmospheric samples are collected in midget impingers loaded with 10 mL of reagent prepared as described above. A calibrated sample rate of 1 L/min is used. A known volume of about 50 L is drawn through the impingers for a range of 0.04-1 ppm phosgene. (For ranges above 1 ppm, 25 L should be drawn.) The final solution volume is recorded and the sample transferred to a small polyethylene or glass bottle or directly to the cuvette. The red color formed is stable for at least 4 hr, but should be measured within 9 hr. A 10%-15% loss in color density may occur after 8 hr.

Analysis of samples. The sample is transferred to a cuvette for the photometric measurement of 475 nm. The photometer should be zeroed (at 100% T or O absorbance) with a matched cuvette containing unreacted reagent at the 475 nm wavelength. The absorbance (or transmittance) of the sample is measured with the photometer at the 475 nm wavelength.

Calibration and Standards

Phosgene is highly reactive, and standards are difficult to prepare. However, dynamic standards may be prepared from a flow dilution system using a standard cylinder of 1000 ppm phosgene in nitrogen. (Lower concentrations of phosgene in cylinders are generally unstable. In any case, the standard cylinder concentration should be checked periodically by mass spectrometric or other methods.) A double dilution system with $COCl_2$-free air to dilute the phosgene about 1000:1 should be used. Permeation type calibration technique can be employed if phosgene standard permeation tubes are available.

Calculations

A calibration curve should be prepared of absorbance at 475 nm versus the phosgene concentration in air with 50 L of gas sample. For a gas sample of volume X, the phosgene concentration as determined by the calibration curve should be multiplied by 50/X for the corrected value.

References

1. Analytical Guides, American Industrial Hygiene Association, Akron, Ohio, 1969.
2. Linch, A.L., S.S. Ford, K.A. Kubitz, and M.R. DeBrunner. Phosgene in air—development of improved detector procedures. Amer. Ind. Hyg. Assoc. J. 26: 465, 1965.
3. Noweir, M.H. and E.A. Pfitzer. The Determination of Phosgene in Air, unpublished paper presented at the American Industrial Hygiene Conference in Philadelphia, PA, on April 28, 1964.

PHOSPHINE

ACGIH TLV: 0.3 ppm (0.4 mg/m³)
OSHA Standard: 0.3 ppm

PH_3

Method	Sampling Duration	Sampling Location	Useful Range (ppm)	System Cost ($)	Test Cost ($)	Manufacturer
DT	Grab	Portable	0.1-3000	150	2.80	Nat'l Draeger
DT	Grab	Portable	0.25-700	165	2.30	Matheson
DT	Grab	Portable	0.15-1000	180	2.25	Bendix
CS	Grab	Portable	0.02-0.5	2465	0.80	MDA
PI	Cont	Portable	0.1-600	4295	0	Airco
PI	Cont	Fixed	0.2-20	4950	0	HNU
PI	Cont	Fixed	0.1-600	6916	0	Airco
GC	1 min	Fixed	0.007-0.75	43,000	0	Airco

NIOSH METHOD NO. S332

Principle

A known volume of air is drawn through a tube containing mercuric cyanide impregnated silica gel to trap the phosphine. The phosphorus is extracted and oxidized to phosphate using a hot, acidic permanganate solution. The extracted sample is analyzed for phosphate by formation of the phosphomolybdate complex, extraction into a mixture of isobutanol and toluene and reduction using stannous choride. The absorbance of the reduced phosphomolybdate complex is measured at 625 nm.

Range and Sensitivity

This method was validated over the range of 0.195-0.877 mg/m³ at an atmospheric temperature and pressure of 19.0° C. and 765.3 mm Hg, respectively, using a 16 L sample. The method may be capable of measuring smaller amounts if the desorption efficiency is adequate. Desorption efficiency must be determined over the range used. The upper limit of the range of the method depends on the adsorptive capacity of the mercuric cyanide treated silica gel. This capacity may vary with the concentration of phosphine and other substances in the air. When an atmosphere at 90% relative humidity containing 0.957 mg/m³ of phosphine was sampled at a flow rate of 0.2 L/min breakthrough was determined to occur at a sampling volume of 20.75 L (capacity=19.86 µg PH_3). To minimize the probability of overloading the sampling tube, the sample size recommended is less than two-thirds the 5% breakthrough capacity at over 80% RH at twice the OSHA standard.

Sampling at 0.2 L/min for 80 min will yield a 16 L sample with 6.4 µg of PH_3 collected at the OSHA standard level. The sensitivity of the method obtained from the slope of the absorbance vs. µg of phosphine calibration curve is 0.0524 absorbance units/µg of PH_3. The detection limit of the method determined from twice the standard deviation for the absorbance of 6 blank treated silica gel tubes corresponds to 0.19 µg PH_3 (or 0.0119 mg/m³ for a 16 L sample).

Interferences

When two or more compounds are known or suspected to be present in the air, such information, including the suspected identities, should be transmitted with the sample. The colorimetric determination of phosphate is subject to interference by any species which also forms a molybdate complex which is extractable into the isobutanol-toluene mix and absorbs at similar wavelengths.

Any phosphorus compound which is retained by the mercuric cyanide coated silica gel tube and oxidized to phosphate by hot aqueous permanganate solution will be a major interference. Possible interfering species include PCl_3 and PCl_5 vapors and organic phosphorus compounds. Particulate H_3PO_4, P_4O_{10} and P_4S_{10} are also possible interferents unless a prefilter is used in conjunction with the sorbent tube. Although a prefilter has not been tested with the method, its use is recommended. If the possibility of interferences do exist, modification of the analytical procedure must be made to circumvent the problems or an alternative procedure should be used.

Precision and Accuracy

The coefficient of variation for the total sampling method in the range of 0.195-0.877 mg/m^3 was 0.0908. This value corresponds to a standard deviation of 0.0363 mg/m^3 at the OSHA standard level. Statistical information and details of the validation and experimental test procedures can be found in references 1 and 2.

On the average, the concentration obtained at the OSHA standard level using the overall sampling and analytical procedure was 0.2% higher than the average taken concentration for a limited number of laboratory experiments. Any difference between the found and taken concentrations may not represent a bias in the sampling and analytical method but rather a random variation from the experimentally determined "true" concentration. Also, collected samples, stored for at least 7 days, are stable, thus no recovery corrections should be applied to the final result.

Advantages and Disadvantages

The sampling device is small, portable and involves no liquids. The precision of the method is affected by the reproducibility of the pressure drop across the tubes. This drop will affect the flow rate and cause the volume to be imprecise, because the pump is usually calibrated for only one tube. The analytical method requires measurement of the absorbance of the phosphomolybdate complex 1 min after reduction with stannous chloride. This may present an inconvenience since the samples need to be handled individually. A disadvantage of the method is that the amount of sample which can be collected is limited by the number of μg of phosphine that the tube will hold before overloading. When the amount of phosphine found on the backup section exceeds 25% of that on the front, the probability of sample loss exists.

Apparatus

Calibrated personal sampling pump. Whose flow can be determined within 5% at the recommended flow rate. The pump must be calibrated with a representative tube in the line.

Treated silica gel tube. Glass tube with both ends flame sealed, 12 cm long with 6 mm O.D. and 4 mm I.D., containing 2 sections of treated silica gel (45/60 mesh, SKC, Inc.). The absorbing section contains 300 mg of the treated silica gel, the backup section 150 mg. A small wad of silylated glass wool is also placed between the front adsorbing section and the backup section; a plug of silylated glass wool is also placed in the front of the adsorbing section and at the end of the backup section. The pressure drop across the tube must be less than 2 inches of mercury at a flow rate of 0.2 L/min.

Procedure for coating the silica gel. Dry 100 g of silica gel at 90° C. for 2 hours. Prepare a 2% w/v mercuric cyanide solution in water (2 g $Hg(CN)_2$ in 100 mL H_2O). Add the dried silica gel to the mercuric cyanide solution and let set for 15 min with occasional stirring. Drain the excess mercuric cyanide solution and dry the remaining silica gel at 90° C. for 3 hours. Cool the silica gel to room temperature in a covered beaker. Expose the silica gel to a humid atmosphere (over 80% RH) for 24 hours.

Spectrophotometer. Capable of measuring absorbance or transmittance at 625 nm.

Two matched 5 cm silica cells. With tight-fitting caps.

Separatory funnel. 125 mL.

Beakers. 50 L.

Volumetric flasks. 10, 25, 100 and 1000 mL.

Pipets. 0.2, 10 and 25 mL and other convenient sizes to make standard dilutions.

Graduated cylinders. 10 mL.

Water bath. Maintained at 65-70° C.

Syringes. 0.5 and 1.0 mL.

Balance.

Barometer.

Thermometer.

Stopwatch.

Reagents

All reagents should be ACS reagent grade or better.

Water, distilled or deionized.

Potassium dihydrogen phosphate, anhydrous.

Sulfuric acid, concentrated.

Ammonium molybdate.

Ferrous ammonium sulfate.

Potassium permanganate.

Stannous chloride.

Glycerol.

Toluene.

Isobutanol.

Methanol.

Standard phosphate solution. Prepare by dissolving 200 mg of KH_2PO_4 in 1 L of distilled water. (1.00 mL=49.94 µg PH_3).

Molybdate solution. Prepare by dissolving 49.4 g of $(NH_4)_6Mo_7O_{24}.4H_2O$ and 112 mL of concentrated H_2SO_4 in water to a total volume of 1 L.

Toluene-isobutanol solvent. Mix equal volumes of toluene and isobutyl alcohol.

Alcoholic sulfuric acid solution. Add 50 mL of concentrated H_2SO_4 to 950 mL of methyl alcohol.

Ferrous solution. Prepare by dissolving 7.9 g of $Fe(NH_4)_2(SO_4)_2.6H_2O$ and 1 mL of concentrated H_2SO_4 in water with a total volume of 100 mL.

Stannous chloride reagent. Prepare by dissolving 0.4 g of $SnCl_2.2H_2O$ in 50 mL glycerol (heat to dissolve).

Acidic permanganate reagent. Prepare by dissolving 0.316 g of $KMnO_4$ and 6 mL of concentrated H_2SO_4 in a total volume of 1 L H_2O.

Procedure

Cleaning of equipment. Before use, all glassware should be initially soaked in a mild detergent solution to remove any residual grease or chemicals. After initial cleaning, the glassware should be thoroughly rinsed with warm tap water, 6 M nitric acid, tap water, distilled water in that order and then dried.

Calibration of personal sampling pumps. Each personal sampling pump must be calibrated with a representative sampling tube in the line. This will minimize errors associated with uncertainties in the sample volume collected.

Collection and shipping of samples. Immediately before sampling, break the two ends of the silica gel tubes to provide an opening at least one-half the internal diameter of the tube (2 mm). The treated silica gel tubes should be placed in a vertical direction during sampling to minimize channeling through the charcoal. Air being sampled should not be passed through any hose or tubing before entering the tube.

A sample size of 16 L is recommended. Sample at a known flow rate between 0.2 and 0.01 L/min. The flow rate should be known with an accuracy of at least 5%. The temperature and pressure of the atmosphere being sampled should be recorded. If pressure reading is not available, record the elevation. The treated silica gel tubes should be labelled appropriately and capped with the supplied plastic caps. Under no circumstances should rubber caps be used.

With each batch or partial batch of 10 samples, submit one treated silica gel tube which had been handled in the same manner as the sample tubes (break, seal, and transport), except that no air is sampled through this tube. This tube should be labeled as a blank. Capped treated silica gel tubes should be packed tightly and padded before they are shipped to minimize tube breakage during shipping.

Preparation of samples. In preparation for analysis, each treated silica gel tube is scored with a file and broken open. The glass wool is removed with care and discarded. The silica gel in the front section is transferred to a 50 mL beaker. The separating section of glass wool is removed and discarded. The backup section of silica gel is transferred to another container. These two section are analyzed separately.

Extraction of the samples. Prior to analysis, 10 mL of the acidic permanganate reagent is pipeted into each beaker containing the silica gel. The extraction is carried out for 90 min at 65-70° C. in a water bath. After extraction the acidic permanganate solution is drained into a 10 mL volumetric leaving the silica gel. The volumetric is made to volume with distilled water. The silica gel is washed twice with 3 mL portions of distilled water and the contents drained into another 10 mL volumetric containing 1 mL of ferrous solution. The flask is made up to volume with distilled water and mixed thoroughly.

Spectrophotometer operation. Turn on the spectrophotometer and allow sufficient time for warmup. Follow the instrument manufacturer's recommendations for specific operating parameters. Adjust the wavelength to 625 nm and set the zero and 100% transmittance scale using 5 cm cells filled with distilled water. Check these settings prior to making any measurement to check on instrument drift.

Analytical procedure. Add the contents of both 10 mL volumetric flasks (extract and washings) to a 125 mL separatory funnel. Add 7.5 mL of molybdate reagent and 25 mL of toluene-isobutanol solvent. Shake for 60 seconds. Allow 60 seconds for the aqueous and nonaqueous layers to separate and discard the lower (aqueous layer). Pipet 10 mL of the nonaqueous layer into a 25 mL volumetric containing 10 mL of the alcoholic sulfuric acid solution.

The next steps must be performed within 1 min. Add 0.5 mL (25 drops) of stannous chloride reagent and make to volume using alcoholic sulfuric acid solution. Mix thoroughly. Transfer the sample into 5 cm cells and stopper immediately. Measure the absorbance or transmittance at 625 nm using water as a blank.

Calibration and Standards

Add 10 mL of acidic permanganate solution, 1 mL of ferrous reagent to the separatory funnel. Add 20-400 µL of the standard phosphorus solution to cover the range of 1-10 µg of PH_3. Add 8-9 mL of H_2O to make the total volume of the permanganate solution, ferrous solution, phosphorus solution and water to 20 mL. Prepare at least 6 calibration standards. A blank containing no phosphorus should also be analyzed.

Proceed as described above under Analytical Procedure. Prepare a calibration curve by plotting the absorbance of the standards after subtraction of the blank versus the amount of each standard in µg of PH_3 added on linear graph paper.

Calculations

Correction for the blank (obtained by extraction and analysis of the treated silica gel tube marked blank) must be made for each sample:

$Ac = As - Ab$

where:

Ac = corrected absorbance
As = sample absorbance
Ab = blank absorbance

A similar procedure is followed for the backup tube. The amount of phosphine present in the front tube, corrected for the blank, is found by reading the amount corresponding to Ac from the standard curve. The amount of phosphine found in the backup tube is similarly determined. Add the amounts present in the front and backup tubes for the same sample to determine the total weight in the sample.

Determine the volume of air sampled at ambient conditions based on the appropriate information, such as flow rate (L/min) multiplied by sampling time (min). If a pump using a rotameter for flow rate control was used for sample collection, a pressure and temperature correction must be made for the indicated flow rate. The expression for this correction is:

Corrected volume = $ft(P_1T_2/P_2T_1)^{1/2}$

where:

f = flow rate sampled
t = sampling time
P_1 = pressure during calibration of sampling pump (mm Hg)
P_2 = pressure of air sampled (mm Hg)
T_1 = temperature during calibration of sampling pump (°K)
T_2 = temperature of air sampled (°K)

The concentration of the analyte in the air sampled can be expressed in mg/m^3, which is numerically equal to µg/L, by:

mg/m^3 = total mg ÷ air volume sampled in L

Another method of expressing concentration is ppm (corrected to standard conditions of 25° C. and 760 mm Hg):

$$\text{ppm} = (mg/m^3)(24.45)(760)(T + 273) \div (34.00)(P)(298)$$

where:

P = pressure (mm Hg) of air sampled
T = temperature (°C.) of air sampled
24.45 = molar volume (L/mol) at 25° C. and 760 mm Hg
34.00 = molecular weight of phosphine
760 = standard pressure (mm Hg)
298 = standard temperature (°K)

References

1. Documentation of NIOSH Validation Tests. National Institute for Occupational Safety and Health, Cincinnati, Ohio (DHEW-NIOSH-Publication No. 77-185), 1977. Available from Superintendent of Documents, Washington D.C., Order No. 017-33-00231-2.
2. Backup Data Report for Phosphine, S332. Prepared under NIOSH Contract No. 210-76-0123, March 17, 1978.

PROPANE

C_3H_8

ACGIH TLV: simple asphyxiant
OSHA Standard: 1000 ppm (1800 mg/m^3)

Method	Sampling Duration	Sampling Location	Useful Range (ppm)	System Cost ($)	Test Cost ($)	Manufacturer
DT	Grab	Portable	5000-13,000	150	2.60	Nat'l Draeger
IR	Cont	Portable	20-2000	4374	0	Foxboro
PI	Cont	Fixed	20-2000	4950	0	HNU
GC	1 min	Portable	1-10^6	10,000	0	Microsensor

NIOSH METHOD NO. S93

(Use as described for liquified petroleum gas, with propane as the calibration gas)

n-PROPYL ACETATE

ACGIH TLV: 200 ppm (840 mg/m³)
OSHA Standard: 200 ppm

$CH_3COOC_3H_7$

Method	Sampling Duration	Sampling Location	Useful Range (ppm)	System Cost ($)	Test Cost ($)	Manufacturer
DT	Grab	Portable	600-9000	180	2.25	Bendix
IR	Cont	Portable	4-400	4374	0	Foxboro
PI	Cont	Fixed	2-200	4950	0	HNU
GC	1 min	Portable	$1-10^6$	10,000	0	Microsensor

NIOSH METHOD NO. P&CAM 127

(Use as described for acetone, with 70° C. column temperature)

n-PROPYL ALCOHOL

Synonyms: n-propanol; 1-propanol
ACGIH TLV: 200 ppm (500 mg/m³) - Skin
OSHA Standard: 200 ppm

C_3H_7OH

Method	Sampling Duration	Sampling Location	Useful Range (ppm)	System Cost ($)	Test Cost ($)	Manufacturer
DT	Grab	Portable	100-3000	150	2.70	Nat'l Draeger
DT	Grab	Portable	200-8000	180	2.25	Bendix
IR	Cont	Portable	4-400	4374	0	Foxboro
PI	Cont	Fixed	2-200	4950	0	HNU

NIOSH METHOD NO. S66

(Use as described for n-butyl alcohol, with 70° C. column temperature)

PROPYLENE DICHLORIDE

Synonyms: 1,2-dichloropropane; propylene chloride $CH_2ClCHClCH_3$
ACGIH TLV: 75 ppm (350 mg/m^3)
OSHA Standard: 75 ppm

Method	Sampling Duration	Sampling Location	Useful Range (ppm)	System Cost ($)	Test Cost ($)	Manufacturer
IR	Cont	Portable	1.5-150	4374	0	Foxboro
GC	1 min	Portable	1-10^6	10,000	0	Microsensor

NIOSH METHOD NO. P&CAM 127

(Use as described for acetone, with 80° C. column temperature)

PROPYLENE OXIDE

Synonyms: propene oxide; methyloxirane; 1,2-epoxypropane OCH_2CHCH_3
ACGIH TLV: 20 ppm (50 mg/m^3)
OSHA Standard: 100 ppm (240 mg/m^3)

Method	Sampling Duration	Sampling Location	Useful Range (ppm)	System Cost ($)	Test Cost ($)	Manufacturer
DT	Grab	Portable	3000-36,000	180	2.25	Bendix
IR	Cont	Portable	2-200	4374	0	Foxboro
PI	Cont	Fixed	0.2-20	4950	0	HNU
GC	1 min	Portable	1-10^6	10,000	0	Microsensor

NIOSH METHOD NO. S112

(Use as described for chloroprene, with 145° C. column temperature)

n-PROPYL NITRATE

ACGIH TLV: 25 ppm (105 mg/m^3)
OSHA Standard: 25 ppm

$C_3H_7NO_3$

Method	Sampling Duration	Sampling Location	Useful Range (ppm)	System Cost ($)	Test Cost ($)	Manufacturer
IR	Cont	Portable	0.5-50	4374	0	Foxboro

NIOSH METHOD NO. P&CAM 127

(Use as described for acetone, with 85° C. column temperature)

PYRIDINE

ACGIH TLV: 5 ppm (15 mg/m^3)
OSHA Standard: 5 ppm

C_5H_5N

Method	Sampling Duration	Sampling Location	Useful Range (ppm)	System Cost ($)	Test Cost ($)	Manufacturer
DT	Grab	Portable	5	150	6.00	Nat'l Draeger
DT	Grab	Portable	0.4-35	180	2.25	Bendix
IR	Cont	Portable	0.4-20	4374	0	Foxboro
PI	Cont	Fixed	0.2-20	4950	0	HNU

NIOSH METHOD NO. S161

Principle

A known volume of air is drawn through a charcoal tube to trap the organic vapors present. The charcoal in the tube is transferred to a small, stoppered sample container, and the analyte is desorbed with methylene chloride. An aliquot of the desorbed sample is injected into a gas chromatograph. The area of the resulting peak is determined and compared with areas obtained for standards.

Range and Sensitivity

This method was validated over the range of 7.59-30.4 mg/m^3 at an atmospheric temperature and pressure of 22° C. and 763 mm Hg, using a 100 L sample. Under the conditions of sample size (100 L) the probable useful range of this method is 1.5-45 mg/m^3 at a detector sensitivity that gives nearly full deflection on the strip chart recorder for a 4.5 mg sample. The method is capable of measuring much smaller amounts if the desorption efficiency is adequate. Desorption efficiency must be determined over the range used.

The upper limit of the range of the method is dependent on the adsorptive capacity of the charcoal tube. This capacity varies with the concentrations of pyridine and other substances in the air. The first section of the charcoal tube was found to hold at least 6.8 mg of pyridine when a test atmosphere containing 30.4 mg/m^3 of pyridine in air was sampled at 0.93 L/min for 240 min; at that time the concentration of pyridine in the effluent was less than 5% of that in the influent. (The charcoal tube consists of two sections of activated charcoal separated by a section of urethane foam.) If a particular atmosphere is suspected of containing a large amount of contaminant, a smaller sampling volume should be taken.

Interferences

When the amount of water in the air is so great that condensation actually occurs in the tube, organic vapors will not be trapped efficiently. Preliminary experiments using toluene indicated that high humidity severely decreases that breakthrough volume.

When two or more compounds are known or suspected to be present in the air, such information, including their suspected identities, should be transmitted with the sample. Since methylene chloride is used rather than carbon disulfide to desorb pyridine from the charcoal, it would not be possible to measure methylene chloride in the sample with this desorbing solvent. If it is suspected that methylene choride is present, a separate sample should be collected for methylene chloride analysis.

It must be emphasized that any compound which has the same retention time as the analyte at the operating conditions described in this method is an interference. Retention time data on a single column cannot be considered proof of chemical identity. If the possibility of interference exists, separation conditions (column packing, temperature, etc.) must be changed to circumvent the problem.

Precision and Accuracy

The coefficient of variation for the total analytical and sampling method in the range of 7.59-30.4 mg/m^3 was 0.059. This value corresponds to a standard deviation of 0.9 mg/m^3 at the OSHA standard level. Statistical information and details of the validation and experimental test procedures can be found in reference 2.

On the average, the concentrations obtained at the OSHA standard level using the overall sampling and analytical method were 6.3% higher than the "true" concentrations for a limited number of laboratory experiments. Any difference between the "found" and "true" concentrations may not represent a bias in the sampling and analytical method, but rather a random variation from the experimentally determined "true" concentration. Therefore, no recovery correction should be applied to the final result.

Advantages and Disadvantages

The sampling device is small, portable, and involves no liquids. Interferences are minimal, and most of those which do occur can be eliminated by altering chromatographic conditions. The tubes are analyzed by means of a quick, instrumental method. The method can also be used for the simultaneous analysis of two or more compounds suspected to be present in the same sample by simply changing gas chromatographic conditions from isothermal to a temperature-programmed mode of operation.

One disadvantage of the method is that the amount of sample which can be taken is limited by the number of mg that the tube will hold before overloading. When the sample value obtained for the backup section of the charcoal tube exceeds 25% of that found in the front section, the possibility of sample loss exists. Furthermore, the precision of the method is limited by the reproducibility of the pressure drop across the tubes. This drop will affect the flow rate and cause the volume to be imprecise, because the pump is usually calibrated for one tube only.

Apparatus

Calibrated personal sampling pump. Whose flow can be determined within 5% at the recommended flow rate.

Charcoal tubes. Glass tube with both ends flame sealed, 7 cm long with 6 mm O.D. and 4 mm I.D., containing 2 sections of 20/40 mesh activated charcoal separated by a 2 mm portion of urethane foam. The activated charcoal is prepared from coconut shells and is fired at 600° C. prior to packing. The adsorbing section contains 100 mg of charcoal, the backup section 50 mg. A 3 mm portion of urethane foam is placed between the outlet end of the tube and the backup section. A plug of silylated glass wool is placed in front of the adsorbing section. The pressure drop across the tube must be less than one inch of mercury at a flow rate of 1 L/min.

Gas chromatograph. Equipped with a flame-ionization detector.

Column. 10 ft × 1/8 in stainless steel, packed with 5% Carbowax 20 M on 80/100 mesh, acid washed DMCS Chromosorb W.

Electronic integrator. Or some other suitable method for measuring peak area.

Sample containers. 2 mL, with glass stoppers or Teflon-lined caps.

Syringes. 10 μL, and other convenient sizes for making standards.

Pipets. 1.0 mL delivery pipets.

Volumetric flasks. 10 mL or convenient sizes for making standard solutions.

Reagents

Chromatographic quality methylene chloride.

Pyridine, reagent grade.

Purified nitrogen.

Prepurified hydrogen.

Filtered compressed air.

Procedure

Cleaning of equipment. All glassware used for the laboratory analysis should be detergent washed and thoroughly rinsed with tap water and distilled water.

Calibration of personal pumps. Each personal pump must be calibrated with a representative charcoal tube in the line. This will minimize errors associated with uncertainties in the sample volume collected.

Collection and shipping of samples. Immediately before sampling, break the ends of the tube to provide an opening at least one-half the internal diameter of the tube (2 mm). The smaller section of charcoal is used as a backup and should be positioned nearest the sampling pump. The charcoal tube should be placed in a vertical direction during sampling to minimize channeling through the charcoal. Air being sampled should not be passed through any hose or tubing before entering the charcoal tube.

A maximum sample size of 100 L is recommended. Sample at a flow of 1.0 L/min or less. The flow rate should be known with an accuracy of at least 5%. The temperature and pressure of the atmosphere being sampled should be recorded. If pressure reading is not available, record the elevation. The charcoal tubes should be capped with the supplied plastic caps immediately after sampling. Under no circumstances should rubber caps be used.

One tube should be handled in the same manner as the sample tubes (break, seal, and transport), except that no air is sampled through this tube. This tube should be labeled as a blank. Capped tubes should be packed tightly and padded before they are shipped to minimize tube breakage during shipping. A sample of the bulk material should be submitted to the laboratory in a glass container with a Teflon-lined cap. This sample should not be transported in the same container as the charcoal tubes.

Preparation of samples. In preparation for analysis, each charcoal tube is scored with a file in front of the first section of charcoal and broken open. The glass wool is removed and discarded. The charcoal in the first (larger) section is transferred to a 2 mL stoppered sample container. The separating section of foam is removed and discarded; the second section is transferred to another stoppered container. These two sections are analyzed separately.

Desorption of samples. Prior to analysis, 1.0 mL of methylene chloride is pipetted into each sample container. Desorption should be done for 30 min. Tests indicate that this is adequate if the sample is agitated occasionally during this period.

GC conditions. The typical operating conditions for the gas chromatograph are:

50 mL/min (60 psig) nitrogen carrier gas flow
65 mL/min (24 psig) hydrogen gas flow to detector
500 mL/min (50 psig) air flow to detector
260° C. injector temperature
285° C. detector temperature
140° C. column temperature

Injection. The first step in the analysis is the injection of the sample into the gas chromatograph. To eliminate difficulties arising from blow back or distillation within the syringe needle, one should employ the solvent flush injection technique. The 10 µL syringe is first flushed with solvent several times to wet the barrel and plunger. Three µL of solvent are drawn into the syringe to increase the accuracy and reproducibility of the injected sample volume. The needle is removed from the solvent, and the plunger is pulled back about 0.2 µL to separate the solvent flush from the sample with a pocket of air to be used as a marker. The needle is then immersed in the sample, and a 5 µL aliquot is withdrawn, taking into consideration the volume of the needle, since the sample in the needle will be completely injected. After the needle is removed from the sample and prior to injection, the plunger is pulled back 1.2 µL to minimize evaporation of the sample from the tip of the needle. Observe that the sample occupies 4.9-5.0 µL in the barrel of the syringe. Duplicate injections of each sample and standard should be made. No more than a 3% difference in area is to be expected.

Measurement of area. The area of the sample peak is measured by an electronic integrator or some other suitable form of area measurement, and preliminary results are read from a standard curve prepared as discussed below.

Determination of desorption efficiency. The desorption efficiency of a particular compound can vary from one laboratory to another and also from one batch of charcoal to another. Thus, it is necessary to determine at least once the percentage of the specific compound that is removed in the desorption process, provided the same batch of charcoal is used.

Activated charcoal equivalent to the amount in the first section of the sampling tube (100 mg) is measured into a 2.5 in, 4 mm I.D. glass tube, flame sealed at one end. This charcoal must be from the same batch as that used in obtaining the samples and can be obtained from unused charcoal tubes. The open end is capped with Parafilm. A known amount of hexane solution of pyridine containing 393 mg/mL is injected directly into the activated charcoal with a microliter syringe, and the tube is capped with more Parafilm. The amount injected is equivalent to that present in a 100 L sample at the selected level.

Six tubes at each of three levels (0.5X, 1X, and 2X the standard) are prepared in this manner and allowed to stand for at least overnight to assure complete adsorption of the analyte onto the charcoal. These tubes are referred to as the samples. A parallel blank tube should be treated in the same manner except that no sample is added to it. The sample and blank tubes are desorbed and analyzed in exactly the same manner as the sampling tube described above.

Two or three standards are prepared by injecting the same volume of compound into 1.0 mL of methylene chloride with the same syringe used in the preparation of the samples. These are analyzed with the samples. The desorption efficiency (D.E.) equals the average weight in mg recovered from the tube divided by the weight in mg added to the tube, or:

D.E. = average wt recovered (mg) ÷ wt added (mg)

The desorption efficiency is dependent on the amount of analyte collected on the charcoal. Plot the desorption efficiency versus the weight of analyte found. The curve is used to correct for adsorption losses.

Calibration and Standards

It is convenient to express concentration of standards in terms of mg/1.0 mL methylene chloride, because samples are desorbed in this amount of methylene chloride. The density of the analyte is used to convert mg into μL for easy measurement with a microliter syringe. A series of standards, varying in concentration over the range of interest, is prepared and analyzed under the same GC conditions and during the same time period as the unknown samples. Curves are established by plotting concentration in mg/1.0 mL versus peak area. Note: since no internal standard is used in the method, standard solutions must be analyzed at the same time that the sample analysis is done. This will minimize the effect of known day-to-day variations and variations during the same day of the FID response.

Calculations

Read the weight, in mg, corresponding to each peak area from the standard curve. No volume corrections are needed, because the standard curve is based on mg/1.0 mL methylene chloride and the volume of sample injected is identical to the volume of the standards injected. Corrections for the blank must be made for each sample:

mg = mg sample − mg blank

where:

mg sample = mg found in front section of sample tube
mg blank = mg found in front section of blank tube

A similar procedure is followed for the backup sections. Add the weights found in the front and backup sections to get the total weight in the sample. Read the desorption efficiency from the curve for the amount found in the front section. Divide the total weight by this desorption efficiency to obtain the corrected mg/sample:

Corrected mg/sample = total weight ÷ D.E.

The concentration of the analyte in the air sampled can be expressed in mg/m^3:

mg/m³ = (corrected mg)(1000 L/m³) ÷ air volume sampled in L

Another method of expressing concentration is ppm:

ppm = (mg/m³)(24.45)(760)(T + 273) ÷ (M.W.)(P)(298)

where:

P = pressure (mm Hg) of air sampled
T = temperature (°C.) of air sampled
24.45 = molar volume (L/mol) at 25° C. and 760 mm Hg
M.W. = molecular weight (g/mol) of analyte
760 = standard pressure (mm Hg)
298 = standard temperature (°K)

References

1. White, L.D. et al. A convenient optimized method for the analysis of selected solvent vapors in the industrial atmosphere. Amer. Ind. Hyg. Assoc. J. 31: 225, 1970.
2. Documentation of NIOSH Validation Tests. NIOSH Contract No. CDC-99-74-45.

RHODIUM DUST OR FUME

ACGIH TLV: 1 mg/m³ (metal); 0.01 mg/m³ (soluble salts)
OSHA Standard: 0.1 mg/m³ (metal); 0.001 mg/m³ (soluble salts)

Rh

Method	Sampling Duration	Sampling Location	Useful Range (mg/m³)	System Cost ($)	Test Cost ($)	Manufacturer
LS	Cont	Personal	0.01-100	1695	0	GCA
LS	Cont	Portable	0.001-100	2785	0	MDA
LS	Cont	Fixed	0.001-200	3990	0	GCA
BA	1,4 min	Portable	0.02-150	4890	0	GCA

NIOSH METHOD NO. S188

Principle

Sample-containing filters are wet ashed using nitric acid to destroy the organic matrix; rhodium and its compounds are then solubilized in a hydrochloric acid solution maintained at a pH of 1; potassium bisulfate is added to eliminate interferences by other common cations. The solutions of samples and standards are aspirated into the oxidizing air-acetylene flame of an atomic absorption (AA) spectrophotometer. A hollow cathode lamp for rhodium is used to provide a characteristic rhodium line at 343.5 nm. The absorbance is proportional to the rhodium concentration within a limited concentration range.

Range and Sensitivity

This method was validated over the range of 0.057-0.21 mg/m³ at an atmospheric temperature and pressure of 24° C. and 756 mm Hg, using a 720 L sample. Under the conditions of sample size (720 L), the linear working range of the method is estimated to be 0.005-0.21 mg/m³. The method may be extended to higher values by further dilution of the sample solution.

Interferences

Additional 3% potassium bisulfate eliminates the interferences of other common cations, and also anions like nitrates and phosphates in the rhodium assay using an oxidizing air-acetylene flame (reference 4).

Precision and Accuracy

The coefficient of variation for the total analytical and sampling method in the range of 0.057-0.21 mg/m³ was 0.079. This value corresponds to a 0.0079 mg/m³ standard deviation at the OSHA standard level. Statistical information and details of the validation and experimental test procedures can be found in reference 3.

A collection efficiency of 1.00 ± 0.01 was determined for the collection medium; thus, no bias was introduced in the sample collection step. There may be some bias in the analytical method—the average recovery from the filters was 94.7%; the data may be adjusted by this correction factor to eliminate any bias. Thus, coefficient of variation is a satisfactory measure of both accuracy and precision of the sampling and analytical method.

Advantages and Disadvantages

The method is simple.

Apparatus

Filter unit. Consists of the filter media and 37 mm 3 piece cassette filter holder.

Personal sampling pump. A calibrated personal sampling pump whose flow can be determined to an accuracy of 5% at the recommended flow rate. The pump must be calibrated with a filter holder and filter in line.

Thermometer.

Manometer.

Stopwatch.

Mixed cellulose ester membrane filter. 37 mm diameter, 0.8 micron pore size.

Atomic absorption spectrophotometer. With a monochromator having a reciprocal linear dispersion of about 6.5 Angstrom/mm in the ultraviolet region. The instrument must be equipped with an air-acetylene burner head.

Rhodium hollow cathode lamp.

Oxidant. Compressed air.

Fuel. Purified acetylene.

Pressure regulator. Two-stage, for each compressed gas tank used.

Phillips beakers. 125 mL, with watchglass covers.

Pipets. Delivery or graduated, 1, 5, 10 mL.

25 mL volumetric flasks.

Adjustable thermostatically controlled hot plate. Capable of reaching 400° C.

Reagents

All reagents used must be ACS reagent grade or better.

Distilled or deionized water.

Concentrated nitric acid.
Hydrochloric acid, 6 N.
Potassium bisulfate, $KHSO_4$. Prepare a 30% (w/v) solution in water.
Aqueous standard rhodium stock solution, 1000 µg/mL. Commercially available.
Rhodium working standard solution, 20 µg/mL. Prepare by appropriate dilution of above solution. Prepare fresh daily.

Procedure

Cleaning of equipment. Before use all glassware should initially be soaked in a mild detergent solution to remove any residual grease or chemicals. After initial cleaning, the glassware should be thoroughly rinsed with warm tap water, concentrated nitric acid, tap water, and distilled water, in that order, and then dried.

Sampling requirements and shipping of samples. To collect rhodium, metal fume and dust, a personal sampler pump is used to pull air through a cellulose ester membrane filter. The filter holder is held together by tape or a shrinking band. If the middle piece of the filter holder does not fit snugly into the bottom piece of the filter holder, the contaminant will leak around the filter. A piece of flexible tubing is used to connect the filter holder to the pump. Sample at a flow rate of 1.5 L/min with face cap on and small plugs removed. After sampling, replace small plugs.

With each batch of ten samples submit one filter from the same lot of filters which was used for sample collection and which is subjected to exactly the same handling as for the samples except that no air is drawn through it. Label this as a blank. The filter cassettes should be shipped in a suitable container, designed to prevent damage in transit.

Analysis of samples. Open the cassette filter holder and carefully remove the cellulose membrane filter from the holder and cellulose backup pad with the aid of a Millipore filter tweezers and transfer to a 125 mL Phillips beaker.

To destroy the organic filter matrix, treat the sample in each beaker with 2 mL of concentrated nitric acid. Cover each beaker with a watch glass and heat on a hot plate (140° C.) in a fume hood until all the filter is dissolved and the volume is reduced to about one-half mL. Repeat this process two more times using 2 mL of concentrated nitric acid each time. Do not allow the solution to evaporate to dryness.

To ensure complete dissolution of rhodium compounds, digest the resulting nitric acid solution by treating with HCl and heating on a high temperature hot plate (400° C.). This is done by adding 2 mL of 6 N aqueous HCl and evaporating to about 0.5 mL; this HCl addition and evaporation is done 3 times. Do not allow the solution to evaporate to dryness at any point. Cool solutions and add 10 mL of distilled (or deionized) water to each one.

Quantitatively transfer the clear solutions into a 25 mL volumetric flask. Rinse each beaker at least twice with 5 mL portions of distilled water, and quantitatively transfer each rinsing to the solution in the volumetric flask. Add 2.5 mL of the 30% potassium bisulfate solution and dilute to 25 mL with distilled water.

Aspirate the solutions into an oxidizing air acetylene flame and record the absorbance at 343.5 nm. The absorbance is proportional to the analyte concentration in the sample and can be determined from the appropriate calibration curve. When very low metal concentrations are found in the sample, scale expansion can be used to increase instrument response or the sample could be concentrated to some smaller volume such as 10 mL before aspiration. In such a case, one should not use any more water than is necessary to effect a quantitative transfer. Follow instrument manufacturer's recommendation for specific operating parameters. Appropriate filter blanks must be analyzed by the same procedure used for the samples.

Determination of sample recovery. To eliminate any bias in the analytical method, it is necessary to determine the recovery of the analyte. The analyte recovery should be determined in duplicate and should cover the concentration ranges of interest. If the recovery of the analyte is less than 95%, the appropriate correction factor should be used to calculate the "true" value.

A known amount of the analyte, preferably equivalent to the concentration expected in the sample, is added to a representative cellulose membrane filter and air-dried. The analyte is then recovered from the filter and analyzed as described earlier. Duplicate determinations should agree within 5%. The percent recovery equals the average weight in μg recovered from the filter divided by the weight in μg added to the filter, or:

Recovery = (average weight recovered)(100) ÷ weight added

Calibration and Standards

From the 20 μg/mL working standard solution, prepare at least 6 working standards to cover the range from 10-160 μg/25 mL. Absorbance may be a nonlinear function of rhodium concentration above 160 μg/25 mL (reference 1) so the concentration range above 160 μg/25 mL must be adequately covered with standards. All standard solutions are made 0.2 N in HCl and are stored in polyethylene bottles. Since the low concentration and standards may deteriorate, the standard solutions should be made fresh each day.

Proceed as described above, beginning with the addition of 2.5 mL of 30% potassium bisulfate solution. Prepare a calibration curve by plotting on linear graph paper the absorbance versus the concentration of each standard in μg/25 mL. It is advisable to run a set of standards both before and after the analysis of a series of samples to ensure that conditions have not changed.

Calculations

Read the weight, in μg, corresponding to the total absorbance from the standard curve. No volume corrections are needed, because the standard curve is based on μg/25 mL. Corrections for the blank must be made for each sample:

μg = μg sample − μg blank:

where:

μg sample = μg found in sample filter
μg blank = μg found in blank filter

Divide the total weight by the recovery to obtain the corrected μg/sample:

Corrected μg/sample = total wt ÷ recovery

The concentration of the analyte in the air sampled can be expressed in mg/m^3 (μg/L = mg/m^3):

mg/m^3 = corrected μg/air volume sampled (L)

References

1. Analytical Methods for Atomic Absorption Spectrophotometry. The Perkin-Elmer Corp., Norwalk, CT, 1971.
2. Methods for Emission Spectrochemical Analysis. ASTM Committee E-2, Philadelphia, 1971.
3. Documentation of NIOSH Validation Tests. Contract No. CDC-99-74-45.
4. Kallmann, S. and Hobart, E.W. Vital parameters in the determination of rhodium by atomic absorption. Anal. Chem. Acta 51: 120-124, 1970.

SILICA DUST

Synonyms: cristobalite; quartz; tridymite $\qquad$ SiO$_2$
ACGIH TLV: 10 mg/m^3 divided by (% respirable quartz + 2)
OSHA Standard: same
NIOSH Recommendation: 0.05 mg/m^3 (respirable free silica)

Method	Sampling Duration	Sampling Location	Useful Range (mg/m^3)	System Cost ($)	Test Cost ($)	Manufacturer
LS	Cont	Personal	0.01-100	1695	0	GCA
LS	Cont	Portable	0.001-100	2785	0	MDA
LS	Cont	Fixed	0.001-200	3990	0	GCA
BA	1,4 min	Portable	0.02-150	4890	0	GCA

NIOSH METHOD NO. P&CAM 106

Principle
The sample is digested in phosphoric acid to remove the silicates. The remaining crystalline material is dissolved in hydrofluoric acid. Silica is determined colorimetrically as silicomolybdate (420 mu) or as molybdenum blue (820 mu).

Range and Sensitivity
In the silicomolybdate range, concentrations from 0.1-2.5 mg can be detected. In the molybdenum blue range, concentration from less than 5 μg to 140 μg can be detected. The range can be extended by varying the 0.50 g sample size.

Interferences
The phosphate ion reacts with molybdic acid to form a yellow phosphomolybdate complex. It can be eliminated by lowering the pH to 1.2-1.3 with 10 N H$_2$SO$_4$. The ferric ion may consume the reducing agent and cause low results. As much as 1.0 mg will not interfere. Iron in excess of 1.0 mg/sample must be removed by preliminary treatment with 10:1 HCl-HNO$_3$.

Precision and Accuracy
Relative standard deviation on a run of 10 replicate samples was 9.25%. The wide deviation is due to error introduced by solution of some of the quartz in the phosphoric acid or the resistance of some silicates to dissolve. The same 10 samples reflect a mean free silica analysis of 3.36% on a sample with a concentration of 3.33% as determined gravimetrically. The error is less than 1%.

Advantages and Disadvantages
The major advantage lies in the versatility and wide range of the method. With facile pretreatment, dusts, airborne particulates and tissues can be analyzed by this method. If the SiO$_2$ concentration falls below the silicomolybdate detection limit, the simple addition of another reagent increases the sensitivity twenty-fold. The major disadvantage is the loss of free silica by solution during the phosphoric acid digestion. Quartz solubility is a function of the particle size, so the small, fine samples are vulnerable to this source of error. Some silicates may be resistant to phosphoric acid.

Apparatus

Filter unit. Consists of the filter media and appropriate cassette filter holder, either a 2 or 3 piece filter cassette.

Personal sampling pump. This pump must be properly calibrated so the volume of air sampled can be measured as accurately as possible. The pump must be calibrated with a representative filter unit in the line. Flow rate, time, and/or volume must be known.

Thermometer.

Manometer.

Stopwatch.

Membrane filters. 47 mm diameter, mean pore size, 0.45 micron.

Precision heater. 500 watt, 115 volt, Type RH (Precision Scientific Company).

Variable transformer. 750 watt, with built-in voltmeter.

Serological rotator.

Beakers. 250 mL borosilicate glass, Phillips type.

Funnels. Short-stemmed glass, with stems bent.

Crucible tongs. Padded with rubber or Tygon tubing.

Spectrophotometer. With capacity of readings at 420 mu and 820 mu.

Filter funnel, membrane mount, and flask assembly.

Assorted plasticware. 125 mL polyethylene beakers, polyethylene watch glass, stirring rods, 50 mm discs. Polyethylene reagent bottles to store water, boric acid and standard solutions.

Bakelite or Nalgene buret (10 mL).

Constant temperature water bath.

Reagents

All reagents used should be ACS reagent grade or better.

Silica-free water. For all solutions and dilutions. Use distilled deionized water and store in polyethylene carboy.

Hydrofluoric acid, 48%.

Orthophosphoric acid, 85%.

Hydrochloric acid, 1:10.

Boric acid solution, 5%. Dissolve 200 g of boric acid crystals in 4 L of warm, silica free water. Cool. Filter with vacuum through a 0.45 micron membrane filter. Store in a polyethylene container.

Molybdate reagent. Dissolve 50 g of ammonium molybdate tetrahydrate in about 400 mL of silica free water. Acidify with 40 mL of concentrated sulfuric acid. Cool. Dilute to 500 mL. Store in dark.

Sulfuric acid, 10 N. Cautiously add 555 mL of concentrated H_2SO_4 to about 1.3 L of water. Cool. Dilute to 2 L.

Reducing solution. For solution A, dissolve 9 g of sodium bisulfite in 80 mL of water. For solution B, in 10 mL of water dissolve 0.7 g of anhydrous sodium sulfite and 0.15 g of 1-amino-2-naphthol-4-sulfonic acid, in that order. Combine solutions A and B and dilute to 100 mL. Reagent stored in refrigerator is stable for about one month.

Quartz, finely ground and acid washed.

Procedure

Cleaning of equipment. All glass and polyethylene ware used should be washed thoroughly and subjected to a final rinse in silica free water.

Collection and shipping of samples. Air samples should be collected on standard 37 or 47 mm cellulose membrane filters as follows. Assemble the filter unit by mounting the filter discs in the filter cassette. Connect the exit end of the filter unit to the pump with a short piece of flexible tubing. Turn on pump to begin sample collection. The flow rate, times, and/or volume must be measured as accurately as possible. The sample should be taken at a flow rate of 2 L/min. A minimum sample of 100 L should be collected. Larger sample volumes are encouraged provided the filters do not become loaded with dust to the point that loose material might fall off or the filter become plugged.

The sample cassettes should be shipped in a suitable container designed by NIOSH to minimize contamination and to prevent damage in transit. Care must be taken during storage and shipping that no part of the sample is dislodged from the filter nor that the sample surface be disturbed in any way. Loss of sample from heavy deposits on the filter may be prevented by mounting a clean filter in the cassettes on top of the sample filter. With each batch of samples, one filter, labelled as a blank, must be submitted. No air should be drawn through this filter. Bulk or rafter samples should be submitted in quantities equaling or exceeding 0.5 g.

Analysis of samples. A weighed sample containing no more than 2.5 mg SiO_2 or a filter membrane on which airborne particulates have been collected is placed in a clean 250 mL Phillips borosilicate glass beaker. Redistilled nitric acid (3-4 mL) is added and the sample heated to absence of brown fumes and dryness. The process is repeated until a white residue remains. (The filter should be consumed.) A blank should be carried through all steps of the analysis. If the sample is received on a membrane filter, an untreated membrane of the same type should be ashed. If a bulk sample is used, a beaker containing only the nitric or perchloric acid is started.

When polyvinyl chloride membranes have been used for sampling, nitric acid ashing is inadequate. Two mL of $HClO_4$ are added and heated slowly to just short of dryness. Ashing should be complete at this point. If necessary a second portion of $HClO_4$ can be added. Add 25 mL of 85% phosphoric acid to the beaker. Start a reagent blank from this point, using only phosphoric acid. Cover the beaker with a bent-stem funnel. In a hood preheat the precision heater for at least 45 min at 70 volts (about 240° C.). Heat the sample for exactly 8 min, swirling it by the action of the serological rotator.

Remove the beaker from the heater by grasping it with padded crucible tongs and swirl it vigorously for 1 min. Allow the beaker to cool and add approximately 125 mL of water at 60-70° C. Swirl to mix the syrupy phosphoric mixture with the water. Filter the sample with suction through a membrane filter. A millipore disc, 47 mm in diameter with mean pore size of 45 micron, is suitable. Wash thoroughly with 1:10 HCl. Place the membrane flat in the bottom of a 150 mL polyethylene beaker and add 0.5 mL of 48% HF to the membrane surface. Float a thin polyethylene disc of about 50 mm diameter over the membrane and cover the beaker. Allow it to stand for 30 min.

Add 25 mL of water and 50 mL of boric acid solution. Stir well, cover. Heat the solution in a 40° C. water bath for at least 10 min. Add 4 mL of molybdic acid reagent while stirring, staggering the addition at 2 min intervals between samples. Time with a stopwatch from the beginning of the addiion to the first sample. Twenty min after the first addition, add 20 mL of 10 N sulfuric acid, and stir thoroughly. If any yellow color persists, read within 2 min after acidification in a spectrophotometer at 420 mu against distilled water. Subtract blank. If a colorless solution results, allow it to stand for 2-5 min and add 1 mL of 1-amino-2-naphthol-4-sulfonic acid reagent. Mix and read after 20 min at 820 mu against distilled water. Subtract blank. This color is stable for several hours.

Calibration and Standards

A 0.5 mg/mL SiO_2 stock standard is made by dissolving 250 mg of finely ground, acid washed quartz in 10 mL of 48% hydrofluoric acid. Solution rate is slow and it may need to stand overnight. Dilute to 500 mL. This standard is stable indefinitely if stored in a polyethylene container. A calibration curve in the silicomolybdate color range is made by diluting 1, 2, 3, 4, 5, and 6 mL aliquots of the stock standard to 25 mL in polyethylene beakers and proceeding as above, beginning with the addition of 25 mL of water and 50 mL of boric acid solution. Absorbance against mg of SiO_2 is plotted.

A standard usable in the molybdenum blue range is made by preparing a 1:25 dilution of the stock standard described above. This standard will contain 20 μg SiO_2/mL. Dilute standards similar to those described above are prepared for use in plotting the calibration curve. The upper limit for this curve is 140 μg.

Calculations

The mg of SiO_2/sample is read from the appropriate calibration curve:

mg SiO_2/g of sample = mg SiO_2 in sample ÷ sample weight in g

The μg SiO_2/sample is read from the appropriate calibration curve:

μg SiO_2/g of sample = μg SiO_2 in sample ÷ sample weight in g

% SiO_2 = (μg SiO_2)(0.0001) ÷ g of sample

References
1. Talvitie, N.A. Determination of quartz in the presence of silicates using phosphoric acid. Anal. Chem. 23: 623, 1951.
2. Talvitie, N.A., and F. Hyslop. Colorimetric determination of siliceous atmospheric contaminants. Amer. Ind. Hyg. Asso. J. 19: 54, 1958.
3. Talvitie, N.A. Determination of free silica: gravimetric and spectrophotometric procedure applicable to air-borne and settled dust. Amer. Ind. Hyg. Asso. J. 25: 169, 1964.

STODDARD SOLVENT

Synonym: hydrocarbons (159-176° C. b.p.)
ACGIH TLV: 100 ppm (525 mg/m^3)
OSHA Standard: 500 ppm (2950 mg/m^3)

Method	Sampling Duration	Sampling Location	Useful Range (ppm)	System Cost ($)	Test Cost ($)	Manufacturer
IR	Cont	Portable	10-1000	4374	0	Foxboro

NIOSH METHOD NO. P&CAM 127

(Use as described for acetone, with 100° C. column temperature and undecane as internal standard)

STYRENE

Synonyms: vinyl benzene; phenylethylene; cinnamene $C_6H_5CH=CH_2$
ACGIH TLV: 50 ppm (215 mg/m^3)
OSHA Standard: pm (425 mg/m^3)

Method	Sampling Duration	Sampling Location	Useful Range (ppm)	System Cost ($)	Test Cost ($)	Manufacturer
DT	Grab	Portable	10-5000	150	2.70	Nat'l Draeger
DT	Grab	Portable	5-300	165	1.70	Matheson
DT	Grab	Portable	2-1000	180	2.25	Bendix
DT	2 hr	Personal	10-250	850	3.10	Nat'l Draeger
IR	Cont	Portable	2-200	4374	0	Foxboro
PI	Cont	Fixed	2-200	4950	0	HNU
GC	1 min	Portable	1-10^6	10,000	0	Microsensor

NIOSH METHOD NO. P&CAM 127

(See acetone)

SULFUR DIOXIDE

ACGIH TLV: 2 ppm (5 mg/m³) SO_2
OSHA Standard: 5 ppm (13 mg/m³)
NIOSH Recommendation: 0.5 ppm (1.3 mg/m³)

Method	Sampling Duration	Sampling Location	Useful Range (ppm)	System Cost ($)	Test Cost ($)	Manufacturer
DT	Grab	Portable	0.1-5000	150	2.60	Nat'l Draeger
DT	Grab	Portable	1-30,000	165	1.90	Matheson
DT	Grab	Portable	0.5-80,000	180	2.25	Bendix
DT	4 hr	Personal	1-50	850	3.10	Nat'l Draeger
ES	1 min	Personal	0.05-5	1765	0	Interscan
ES	Cont	Portable	0.01-1(50)	2211	0	Interscan
CS	Grab	Portable	1-20	2465	0.80	MDA
ES	Cont	Fixed	0.05-5(50)	2625	0	Interscan
ES	Cont	Port/Fixed	0.05-5(50)	3343	0	Interscan
IR	Cont	Portable	1-20	4374	0	Foxboro
CS	Cont	Fixed	1-20	4950	Variable	MDA
CM	Cont	Port/Fixed	0.002-10	4995	Variable	CEA
GC	1 min	Portable	$1-10^6$	10,000	0	Microsensor

NIOSH METHOD NO. S308

Principle

A known volume of air is drawn through a midget bubbler containing hydrogen peroxide. Sulfur dioxide vapor is collected and oxidized to sulfuric acid. Isopropyl alcohol is added to the contents in the bubbler, and the pH of the sample is adjusted with dilute perchloric acid. The resulting solution is titrated with 0.005 M barium perchlorate using Thorin as the indicator. There is a sharp color change from yellow or yellow-orange to pink when the end point is reached.

Range and Sensitivity

This method was validated over the range of 6.55-26.8 mg/m³ at an atmospheric temperature and pressure of 21° C. and 757 mm Hg, using a 90 L sample. The upper limit of the range of the method is dependent on the capacity of the midget bubbler. If higher concentrations than those tested are to be sampled, smaller sample volumes should be used.

Interferences

The presence and suspected identities of other air contaminants must be recorded and reported to the laboratory. Volatile phosphates will be a significant interference. Volatile metals can be converted to metallic ions in the bubbler. The presence of these and other air contaminants must be noted so that the analyst can pretreat the sample. Particulate contaminants such as sulfate, sulfuric acid, and metals are removed by the prefilter.

Metal particulates will also be removed using the 0.8 micron cellulose ester membrane filter. Volatile metal interferences that are converted to metallic ions in the midget bubbler can be removed by passing the sample through a cation exchange resin. Concentrations of phosphate ions greater than the sulfate ion concentration will cause appreciable interference. Particulate phosphates will be removed using the 0.8 micron cellulose ester membrane filter connected in front of the midget bubbler. Phosphates can also be removed by precipitation with magnesium carbonate.

Precision and Accuracy

The coefficient of variation for the total analytical and sampling method in the range of 6.55-26.8 mg/m^3 was 0.054. This value corresponds to a standard deviation of 0.70 mg/m^3 at the OSHA standard level. Statistical information can be found in reference 1. Details of the test procedures can be found in reference 2.

On the average the concentrations obtained in the laboratory validation study at 0.5X, 1X, and 2X the OSHA standard level were 1.5% lower than the "true" concentrations for 18 samples. Any difference between the "found" and "true" concentrations may not represent a bias in the sampling and analytical method, but rather a random variation from the experimentally determined "true" concentration. Therefore, the coefficient of variation is a good measure of the accuracy of the method since the recoveries, storage stability, and collection efficiency were good. Storage stability studies on samples collected from a test atmosphere at a concentration of 12.90 mg/m^3 indicate that collected samples are stable for at least one week.

Advantages and Disadvantages

Collected samples are analyzed by means of a quick and simple method. The sulfuric acid formed is stable and nonvolatile, making this manner of collection of sulfur dioxide desirable. A disadvantage of the method is the awkwardness in using midget bubblers for collecting personal samples. If the worker's job performance requires much body movement, loss of the collection solution during sampling may occur.

The bubblers are more difficult to ship than adsorption tubes or filters due to possible breakage and leakage of the bubblers during shipping. The precision of the method is limited by the reproducibility of the pressure drop across the prefilter and bubbler. This drop will affect the flow rate and cause the volume to be imprecise, because the pump is usually calibrated for one filter/bubbler combination only.

Apparatus

Prefilter unit. The prefilter unit, which is used to remove particulate interferences, consists of a 37 mm diameter cellulose ester membrane filter (millipore type AA or equivalent) with a pore size of 0.80 micron, contained in a 37 mm two piece cassette filter holder. The filter is supported by a stainless steel screen.

Glass midget bubbler. Contains the collecting medium.

Personal sampling pump. A calibrated personal sampling pump whose flow rate can be determined to an accuracy of 5%. The sampling pump is protected from splashover or solvent condensation by a trap. The trap is a midget bubbler or impinger with the stem broken off which is used to collect spillage. The trap is attached to the pump with a metal holder. The outlet of the trap is connected to the pump by flexible tubing.

Thermometer.

Manometer.

Volumetric flasks. Convenient sizes for preparing standard solutions.

Beakers. 250 mL.

Pipets. Convenient sizes for preparing standard solutions and for measuring the collection medium.

Burette. A burette of 10 mL capacity, graduated in 0.05 mL.
Daylight fluorescent lamp. To aid in identifying the end point.

Reagents

Isopropanol, reagent grade.

Barium perchlorate, 0.005 M. Dissolve 2.0 barium perchlorate trihydrate in 150 mL of distilled water and add 850 mL of isopropanol. Adjust the pH to about 3.5 with perchoric acid. Standardize against 0.005 M sulfuric acid.

Thorin. Prepare a 0.1-0.2% solution in distilled water.

Standard sulfate solution. Prepare a 0.005 M solution of sulfuric acid and standardize by titration with 0.02 M sodium hydroxide.

Perchloric acid, 1.8%. Dilute 25 mL of reagent grade perchloric acid (70-72%) to 1 L with distilled water.

Hydrogen peroxide, 0.3 N. Dilute 17 mL of 30% hydrogen peroxide solution to 1 L with distilled water.

pH paper, Hydrion.

Procedure

Cleaning of equipment. All glassware used for the laboratory analysis should be detergent washed and thoroughly rinsed with tap water and distilled water.

Calibration of personal sampling pumps. Each personal sampling pump should be calibrated with a representative filter cassette, bubbler, and splashover tube in the line to minimize errors associated with uncertainties in the volume sampled.

Collection and shipping of samples. Assemble the filter in the two piece filter cassette holder and close firmly. The filter is backed up by a stainless steel screen rather than a filter pad. Secure the cassette holder together with tape or shrinkable band. Pipet 15 mL of 0.3 N hydrogen peroxide into each midget bubbler. Remove the filter holder's plugs and attach the outlet of the filter holder to the inlet arm of the midget bubbler with a short piece of flexible tubing. The outlet of the midget bubbler is attached to the pump's inlet or a trap which may be used to protect the pump during personal sampling. The trap is a midget impinger or bubbler with the stem broken off which is used to collect spillage. The trap is attached to the pump with a metal holder. The outlet of the trap is connected to the pump by flexible tubing.

Air being sampled should not pass through any hose or tubing before entering the filter cassette. At levels corresponding to the standard, a sample size of 90 L is recommended. Sample at a known flow rate between 1.5 and 1.0 L/min. At levels outside of the range over which the method was validated, adjust the sample volume to collect between 0.6 and 2.3 mg of sulfur dioxide. Turn the pump on and begin sample collection. Since it is possible for a filter to become plugged by heavy particulate loading or by the presence of oil mists or other liquids in the air, the pump rotameter should be observed frequently, and the sampling should be terminated at any evidence of a problem.

Terminate sampling at the predetermined time and record sample flow rate, collection time, and ambient temperature and pressure. If pressure reading is not available, record the elevation. Also record the type of sampling pump used. Remove the bubbler stem and tap the stem gently against the inside wall of the bubbler bottle to recover as much of the sampling solution as possible. Rinse the bubbler stem with 1-2 mL of unused 0.3 N hydrogen peroxide, and add the wash to the bubbler bottle. Replace the bubbler stem. The inlet and outlet of the bubbler stem should be sealed by connecting a piece of Teflon tubing between them or inserting Teflon plugs in the inlet and outlet. Do not seal with rubber. The standard taper joint of the bubbler should be taped securely to prevent leakage during shipping.

The filter should be removed from the cassette filter holder and discarded. The cassette holders and stainless steel screens should be cleaned and saved for future use. Care should be taken to minimize spillage or loss of sample by evaporation at all times. With each batch of ten samples submit one bubbler containing 15 mL of 0.3 N hydrogen peroxide prepared from the same stopper as that used for sample collection. This bubbler must be subjected to exactly the same handling as the samples except that no air is drawn through it. Label this bubbler as the blank.

Analysis of samples. Empty the contents of the bubbler into a 250 mL beaker, using 2 mL of 0.3 N hydrogen peroxide to rinse the bubbler flask. Add rinse to the beaker. Add 100 mL isopropanol to the beaker. Adjust the pH of the sample to 3.5 with 1.8% perchloric acid. Add 8-10 drops of Thorin indicator, and titrate the sample with barium perchlorate to a pink colored end point. Analyze the standard and absorbing solution blank in the same manner.

Calibration and Standards

The barium perchlorate solution is standardized by titrating a 5 mL aliquot of the standardized 0.005 M sulfuric acid to the end point with Thorin as the indicator. The molarity (M) of barium perchlorate is calculated as follows:

$$M = (\text{mL of } H_2SO_4)(\text{molarity of } H_2SO_4) \div \text{mL of } Ba(ClO_4)_2 \text{ used}$$

The molarity of $Ba(ClO_4)_2$ should be checked periodically.

Calculations

The following reactions are the basis for this analytical method:

$$SO_2 + H_2O_2 \rightarrow H_2SO_4$$

$$H_2SO_4 + Ba(ClO_4)_2 \rightarrow BaSO_4 + 2HClO_4$$

The mg of SO_2 can be calculated as follows:

$$\text{mg } SO_2 = (\text{molarity of } Ba(ClO_4)_2)(\text{mL of } Ba(ClO_4)_2)(64)$$

where:

64 = molecular weight of SO_2

Corrections for the blank must be made for each sample:

$$\text{mg} = \text{mg sample} - \text{mg blank}$$

where:

mg sample = mg SO_2 found in sample bubbler
mg blank = mg SO_2 found in blank bubbler

For personal sampling pumps with rotameters only, the following air volume correction should be made:

$$\text{Corrected volume} = ft(P_1T_2/P_2T_1)^{1/2}$$

where:

    f = flow rate sampled
    t = sampling time
    P_1 = pressure during calibration of sampling pump (mm Hg)
    P_2 = pressure of air sampled (mm Hg)
    T_1 = temperature during calibration of sampling pump (°K)
    T_2 = temperature of air sampled (°K)

The concentration of sulfur dioxide in the air sample can be expressed in mg/m^3:

$$mg/m^3 = (mg)(1000\ L/m^3) \div \text{corr. air volume sampled in L}$$

Another method of expressing concentration is ppm:

$$ppm = (mg/m^3)(24.45)(760)(T + 273) \div (M.W.)(P)(298)$$

where:

    P = pressure (mm Hg) of air sampled
    T = temperature (°C.) of air sampled
    24.45 = molar volume (L/mol) at 25° C. and 760 mm Hg
    M.W. = molecular weight (g/mol) of sulfur dioxide
    760 = standard pressure (mm Hg)
    298 = standard temperature (°K)

References

1. Documentation of NIOSH Validation Tests. Contract No. CDC-99-74-45.
2. Backup Data Report for Sulfur Dioxide. Prepared under NIOSH Contract No. 210-76-0123.
3. Fritz, J.S. and S.S. Yamamura. Anal. Chem. 27: 1461, 1955.

SULFUR HEXAFLUORIDE

ACGIH TLV: 1000 ppm (6000 mg/m³)
OSHA Standard: 1000 ppm

SF_6

Method	Sampling Duration	Sampling Location	Useful Range (ppm)	System Cost ($)	Test Cost ($)	Manufacturer
IR	Cont	Portable	20-2000	4374	0	Foxboro

NIOSH METHOD NO. S244

Principle
An air sample is pumped into a gas sampling bag with a personal sampling pump. The sulfur hexafluoride content of the sample is determined by gas chromatography using thermal conductivity detection.

Range and Sensitivity
This method was validated over the range of 500-2010 ppm at an atmospheric temperature of 21° C. an atmospheric pressure of 760 mm Hg using a 3 L sample volume. Under the instrumental conditions used in the validation study, a 5 mL injection of 2000 ppm sulfur hexafluoride standard resulted in a peak whose height was 100% of full scale on a 1 mv recorder. The amplifier of the gas chromatograph was set on range 1 and attenuation 2. The limit of detection is estimated at less than 1.5 μg sulfur hexafluoride. Using a 5 mL gas sampling loop, this corresponds to a level of approximately 50 ppm.

Interferences
When two or more compounds are known or suspected to be present in the air, such information, including their suspected identities, should be transmitted with the sample. It must be emphasized that any compound which has the same retention time as the analyte at the operating conditions described in this method is an interference. Retention time data on a single column cannot be considered as proof of chemical identity.

Precision and Accuracy
The coefficient of variation for the total analytical and sampling method in the range of 500-2010 ppm was 0.031. This value corresponds to a standard deviation of 31 ppm at the OSHA standard level. Statistical information can be found in reference 1. Details of the test procedure can be found in reference 2.

In validation experiments, this method was found to be capable of coming within 25% of the "true" value on the average 95% of the time over the validation range. The average of the concentrations obtained at 0.5, 1, and 2 times the OSHA environmental limit were 4.4% lower than the dynamically generated test concentration (n=18). In storage stability studies, the mean of samples analyzed after 7 days were within 0.2% of the mean of samples analyzed immediately after collection. Experiments performed in the validation study are described in reference 2.

Advantages and Disadvantages
The sampling device is portable and involves no liquids. Interferences are minimal, and most of those which do occur can be eliminated by altering chromatographic conditions. The samples in bags are analyzed by means of a quick instrumental method. One disadvantage of the method is that the gas sampling bag is rather bulky and may be punctured during sampling and shipping. It is difficult to ship the samples by air.

Apparatus
Personal sampling pump. A personal sampling pump capable of filling a gas sampling bag at approximately 0.05 L/min is required. Each personal pump should be calibrated to within 5%. Although sample volume is not used to determine sample concentration, the pump should be calibrated to make certain that the collected sample represents a time-weighed average concentration and to avoid over filling of the bags; i.e., a maximum sampling time can be determined based on the flow rate and sample volume which is less than 80% of the volume of the bag.

The personal sampling pump must be fitted with an outlet port so it is capable of filling a bag. To ensure a leak-free apparatus, adjust the pump so that it delivers at the proper flow rate, and attach the pump outlet to a water manometer with a short piece of flexible tubing. Turn the pump on and observe the water level difference; it should push at least 30 cm of water. If it does not, the pump is incapable of filling the sampling bag and cannot be used.

Gas sampling bag. Five L capacity, five-layer sampling bags manufactured by Calibrated Instruments, Inc. (Ardsley, NY 10502) were found to be satisfactory for sample collection and storage for at least 7 days. This bag is fitted with a metal valve and hose bib. The valves used in validation studies were found to leak when in the open position. It is necessary to wrap the valve stem connection with Teflon tape or Parafilm to ensure a leak-free connection. For the preparation of calibration standards in the laboratory, Saran or Tedlar bags could be used.

Gas chromatograph. Equipped with thermal conductivity detector and 5 mL gas sampling loop.

Column. 6 ft × 1/8 in stainless steel, packed with 30/60 mesh Molecular Sieve 13X.

Area integrator. An electronic integrator or some other suitable method for measuring peak areas.

Gas tight syringes. Convenient sizes for preparing standards.

Regulator for compressed air. Capable of metering gas at approximately 1 L/min. The gas line from the regulator should be equipped with a septum-tee for standard preparation.

Water manometer.

Thermometer.

Reagents

All reagents used must be ACS reagent grade or better.

Sulfur hexafluoride, 100%.

Helium, purified.

Air, filtered, compressed.

Procedure

Cleaning of sampling bags and checking for leaks. The bags are cleaned by opening the valve and bleeding out the air sample. The use of a vacuum pump is recommended although this procedure can be carried out by manually flattening the bags. The bags are then flushed with air and evacuated. This procedure is repeated at least twice. Bags should be checked for leaks by filling the bag with air until taut, sealing and applying gentle pressure to the bag. Check for any discernable leaks and any volume changes or slackening of the bag, especially along seams and in the valve stem, for at least a 1 hour period.

Collection and shipping of samples. Immediately before sampling, attach a small piece of Teflon tubing to the hose bib of the five layer gas sampling bag. Rubber tubing should not be used. Unscrew the valve fitting and attach the tubing to the outlet of the sampling pump. Make sure that all connections are tight and leak free. The bag valve must be fully opened during sampling. Air being sampled will pass through the pump and tubing before entering the sampling bag, since a push type pump is required. No tubing is attached to the inlet of the pump.

A sample size of 3-4 L is recommended. Sample at a flow rate of 0.05 L/min or less, but not less than 0.01 L/min. The flow rate should be known with an accuracy of at least 5%. Set the flow rate as accurately as possible using the manufacturer's directions. Although the volume of sample collected is not used in determining the concentration, it is necessary to keep the volume to 80% or less of the bag's capacity. Observe the bag frequently to ensure that it is filling properly.

The temperature and pressure of the atmosphere being sampled should be recorded. If pressure reading is not available, record the elevation. Also record sampling time, flow rate, and type of sampling pump used. The gas sampling bag should be labeled appropriately and sealed tightly.

Gas sampling bags should be packed loosely and padded before they are shipped to minimize the danger of their being punctured during shipping. Do not ship the bags by air, unless they are stored in a pressurized cabin.

GC conditions. The typical operating conditions for the gas chromatograph are:

30 mL/min helium carrier gas flow
100° C. column temperature
125° C. detector temperature
injector port at ambient conditions

GC analysis. Attach the gas sampling bag to the sample loop of the gas chromatograph with a short piece of flexible tubing. Open the valve of the bag and fill the loop by using a vacuum pump or manually applying pressure to the sample bag. Allow the loop to attain atmospheric pressure, and inject the sample. Sulfur hexafluoride elutes after oxygen, nitrogen, and carbon dioxide. Duplicate injections of each sample and standard should be made. No more than a 3% difference in area is to be expected. A retention time of approximately 4 min is to be expected using the above conditions.

Measurement of area. The area of the sample peak is measured by an electronic integrator or some other suitable form of area measurement, and the results are read from a standard curve.

Calibration and Standards

A series of standards, varying in concentration over the range of 100-3000 ppm is prepared and analyzed under the same GC conditions and during the same time period as the unknown samples. Curves are established by plotting concentration in ppm versus peak area.

Completely evacuate and flush several times with air a 5 L gas sampling bag, preferably with the aid of a vacuum pump. Using a calibrated source of air equipped with a septum-tee, meter 3-4 L of air into the bag. Inject appropriate aliquots of sulfur hexafluoride via a gas tight syringe through the septum. Knead the bag to ensure adequate mixing. Prepare at least 5 working standards to cover the range of 100-3000 ppm. The concentration of the bag in ppm equals the volume of sulfur hexafluoride in mL divided by the amount of air in L × 1000:

ppm = (mL of sulfur hexafluoride)(1000) ÷ volume of air in L

Calculations

Read the concentration in ppm, corresponding to each peak area from the standard curve. Another method of expressing concentration is mg/m^3:

$$mg/m^3 = (ppm)(M.W.)(P)(298) \div (24.45)(760)(T + 273)$$

where:

P = pressure (mm Hg) of air sampled
T = temperature (°C.) of air sampled
24.45 = molar volume (L/mol) at 25° C. and 760 mm Hg
M.W. = molecular weight, 146.07 g/mol
760 = standard pressure (mm Hg)
298 = standard temperature (°K)

SULFURIC ACID

ACGIH TLV: 1 mg/m³
OSHA Standard: 1 mg/m³

H_2SO_4

Method	Sampling Duration	Sampling Location	Useful Range (mg/m³)	System Cost ($)	Test Cost ($)	Manufacturer
DT	Grab	Portable	1-5	150	2.80	Nat'l Draeger

NIOSH METHOD NO. P&CAM 187

Principle

Sulfuric acid mist in air is absorbed in 10-15 mL of water in a midget impinger. The sulfate in the sample solution is precipitated as barium sulfate and the turbidity of the suspension is measured at 420 nm with a spectrophotometer.

Range and Sensitivity

As little as 10 µg of sulfuric acid in 10 mL of impinger bubbler sample solution can be determined. This lower limit of the range corresponds to a concentration of 0.1 mg/m³ of sulfuric acid in a 100 L sample of air. When the amount of sulfuric acid in 10 mL of impinger bubbler sample solution exceeds 0.4 mg, the suspension of barium sulfate may lose stability and the accuracy of the method may decrease. This upper limit of the range corresponds to a concentration of 4.0 mg/m³ of sulfuric acid in a 100 L sample of air. Larger concentrations of sulfuric acid may be determined by appropriate dilution of the impinger bubbler sample solution.

Interferences

Colored or suspended matter in large concentrations will interfere. Filtration of the sample solution may remove some suspended matter. If the concentration of colored or suspended matter is small in comparison with that of sulfate, a correction may be made by subtracting the turbidity of the sample solution without added barium chloride from that of the barium sulfate suspension. Silica in concentrations above 500 mg/L in the sample solution will interfere. Large concentrations of organic materials in the sample solution may prevent the satisfactory precipitation of barium sulfate. Sulfate salts, if present as particulate material in the sample air, will be reported as sulfuric acid.

Precision and Accuracy

In the application of the turbidimetric method to water analysis, a coefficient of variation of 0.09 was found for a synthetic water sample containing a sulfate concentration of 0.26 mg/mL; when this synthetic water sample was appropriately diluted for analysis, the concentration of sulfate obtained corresponded approximately to that which would have been obtained from a 100 L sample of air containing 2.6 mg/m^3 of H_2SO_4. The relative error was approximately 2% (reference 2). For the determination of sulfuric acid mist in air, where sampling errors may be significant, the accuracy and precision probably are poorer than for water analysis.

Advantages and Disadvantages

The method is simple to perform and relatively fast. It requires no elaborate preparation of reagents and no complex equipment other than a spectrophotometer. The method is reasonably specific since no common anions other than sulfate form an insoluble barium salt in acid solution. The interferences discussed above are not likely to be encountered in air in concentrations that would cause serious difficulty in the determination of sulfuric acid mist. Conditions for the precipitation of barium sulfate must be carefully controlled to obtain good reproducibility.

Apparatus

Personal sampling pump. A personal sampling pump capable of sampling accurately at 1.5 L/min is required. The pump should be calibrated with a midget impinger containing 10 mL of water in the intake line. A wet or dry test meter or a glass rotameter capable of measuring the required flow rate to within 5% may be used in the calibration.

Midget impinger. Glass midget impinger of a size to contain 25 mL of solution are required. Provision should be made for connection to the pump and for appropriate mounting during sampling.

Sample containers. Bottles (25 mL) are needed for shipping samples to the laboratory. Alternatively, the samples may be shipped in the midget impingers, if suitable closures are available.

Magnetic stirrer. The stirrer should preferably be equipped with a timing device. If it is not, a stopwatch may be used instead. The stirring speed should not vary appreciably.

Spectrophotometer. With absorption cells, for use at 420 nm.

Volumetric flasks and pipets.

Reagents

Distilled water for collection of samples.

SulfaVer IV powder. Commercially available from Hach Chemical Company, or the equivalent. This reagent is supplied in a prepackaged "Powder Pillow" in an amount suitable for a single determination (reference 1).

Standard solution of sulfate (0.1 mg/mL). Dissolve 0.1479 g of ACS reagent grade anhydrous sodium sulfate in distilled water and dilute to 1 L.

Procedure

Cleaning of equipment. Wash all glassware with detergent solution and then rinse thoroughly with tap water and distilled water. After each use the absorption cells should be cleaned with detergent and a brush to prevent the accumulation of a deposit on the inside walls of the cell.

Collection and shipping of samples. Place approximately 10 mL of distilled water in a midget impinger and connect the impinger to the pump. Pump air sample through the impinger at the rate of 1.5 L/min for at least 60 min. Measure the flow rate and time as accurately as possible. If sulfuric acid concentrations greater than 4 mg/m³ are expected, sample for a shorter length of time. Alternatively, the sample solution may be diluted in the laboratory. One of every ten samples collected as directed above should be labeled as a blank. The absorbance of this sample will be measured without the addition of SulfaVer IV reagent to allow a correction to be made for color and suspended matter. The samples should be packed for shipment to the laboratory in a manner that will prevent loss or damage in transit.

Analysis of samples. Adjust the volume of the sample solution to 25 mL in a volumetric flask. Transfer the solution to a 100 mL Erlenmeyer flask and place the flask on the magnetic stirrer. Add one package ("Powder Pillow") of SulfaVer IV reagent to the sample solution. The exact amount of SulfaVer IV powder added is not critical. However, the same amount must be added to all samples. The amount added must be in excess of the amount required to precipitate all of the sulfate as barium sulfate.

Start the magnetic stirrer immediately upon addition of the SulfaVer IV powder. Stir for exactly 1 min. The speed of stirring is not critical, but it should be constant for each set of samples and standards and should be adjusted to about the maximum rate at which no splashing occurs. Immediately after the stirring has been stopped, fill an absorption cell and measure the turbidity at 420 nm at 30 sec intervals for 4 min. Record the maximum turbidity (or absorbance reading) observed during the 4 min period. The maximum turbidity usually occurs within 2 min and the readings remain constant thereafter for 3-10 min. Dilute the blank sample to 25 mL and measure its turbidity (absorbance). Subtract this turbidity value from the sample values to correct for color and suspended matter.

Calibration and Standards

Pipet 0.5, 1.0, 1.5, 2.0, 2.5, 3.0, 3.5, and 4.0 mL of the standard sulfate solution into 25 mL volumetric flasks and dilute to the mark. Transfer the standards to 100 mL Erlenmeyer flasks and follow the procedure outlined in section above. Check the calibration curve periodically by running a standard every 3 or 4 samples.

Calculations

The concentration of sulfuric acid mist in air is calculated by the following equation:

$$\text{mg/m}^3 = (\text{mg H}_2\text{SO}_4 \text{ in sample})(1000 \text{ L/m}^3) \div \text{volume of sample in L}$$

References

1. *Hach Water Analysis Handbook,* Hach Chemical Company, Ames, Iowa, 1974.
2. *Standard Methods for the Examination of Water and Wastewater,* 13th ed., American Public Health Association, Washington, D.C., 1971, pp. 334-335.

SULFURYL FLUORIDE

ACGIH TLV: 5 ppm (20 mg/m³)
OSHA Standard: 5 ppm

SO_2F_2

Method	Sampling Duration	Sampling Location	Useful Range (ppm)	System Cost ($)	Test Cost ($)	Manufacturer
IR	Cont	Portable	0.1-10	4374	0	Foxboro

NIOSH METHOD NO. S245

Principle
An air sample is pumped into a gas sampling bag with a personal sampling pump. The sulfuryl fluoride content of the sample is determined by gas chromatography using flame photometric detection in the sulfur mode.

Range and Sensitivity
This method was validated over the range of 2.54-10.29 ppm at an atmospheric temperature of 20° C. and atmospheric pressure of 760 mm Hg using a 3 L sample volume. Under the instrumental conditions used in the validation study, a 2 mL injection of a 3 ppm sulfuryl fluoride standard resulted in a peak whose height was 64% of full scale on a 1 mv recorder. The amplifier of the gas chromatograph was set on range 10^4 and attenuation 1. The limit of detection is estimated at less than 0.8 ng sulfuryl fluoride. Using a 2.5 mL gas sampling loop, this corresponds to a level of approximately 0.1 ppm.

Interferences
When two or more compounds are known or suspected to be present in the air, such information, including their suspected identities, should be transmitted with the sample. It must be emphasized that any sulfur compound which has the same retention time as the analyte at the operating conditions described in this method is an interference. Retention time data on a single column cannot be considered as proof of chemical identity.

Precision and Accuracy
The coefficient of variation for the total analytical and sampling method in the range of 2.54-10.29 ppm was 0.025. This value corresponds to a standard deviation of 0.1 ppm at the OSHA standard level. Statistical information can be found in reference 1. Details of the test procedure can be found in reference 2.

In validation experiments, this method was found to be capable of coming within 25% of the "true" value on the average of 95% of the time over the validation range. The average of the concentrations obtained at 0.5, 1, and 2 times the OSHA environmental limit were 2.0% lower than the dynamically generated test concentrations (n=18). In storage stability studies, the mean of samples analyzed after 7 days was within 0.6% of the mean of samples analyzed immediately after collection. Experiments performed in the validation study are described in reference 2.

Advantages and Disadvantages
The sampling device is portable and involves no liquids. Interferences are minimal, and most of those which do occur can be eliminated by altering chromatographic conditions. The samples in bags are analyzed by means of a quick instrumental method. One disadvantage of the method is that the gas sampling bag is rather bulky and may be punctured during sampling and shipping. It is difficult to ship the samples by air.

Apparatus

Personal sampling pump. A personal sampling pump capable of filling a gas sampling bag at approximately 0.05 L/min is required. Each personal pump should be calibrated to within 5%. Although sample volume is not used to determine sample concentration, the pump should be calibrated to make certain that the collected sample represents a time-weighted average concentration and to avoid over filling of the bags, i.e., a maximum sampling time can be determined based on the flow rate and sample volume which is less than 80% of the volume of the bag.

The personal sampling pump must be fitted with an outlet port so it is capable of filling a bag. To ensure a leak-free apparatus, adjust the pump so that it delivers at the proper flow rate, and attach the pump outlet to a water manometer with a short piece of flexible tubing. Turn the pump on and observe the water level difference; it should push at least 30 cm of water. If it does not, the pump is incapable of filling the sampling bag and cannot be used.

Gas sampling bag. Five liter capacity, five-layer sampling bags (Calibrated Instruments, Ardsley, NY 10502) were found to be satisfactory for sample collection and storage for at least 7 days. This bag is fitted with a metal valve and hose bib. The valves used in validation studies were found to leak when in the open position. It is necessary to wrap the valve stem connection with Teflon tape or Parafilm to ensure a leak-free connection. For the preparation of calibration standards in the laboratory, Saran or Tedlar bags can be used.

Gas chromatograph. Equipped with a flame photometric detector in the sulfur mode and 2 mL gas sampling loop.

Column. 0.5 m × 2 mm I.D. glass, packed with 60/80 mesh Carbosieve S (Supelco, Inc.).

Area integrator. An electronic integrator or some other suitable method for measuring peak areas.

Gas tight syringes. Convenient sizes for preparing standards.

Regulator for compressed air. Capable of metering gas at approximately 1 L/min. The gas line from the regulator should be equipped with a septum-tee for standards preparation.

Water manometer.
Thermometer.
Stopwatch.

Reagents

All reagents used must be ACS reagent grade or better.
Sulfuryl fluoride, 100%.
Helium, purified.
Air, filtered, compressed.
Hydrogen, prepurified, compressed.
Oxygen, purified, compressed.

Procedure

Cleaning of sampling bags and checking for leaks. The bags are cleaned by opening the valve and bleeding out the air sample. The use of a vacuum pump is recommended although this procedure can be carried out by manually flattening the bags. The bags are then flushed with air and evacuated. This procedure is repeated at least twice. Bags should be checked for leaks by filling the bag with air until taut, sealing and applying gentle pressure to the bag. Check for any discernable leaks and any volume changes or slackening of the bag, especially along seams and in the valve stem, for at least a 1 hour period.

Collection and shipping of samples. Immediately before sampling, attach a small piece of Teflon tubing to the valve fitting of the five-layer gas sampling bag. Rubber tubing should not be used. Unscrew the valve fitting and attach the tubing to the outlet of the sampling pump. Make sure that all connections are tight and leak-free. The bag valve must be fully opened during sampling. Air being sampled will pass through the pump and tubing before entering the sampling bag, since a push type pump is required. No tubing is attached to the inlet of the pump.

Sample at a flow rate of 0.05 L/min or less. Set the flow rate as accurately as possible using the manufacturer's directions. Although the volume of sample collected is not used in determining the concentration, it is necessary to keep the volume to 80% or less of the bag's capacity. Observe the bag frequently to ensure that it is filling properly. The temperature and pressure of the atmosphere being sampled should be recorded. If pressure reading is not available, record the elevation. Also record sampling time, flow rate, and type of sampling pump used.

The gas sampling bag should be labeled appropriately and sealed tightly. Gas sampling bags should be packed loosely and padded before they are shipped to minimize the danger of being punctured during shipping. Do not ship the bags by air, unless they are stored in a pressurized cabin.

GC conditions. The typical conditions for the gas chromatograph are:

45 mL/min helium carrier gas flow
150 mL/min hydrogen flow to detector
40 mL/min air flow to detector
20 mL/min oxygen flow to detector
110° C. column temperature
225° C. detector temperature

GC analysis. Attach the gas sampling bag to the sample loop of the gas chromatograph with a short piece of flexible tubing. Open the valve of the bag and fill the loop by using a vacuum pump or manually applying pressure to the sample bag. Allow the loop to attain atmospheric pressure, and inject the sample. Duplicate injections of each sample and standard should be made. No more than a 3% difference in area is to be expected. A retention time of approximately 4 min is to be expected using the above conditions.

Measurement of area. The area of the sample peak is measured by an electronic integrator or some suitable forms of area measurement, and the results are read from a standard curve.

Calibration and Standards

A series of standards varying in concentration over the range of 0.5-15 ppm is prepared and analyzed under the same GC conditions and during the same time period as the unknown samples. Curves are established by plotting concentration in ppm versus peak area. A logarithmic plot may be used to obtain a linear curve for this detector.

Completely evacuate and flush several times with air a 5 L gas sampling bag, preferably with the aid of a vacuum pump. Using a calibrated source of air equipped with a septum-tee, meter 3-4 L of air into the bag. Inject appropriate aliquots of sulfuryl fluoride via a gas tight syringe through the septum. Knead the bag to ensure adequate mixing. Prepare at least 5 working standards to cover the range of 0.5-15 ppm. Analyze these standards as described earlier. The concentration of the bag in ppm equals the volume of sulfuryl fluoride in mL divided by the amount of air in L $\times$ 1000:

ppm = (mL of sulfuryl fluoride)(1000) ÷ volume of air in L

Calculations

Read the concentration in ppm, corresponding to each peak area from the standard curve. Another method of expressing concentration is mg/m^3:

mg/m^3 = (ppm)(M.W.)(P)(298) ÷ (24.45)(760)(T + 273)

where:

P = pressure (mm Hg) of air sampled
T = temperature (°C.) of air sampled
24.45 = molar volume (L/mol) at 25° C. and 760 mm Hg

M.W. = molecular weight, 102.07 g/mol
760 = standard pressure (mm Hg)
298 = standard temperature (°K)

References

1. Documentation of NIOSH Validation Tests. National Institute for Occupational Safety and Health, Cincinnati, Ohio (DHEW-NIOSH Publication #77-185), 1977. Available from Superintendent of Documents, U.S. Government Printing Office, Washington, D.C., Order No. 017-033-00231-2.
2. Backup Data Report for Sulfuryl Fluoride. Prepared under NIOSH Contract No. 210-76-0123.

1,1,2,2-TETRACHLORO-1,2-DIFLUOROETHANE

Synonyms: Freon 112; Refrigerant 112; F-112
ACGIH TLV: 500 ppm (4170 mg/m³)
OSHA Standard: 500 ppm

CCl_2FCCl_2F

Method	Sampling Duration	Sampling Location	Useful Range (ppm)	System Cost ($)	Test Cost ($)	Manufacturer
DT	Cont	Portable	10-1000	4374	0	Foxboro

NIOSH METHOD NO. P&CAM 127

(Use as described for acetone, with 50° C. column temperature)

1,1,2,2-TETRACHLOROETHANE

$CHCl_2CHCl_2$

Synonym: acetylene tetrachloride
ACGIH TLV: 1 ppm (7 mg/m^3) - Skin
OSHA Standard: 5 ppm
NIOSH Recommendation: 1 ppm

Method	Sampling Duration	Sampling Location	Useful Range (ppm)	System Cost ($)	Test Cost ($)	Manufacturer
DT	Grab	Portable	3-75	180	2.25	Bendix
IR	Cont	Portable	0.4-20	4374	0	Foxboro
PI	Cont	Fixed	0.2-20	4950	0	HNU

NIOSH METHOD NO. P&CAM 127

(Use as described for acetone, with 160° C. column temperature)

TETRACHLOROETHYLENE

$CCl_2=CCl_2$

Synonym: perchloroethylene
ACGIH TLV: 50 ppm (335 mg/m^3)
OSHA Standard: 100 ppm (680 mg/m^3)
NIOSH Recommendation: 50 ppm with 100 ppm ceiling (15 min)

Method	Sampling Duration	Sampling Location	Useful Range (ppm)	System Cost ($)	Test Cost ($)	Manufacturer
DT	Grab	Portable	5-14,000	150	2.60	Nat'l Draeger
DT	Grab	Portable	5-300	165	4.60	Matheson
DT	4 hr	Personal	12.5-300	850	3.10	Nat'l Draeger
IR	Cont	Portable	2-200	4374	0	Foxboro
PI	Cont	Fixed	2-200	4950	0	HNU
GC	1 min	Portable	1-10^6	10,000	0	Microsensor

NIOSH METHOD NO. P&CAM 127

(See acetone)

TETRAHYDROFURAN

Synonyms: tetramethylene oxide; THF
ACGIH TLV: 200 ppm (590 mg/m³)
OSHA Standard: 200 ppm

$\overline{OCH_2CH_2CH_2CH_2}$

Method	Sampling Duration	Sampling Location	Useful Range (ppm)	System Cost ($)	Test Cost ($)	Manufacturer
DT	Grab	Portable	100-8000	180	2.25	Bendix
IR	Cont	Portable	4-400	4374	0	Foxboro
PI	Cont	Fixed	2-200	4950	0	HNU
GC	1 min	Portable	1-10⁶	10,000	0	Microsensor

NIOSH METHOD NO. S112

(Use as described for chloroprene, with 185° C. column temperature)

TOLUENE

Synonyms: toluol; methylbenzene
ACGIH TLV: 100 ppm (375 mg/m³) - Skin
OSHA Standard: 200 ppm
NIOSH Recommendation: 100 ppm

$C_6H_5CH_3$

Method	Sampling Duration	Sampling Location	Useful Range (ppm)	System Cost ($)	Test Cost ($)	Manufacturer
DT	Grab	Portable	5-1860	150	2.60	Nat'l Draeger
DT	Grab	Portable	10-500	165	1.70	Matheson
DT	Grab	Portable	10-8500	180	2.25	Bendix
DT	8 hr	Personal	20-4000	850	3.10	Nat'l Draeger
IR	Cont	Portable	4-400	4374	0	Foxboro
PI	Cont	Fixed	2-200	4950	0	HNU
GC	1 min	Portable	1-10⁶	10,000	0	Microsensor

NIOSH METHOD NO. P&CAM 127

(See acetone)

TOLUENE DIISOCYANATE

Synonyms: TDI; toluene-2,4-diisocyanate $CH_3C_6H_3(NCO)_2$
ACGIH TLV: 0.005 ppm (0.04 mg/m^3) ceiling
OSHA Standard: 0.02 ppm ceiling
NIOSH Recommendation: 0.005 ppm TWA; 0.02 ppm ceiling (20 min)

Method	Sampling Duration	Sampling Location	Useful Range (ppm)	System Cost ($)	Test Cost ($)	Manufacturer
DT	Grab	Portable	0.02-2	150	4.80	Nat'l Draeger
CS	Grab	Portable	0.001-0.02	2465	0.80	MDA
CS	Cont	Portable	0.002-0.08	2520	Variable	MDA
CS	Cont	Fixed	0.001-0.02	4950	Variable	MDA
CM	Cont	Port/Fixed	0.001-2	4995	Variable	CEA

NIOSH METHOD NO. P&CAM 141

Principle

TDI is hydrolyzed by the absorbing solution to the corresponding toluenediamine derivative. The diamine is diazotized by the sodium nitrite/sodium bromide solution. The diazo compound is coupled with N-(1-naphthyl)-ethylenediamine to form a colored complex. The amount of colored complex formed is in direct proportion to the amount of TDI present. The amount of colored complex is determined by reading the absorbance of the solution at 550 nm.

Toluenediamine is formed via hydrolysis of TDI on a mole to mole basis. This amine is used in place of the TDI for standards. This accomplishes two things. First, the amine is not as toxic as the TDI. Second, TDI liquifies to a semi-solid at room temperature. Weighing the semi-solid is more difficult than weighing the dry amine. Both compounds have been tested by this method and the results compare favorably. TDI kits based on the Marcali method are commercially available but have not been thoroughly tested to date.

Range and Sensitivity

The range of the standards used is equivalent to 1.0-20.0 μg TDI. In a 20 L air sample, this range converts to 0.007-0.140 ppm. For samples of high concentration whereby absorbance is greater than the limits of the standard curve (1.0-20.0 μg TDI), sample dilution with absorbing solution and re-reading the absorbance extends the upper limit of the range.

It is not known how much TDI would saturate 15 mL of the absorbing solution. It may be possible that, in extremely high concentrations, some of the TDI would not be absorbed by the absorbing solution. Therefore, if a sample is diluted and reread, it could give an erroneously low value. A single bubbler absorbs 95% of the diisocyanate if the air concentration is below 2 ppm. Above 2 ppm, about 90% of the diisocyanate is recovered. At high levels, it is suggested that two impingers in series be used.

Interferences

Any free aromatic amine may give a coupling color and thus may be a positive interference. Methylene-di-(4-phenylisocyanate) (MDI) will form a colored complex in this reaction. However, its color development time is about 1-2 hours compared with 5 min for TDI. Therefore, MDI is not a serious problem.

Precision and Accuracy

This method is a modification of the Marcali method. Its precision and accuracy are unknown. There has been no collaborative testing.

Advantages and Disadvantages

This method is based on the well tested Marcali method. Any free organic amine will interfere. The method cannot be considered specific for TDI.

Apparatus

All glass midget impinger. Contains the absorbing solution or reagent.

Battery operated personal sampling pump. MSA Model G or equivalent. The sampling pump is protected from splashover or water condensation by an adsorption tube loosely packed with a plug of glass wool and inserted between the exit arm of the impinger and the pump.

Integrating volume meter. Such as a dry gas or wet test meter.

Thermometer.

Manometer.

Stopwatch.

Various clips, tubing, spring connectors, and belt. For connecting sampling apparatus to worker being sampled.

Spectrophotometer. Beckman model B or equivalent.

Cells. 1 cm and 5 cm matched quartz cells.

Volumetric flasks. Several of each, glass-stoppered, 50, 100, and 1000 mL.

Balance. Capable of weighing to at least three decimal places, preferably 4 decimal places.

Pipets. 0.5, 1, 15 mL.

Graduated cylinders. 25, 50 mL.

Reagents

All reagents must be made using ACS reagent grade or a better grade.

Double distilled water.

2,4-Toluenediamine.

Hydrochloric acid, concentrated, 11.7 N.

Glacial acetic acid, concentrated, 17.6 N.

Sodium nitrite.

Sodium bromide.

Sodium nitrite solution. Dissolve 3.0 g sodium nitrite and 5.0 g sodium bromide in about 80 mL double distilled water. Adjust volume to 100 mL with double distilled water. This solution is stable for one week if refrigerated.

Sulfamic acid.

Sulfamic acid solution, 10% w/v. Dissolve 10 g sulfamic acid in 100 mL double distilled water.

N-(1-Naphthyl)-ethylenediamine dihydrochloride.

N-(1-Naphthyl)-ethylenediamine solution. Dissolve 50 mg in about 25 mL double distilled water. Add 1 mL concentrated hydrochloric acid and dilute to 50 mL with double distilled water. Solution should be clear and colorless; coloring is due to contamination by free amines, and solution should not be used. The solution is stable for 4 days.

Absorbing solution. Add 35 mL concentrated hydrochloric acid and 22 mL glacial acetic acid to approximately 600 mL double distilled water. Dilute the solution to 1 L with double distilled water. 15 mL is used in each impinger.

Standard solution A. Weigh out 140 mg of 2,4-toluenediamine (equivalent to 200 mg of 2,4-toluenediisocyanate). Dissolve in 660 mL of glacial acetic acid, transfer to a 1 L glass-stoppered volumetric flask, and make up to volume with distilled water.

Standard solution B. Transfer 10 mL of standard solution A to a glass-stoppered 1 L volumetric flask. Add 27.8 mL of glacial acetic acid so that when solution B is diluted to 1 L with distilled water, it will be 0.6 N with respect to acetic acid. This solution contains an equivalent of 2 μg TDI/mL.

Procedure

Cleaning of equipment. Wash all glassware in a hot detergent solution such as Alconox to remove any oil. Rinse well with hot tap water. Rinse well with double distilled water. Repeat this rinse several times.

Collection and shipping of samples. Pipet 15 mL of the absorbing solution into the midget impinger. Connect the impinger (via the absorption tube) to the vacuum pump and the prefilter assembly (if needed) with a short piece of flexible tubing. The minimum amount of tubing necessary should be used between the breathing zone and impinger. The air being sampled should not be passed through any other tubing or other equipment before entering the impinger. Turn on pump to begin sample collection. Care should be taken to measure the flow rate, time and/or volume as accurately as possible. Record the atmospheric pressure and temperature. The sample should be taken at a flow rate of 1 L/min. Sample for 20 min, making the final air volume 20 L.

After sampling, the impinger stem can be removed and cleaned. Tap the stem gently against the inside wall of the impinger bottle to recover as much of the sampling solution as possible. Wash the stem with a small amount (1-2 mL) of unused absorbing solution and add the wash to the impinger. Then the impinger is sealed with a hard, non-reactive stopper (preferably Teflon or glass). Do not seal with rubber. The stoppers on the impingers should be tightly sealed to prevent leakage during shipping. If it is preferred to ship the impingers with the stems in, the outlets of the stem should be sealed with Parafilm or other non-rubber covers, and the ground glass joints should be sealed (i.e., taped) to secure the top tightly.

Care should be taken to minimize spillage or loss by evaporation at all times. Refrigerate samples if analysis cannot be done within a day. Whenever possible, hand delivery of the samples is recommended. Otherwise, special impinger shipping cases designed by NIOSH should be used to ship the samples. A blank impinger should be handled as the other samples (fill, seal and transport), except that no air is sampled through this impinger.

Analysis of samples. Remove bubbler tube, if it is still attached, taking care not to remove any absorber solution. Start reagent blank at this point by adding 15 mL fresh absorbing solution to a clean bubbler cylinder. To each bubbler, add 0.5 mL of 3% sodium nitrite solution, gently agitate, and allow solution to stand for 2 min. Add 1 mL of 10% sulfamic acid solution, agitate vigorously to evolve all the N_2 from solution and allow solution to stand 2 min to destroy all the excess nitrous acid present.

Add 1 mL of 0.1% N-(1-Naphthyl)-ethylenediamine solution. Agitate and allow color to develop. Color will be developed in 5 min. A reddish blue color indicates the presence of TDI. Add double distilled water to adjust the final volume to 20 mL in the bubbler cylinder mix. Transfer each solution to a 1 cm or 5 cm quartz cell. Using the blank, adjust the spectrophotometer to 0 absorbance at 550 nm. Determine the absorbance of each sample at 550 nm.

Calibration and Standards

To each of a series of 8 graduated cylinders add 5 mL of 1.2 N hydrochloric acid. To these cylinders add the following amounts of 0.6 N acetic acid: 10.0, 9.5, 9.0, 8.0, 7.0, 6.0, 5.0, and 0.0

mL, respectively. To the cylinders add standard solution B in the same order as the acetic acid was added: 0.0, 0,5, 1.0, 2.0, 3.0, 4.0, 5.0 and 10 mL, so that the final volume is 15 mL (i.e., 0.0 mL of the standard is added to the 10 mL acetic acid; 0.5 mL of the standard is added to the 9.5 mL acid, etc.). The cylinders now contain the equivalent of 0.0, 1.0, 2.0, 4.0, 6.0, 8.0, 10.0, and 20.0 μg TDI, respectively. The standard containing 0.0 mL standard solution is a blank.

Add 0.5 mL of the 3.0% sodium nitrite reagent to each cylinder. Mix. Allow to stand 2 min. Add 1 mL of the 10% sulfamic acid solution. Mix. Allow to stand for 2 min. Add 1 mL of the N-(1-Naphthyl)-ethylenediamine solution. Mix. Let stand for 5 min. Make up to 20 mL with double distilled water. Transfer each solution to a 1 cm or 5 cm quartz cell. Using the blank, adjust the spectrophotometer to 0 absorbance at 550 nm. Determine the absorbance of each standard at 550 nm. A standard curve is constructed by plotting the absorbance against μg TDI.

Calculations

Subtract the blank absorbance, if any, from the sample absorbance. From the calibration curve, read the μg TDI corresponding to absorbance of the sample. Calculate the concentration of TDI in air in ppm, defined as μL TDI/L of air:

$$\text{ppm} = (\mu g)(24.45) \div (Vs)(M.W.)$$

where:

ppm = parts per million TDI
μg = μg TDI
Vs = corrected volume of air in L
24.45 = molar volume of an ideal gas at 25° C. and 760 mm Hg
M.W. = molecular weight of TDI, 174.15

Convert the volume of air sampled to standard conditions of 25° C. and 760 mm Hg:

$$Vs = (V)(P)(298) \div (760)(T + 273)$$

where:

Vs = volume of air in L at 25° C. and 760 mm Hg
V = volume of air in L as measured
P = barometric pressure in mm Hg
T = temperature of air (°C.)

References

1. Marcali, K. Microdetermination of toluenediisocyanates in atmosphere. Anal. Chem. 4: 552, 1957.
2. Larkin, R.L. and R.E. Kupel. Microdetermination of toluene diisocyanate using toluenediamine as the primary standard. Amer. Ind. Hyg. Assoc. J. 30: 640, 1969.

o-TOLUIDINE

Synonym: o-methylaniline $CH_3C_6H_4NH_2$
ACGIH TLV: 2 ppm (9 mg/m^3) - Skin, suspected carcinogen
OSHA Standard: 5 ppm (22 mg/m^3)

Method	Sampling Duration	Sampling Location	Useful Range (ppm)	System Cost ($)	Test Cost ($)	Manufacturer
DT	Grab	Portable	2.5-35	180	2.25	Bendix
IR	Cont	Portable	1-20	4374	0	Foxboro
PI	Cont	Fixed	0.2-20	4950	0	HNU

NIOSH METHOD NO. P&CAM 168

(See aniline)

1,1,2-TRICHLOROETHANE

Synonym: vinyl trichloride $CHCl_2CH_2Cl$
ACGIH TLV: 10 ppm (45 mg/m^3) - Skin
OSHA Standard: 10 ppm

Method	Sampling Duration	Sampling Location	Useful Range (ppm)	System Cost ($)	Test Cost ($)	Manufacturer
IR	Cont	Portable	0.6-20	4374	0	Foxboro
PI	Cont	Fixed	20-2000	4950	0	HNU
GC	1 min	Portable	1-10^6	10,000	0	Microsensor

NIOSH METHOD NO. P&CAM 127

(See acetone)

TRICHLOROETHYLENE

Synonym: ethinyl trichloride $\qquad$ CHCl=CCl$_2$
ACGIH TLV: 50 ppm (270 mg/m^3)
OSHA Standard: 100 ppm (535 mg/m^3)
NIOSH Recommendation: 25 ppm (134 mg/m^3)

Method	Sampling Duration	Sampling Location	Useful Range (ppm)	System Cost ($)	Test Cost ($)	Manufacturer
DT	Grab	Portable	2-500	150	2.60	Nat'l Draeger
DT	Grab	Portable	5-300	165	3.80	Matheson
DT	Grab	Portable	2-560	180	2.25	Bendix
DT	4 hr	Personal	5-200	850	3.10	Nat'l Draeger
IR	Cont	Personal	2-200	4374	0	Foxboro
PI	Cont	Fixed	2-200	4950	0	HNU
GC	1 min	Portable	1-10^6	10,000	0	Microsensor

NIOSH METHOD NO. P&CAM 127

(See acetone)

1,2,3-TRICHLOROPROPANE

Synonyms: glycerol trichlorohydrin; allyl trichloride; trichlorohydrin $\qquad$ CH$_2$ClCHClCH$_2$Cl
ACGIH TLV: 50 ppm (300 mg/m^3)
OSHA Standard: 50 ppm

Method	Sampling Duration	Sampling Location	Useful Range (ppm)	System Cost ($)	Test Cost ($)	Manufacturer
IR	Cont	Portable	1-100	4374	0	Foxboro

NIOSH METHOD NO. P&CAM 127

(Use as described for acetone, with 160° C. column temperature)

1,1,2-TRICHLORO-1,2,2-TRIFLUOROETHANE

Synonyms: Freon 113; Refrigerant 113; F-113
ACGIH TLV: 1000 ppm (7600 mg/m³)
OSHA Standard: 1000 ppm

CCl_2FCClF_2

Method	Sampling Duration	Sampling Location	Useful Range (ppm)	System Cost ($)	Test Cost ($)	Manufacturer
IR	Cont	Portable	20-2000	4374	0	Foxboro
GC	1 min	Portable	$1-10^6$	10,000	0	Microsensor

NIOSH METHOD NO. S112

(Use as described for chloroprene, with 150° C. column temperature)

TRIETHYLAMINE

ACGIH TLV: 10 ppm (40 mg/m³)
OSHA Standard: 25 ppm (100 mg/m³)

$N(C_2H_5)_3$

Method	Sampling Duration	Sampling Location	Useful Range (ppm)	System Cost ($)	Test Cost ($)	Manufacturer
DT	Grab	Portable	5-60	150	2.60	Nat'l Draeger
DT	Grab	Portable	3.5-140	180	2.25	Bendix
IR	Cont	Portable	0.6-50	4374	0	Foxboro
PI	Cont	Fixed	2-200	4950	0	HNU

NIOSH METHOD NO. P&CAM 221

(See n-butylamine)

TRIFLUOROMONOBROMOMETHANE

Synonyms: trifluorobromomethane; Freon 13B1; Refrigerant 13B1 CF_3Br
ACGIH TLV: 1000 ppm (6100 mg/m³)
OSHA Standard: 1000 ppm

Method	Sampling Duration	Sampling Location	Useful Range (ppm)	System Cost ($)	Test Cost ($)	Manufacturer
IR	Cont	Portable	20-2000	4374	0	Foxboro

NIOSH METHOD NO. S111

(Use as described for dichlorodifluoromethane, with 160° C. column temperature)

TURPENTINE

ACGIH TLV: 100 ppm (560 mg/m³) hydrocarbons (150-170° C. b.p.)
OSHA Standard: 100 ppm

Method	Sampling Duration	Sampling Location	Useful Range (ppm)	System Cost ($)	Test Cost ($)	Manufacturer
IR	Cont	Portable	2-200	4374	0	Foxboro

NIOSH METHOD NO. P&CAM 127

(Use as described for acetone, with 100° C. column temperature and undecane as internal standard)

VANADIUM PENTOXIDE FUME

V_2O_5

Synonym: vanadium oxide
ACGIH TLV: 0.05 mg/m³
OSHA Standard: 0.05 mg/m³ ceiling
NIOSH Recommendation: 0.05 mg/m³ ceiling (15 min)

Method	Sampling Duration	Sampling Location	Useful Range (mg/m³)	System Cost ($)	Test Cost ($)	Manufacturer
LS	Cont	Personal	0.01-100	1695	0	GCA
LS	Cont	Portable	0.001-100	2785	0	MDA
LS	Cont	Fixed	0.001-200	3990	0	GCA
BA	1,4 min	Portable	0.02-150	4890	0	GCA

NIOSH METHOD NO. P&CAM 290

Principle

Vanadium oxide sample collected on mixed cellulose ester membrane filter is dissolved from the sample-containing filter by heating in a sodium hydroxide solution at 50° C. for 15-30 min. The solutions of samples and standards are analyzed by atomic absorption spectrophotometry in a high temperature graphite atomizer (HGA). A hollow cathode lamp for vanadium is used to provide a characteristic vanadium line at 318.4 nm. The absorbance is proportional to the vanadium concentration.

Range and Sensitivity

The working range of this method, based on a 25 L air sample, is for pyrolytically coated graphite tubes, 0.05-5.0 µg/sample or 0.02-0.2 mg/m³; and for standard graphite tubes, 5.0-10.0 µg/sample or 0.2-0.5 mg/m³. The sensitivity and detection limits for standard graphite tubes and pyrolytically coated graphite tubes are:

	Standard Tube	Pyro Tube
Characteristic conc. (µg/1% abs)	0.220	0.108
Detection limit (pg)	0.380	0.208

Inteferences

The use of an alkaline dissolution method minimizes the potential interference of ferrovanadium dust and other vanadium compounds. Molybdenum is a positive interference in the analysis of vanadium by graphite furnace AAS. This interference may be minimized by using the 306.4 nm analytical line. Vanadium exhibits self-absorption at 318.4 nm. It is necessary to operate the vanadium hollow cathode lamp at low currents to minimize self-absorption (reference 2).

Precision and Accuracy

The relative standard deviation for the analytical method is 1.9% based on the analysis of 18 spiked filters, 6 replicates at 3 concentration levels (0.625, 1.25, and 2.5 µg/filter). Vanadium recovery was 100.6% for the 18 filters (6 replicates at 3 concentration levels of 0.625, 1.25, and

2.5 µg/filter). A collection efficiency of 100% was determined for the collection medium in the validation tests (reference 1). This method is an improvement over method no. S391. The changes made herein should not affect the validation.

Advantages and Disadvantages

The samples collected on filters are analyzed by means of a quick instrumental method. Dissolution with NaOH separates the oxides, particularly V_2O_5, from aluminum and titanium. Analysis in a nitric acid matrix eliminates interferences from aluminum and titanium. The method does not distinguish between vanadium dust and fume.

Apparatus

Filter unit. Consists of the filter media and 37 mm, 2 piece cassette filter holder.

Personal sampling pump. A calibrated personal sampling pump whose flow can be determined to an accuracy of 5% at the recommended flow rate. The pump must be calibrated with a filter holder and filter in the line.

Thermometer.

Barometer.

Stopwatch.

Mixed cellulose ester membrane filter. 37 diameter, 0.8 micron pore size.

Atomic absorption spectrophotometer. With a high temperature graphite atomizer (HGA) and a deuterium arc background corrector.

Vanadium hollow cathode lamp.

Deuterium arc lamp.

Purge gas. Argon (HGA).

Pressure regulators. Two-stage, for each compressed gas tank used.

Glassware, borosilicate. 50 mL beakers with watchglass covers; pipets or automatic pipetors at convenient sizes for preparing standards; 10 mL volumetric flasks.

Water bath maintained at 50° C.

Reagents

All reagents used must be ACS reagent grade or better.

Distilled or deionized water.

0.01 N Sodium hydroxide.

Concentrated nitric acid.

Aqueous standard stock solution. Containing 250 µg/mL of vanadium pentoxide. This solution can be prepared by dissolving 0.250 g V_2O_5 in 1 L of 2% (v/v) nitric acid.

Commercially prepared aqueous stock standard solution. Containing 1000 µg/mL of vanadium.

Vanadium working standard solution, 100 µg/mL. Prepare by dilution of 1 mL of 1000 µg/mL in 10 mL volumetric flask with approximately 2% (v/v) nitric acid.

Procedure

Cleaning of equipment. Before use, all glassware should be initially soaked in a mild detergent solution to remove any residual grease or chemicals. After initial cleanup, the glassware should be thoroughly rinsed with warm tap water, soaked in dilute (1:10) nitric acid for 30 min, rinse with tap water, then deionized water, in that order, and then dried.

Collection and shipping of samples. To collect vanadium oxides, a personal sampler pump is used to pull air through a cellulose ester membrane filter. The filter holder is held together with tape or shrinking band. If the filter holder does not fit snugly together the contaminant will leak around the filter. A piece of flexible tubing is used to connect the filter holder to the pump.

Sample at a flow rate of 1.7 L/min for 15 min with a face cap on and small plugs removed. After sampling, replace small plugs.

With each batch of ten samples submit one filter from the same lot of filters which was used for sample collection and which is subjected to exactly the same handling as for the samples except that no air is drawn through it. Label this as a blank. The filter cassettes should be shipped in a suitable container, designed to prevent damage in transit.

Analysis of samples. Open the cassette filter holder and carefully remove the cellulose membrane filter from the holder and cellulose backup pad with the aid of appropriate tweezers. Transfer the filter to a 50 mL beaker. To insure complete dissolution of vanadium oxide from the filter, add 3 mL of 0.01 N NaOH to the beaker and heat in a water bath at 50° C. for 15-30 min. Cool solutions and quantitatively transfer the clear solutions into a 10 mL volumetric flask containing 250 μL concentrated nitric acid. Rinse each beaker at least twice with 1-2 mL portions of 0.01 N NaOH, and quantitatively transfer each rinsing to the solution in the volumetric flask. Dilute all samples to 10 mL with 0.01 N NaOH.

Inject 20 μL of solution into a high temperature graphite atomizer. Dry at 100° C. for 20 sec, char with ramp at 1300° C. for 10 sec and atomize at 2800° C. for 8 sec. Use the deuterium arc lamp to correct for background noise. Record the absorbance at 318.4 nm. The absorbance is proportional to the sample concentration and can be determined form the appropriate calibration curve if the graphite tube in use gives reproducibility results (reference 4).

Note: The characteristics of the graphite tubes can influence the results drastically. Careful attention must be paid to the response of the standard, i.e., if the graphite tube gives erratic results and nonreproducible absorbance peaks, it must be replaced because results so obtained are not reliable. Also, follow instrument manufacturer's recommendations for specific operating parameters.

In case a calibration curve cannot be used because of the changing characteristics of the graphite tube, it is recommended that samples be frequently alternated with standards which give responses close to that of the sample. The experimental protocol recommended would be as follows: inject a standard solution in duplicate, inject a sample in duplicate, and reinject standard in duplicate...etc. Appropriate filter blanks must be analyzed by the same procedure used for the samples. Reagent blanks must also be analyzed.

Calibration and Standards

From the V_2O_5 working standard solution, prepare at least six calibration standards to cover the vanadium concentration range from 0.05-0.40 μg/mL. Prepare these calibration standards fresh daily and make all dilutions with 2% (v/v) nitric acid. Prepare a calibration curve by plotting on linear graph paper the absorbance versus the concentration of each standard in μg/10 mL. It is advisable to run a set of standards both before and after the analysis of a series of samples to insure that conditions have not changed. In cases where a calibration curve could not be used reliably, determine the appropriate response factor, by analyzing the appropriate calibration standards alternately with the samples. This practice will minimize the effect of observed fluctuations or variations in absorbance and peak width readings during any given day.

Calculations

Read the weight, in μg, corresponding to the total absorbance from the standard curve. No volume corrections are needed, because the standard curve is based on μg/dL. An alternative procedure is to determine the weight in μg corresponding to the absorbance area of the sample by using the appropriate response factor determined from the response of the calibration standard. The concentration of the analyte in the air sampled can be expressed in mg/m^3 (1 μg/L = 1 mg/m^3):

$$C = W \div V$$

where:

C = concentration of vanadium (mg/m³)
W = concentration of vanadium (μg/filter)
V = volume of air sampled (L)

For personal sampling pumps with rotameters only, the following air volume correction should be made:

$$\text{Corrected volume} = ft(P_1 T_2 / P_2 T_1)^{1/2}$$

where:

f = flow rate sampled
t = sampling time
P_1 = pressure during calibration of sampling pump (mm Hg)
P_2 = pressure of air sampled (mm Hg)
T_1 = temperature during calibration of sampling pump (°K)
T_2 = temperature of air sampled (°K)

References

1. Vanadium, V₂O₅ Dust, S391. Backup Data Report prepared under NIOSH Contract No. 210-76-0123.
2. Christian, G. and Feldman, F. *Atomic Absorption Sepctroscopy,* Wiley-Interscience, N.Y. 1970.
3. Memoranda. K.A. Busch, (Chief, Statistical Services, DLCD), to Deputy Director, DLCD, dated 1/6/75, 11/8/74, subject; Statistical Protocol for Analysis of Data from Contract CDC-99-74-45.
4. Analytical Methods for Atomic Absorption Spectrophotometry. The Perkin-Elmer Corp, Norwalk, CT, 1976.
5. Cassinelli, M.E. and A.W. Verstuyft. Evaluation of AAS Methods for the Determination of Vanadium, IMDS, MRB, Technical Report, 1978.

VINYL ACETATE

Synonym: ethenyl ethanoate $\qquad$ CH₃COOCH=CH₂
ACGIH TLV: 10 ppm (30 mg/m³)
OSHA Standard: none
NIOSH Recommendation: 4 ppm (15 mg/m³) ceiling (15 min)

Method	Sampling Duration	Sampling Location	Useful Range (ppm)	System Cost ($)	Test Cost ($)	Manufacturer
DT	Grab	Portable	5-10,000	180	2.25	Bendix
IR	Cont	Portable	0.2-20	4374	0	Foxboro
PI	Cont	Fixed	0.2-20	4950	0	HNU

NIOSH METHOD NO. P&CAM 278

Principle

A known volume of air is drawn through a tube containing a Chromosorb 107 to trap the vinyl acetate present. The tube is thermally desorbed into a 300 mL chamber. An aliquot of the desorbed vapor is injected into a gas chromatograph. The area of the resulting peak is determined and compared with the areas obtained from the injection of standards.

Range and Sensitivity

Sample loading between 2 and 332 μg of vinyl acetate/sampling device are acceptable. Samples have been successfully collected from dynamically generated atmospheres of vinyl acetate over the concentration range of 8.2-206 mg/m^3. The relative humidity of the sampled air was greater than 80%. The slope of a typical calibration curve (integrator response versus mass of vinyl acetate/sample tube) was 687 (v.S)/g. The sensitivity of the flame ionization detector was 4.04 $\times$ 10^{-6} C/g of vinyl acetate. The lowest quantifiable level was determined to be 0.5 μg of vinyl acetate/300 mg bed of Chromosorb 107. At this loading the relative standard deviation of replicate samples was better than 10%.

Interferences

When two or more substances are known or suspected to be present in the air sampled, such information should be transmitted with the sample, because the substances may interfere with the analysis of vinyl acetate. Any substance which has the same retention time as vinyl acetate at the operating conditions described in this method is an interference. Therefore, retention time data on single or multiple columns cannot be considered proof of chemical identity. If the possibility of interference exists, separation conditions (column packing, temperature, carrier flow, detector, etc.) must be changed to circumvent the problem.

Precision and Accuracy

The pooled relative standard deviation of the sampling and analytical method was 8.1%. This reflects the precision of sampling and analysis of 50 samples of vinyl acetate collected with calibrated personal sampling pumps from humid atmospheres (over 80%) over the concentration range of 8-200 mg/m^3. The relative standard deviation of samples collected from atmospheres averaging 8.6 (range 8.2-9.0) mg/m^3 was 9.5%, from atmospheres averaging 24.3 (range 18.6-39.2) mg/m^3 was 8.0%, and from atmospheres averaging 181 (range 159-206) mg/m^3 was 5.9%.

The concentration of the sampled air was independently determined using a gas phase infrared analyzer. The samples were collected from humid air and stored at room temperature. The determination averaged 7% high, 5% high, and 4% low when analyzed on days 1, 7, and 14, respectively. Thus the analyte displayed an 0.8%/day storage loss when stored at room temperature. This loss can be attributed to the analyte degrading in the presence of water, because samples spiked with 18 μg of vinyl acetate in hexane gave 98% recovery when stored for 14 days at room temperature.

The breakthrough volume and therefore capacity of Chromosorb 107 for vinyl acetate decreased with increasing relative humidity. Under the most adverse conditions tested, 83% relative humidity, the breakthrough volume was found to be 4.0 L when an atmosphere of vinyl acetate at 138 mg/m^3 was sampled at 0.125 L/min.

Advantages and Disadvantages

The sampling device is small, portable, and involves no liquids. Many of the interferences can be eliminated by altering chromatographic conditions. The sampling tubes are analyzed using a quick instrumental method and can be reused after the analysis is completed. The precision of the method is limited by the reproducibility of the pressure drop across the sampling tubes. Variations in pressure drop will affect the flow rate. The reported sample volume will be imprecise because the pump is usually calibrated for one tube only.

One disadvantage of the method is that the amount of sample that can be taken is limited by the capacity of the sampling device. When the amount of vinyl acetate found on the backup section exceeds 10% of the amount found on the front section, the possibility of sample loss exists. Migration from the front to the backup section is not a problem because the sections are separated and individually capped immediately after sampling.

Apparatus

Personal sampling pump. Capable of accurate performance at 0.1 L/min. The pump must be calibrated with a representative sampling device in line and must be fully charged prior to being used.

Chromosorb 107 sampling tubes. Individual front and backup tubes (Century Systems Corporation "Flare" tubes or equivalent) are used. The front section is a stainless steel tube 7.3 cm long with 6 mm O.D. by 4 mm I.D., and a 45° flare at one end. The backup section is a chrome plated nickel tube 3 cm long with 6 mm O.D. by 4 mm I.D., a 45° flare at one end and a hose connection at the other end. Each tube has a permanent metal frit in the outlet end of the tube. The front section contains 300 mg of prewashed Chromosorb 107 held in place with a removable metal frit. The backup section contains 50 mg of prewashed Chromosorb 107 held in place with a plug of silylated glass wool. The sampling device is assembled by joining the front and backup section with a nylon nut and fitting. A hollow nylon ferrule is placed between the two sections. The pressure drop across the tubes must be less than 10 inches of water at a flow rate of 0.1 L/min.

The Chromosorb 107 is washed in a Soxhlet extractor for 8 hours with water, 8 hours with methanol and 8 hours with dichloromethane. The sorbent is then dried overnight in a vacuum oven. The tubes are loaded with sorbent and thermally purged for 2 min with helium at 150° C. After cooling in a closed container the ends are capped.

Thermal desorber. Equipped with thermostatted desorbing oven, 300 mL sample reservoir, and a 2 mL gas sampling loop (Century Systems Corporation Programmed Thermal Desorber or equivalent).

Gas chromatograph. Equipped with a flame-ionization detector and electronic integrator.

GC column. 20 ft × 1/8 in O.D., made of silanized stainless steel and packed with 10% FFAP on 80/100 mesh Chromosorb W AW.

Vials. 1.5 mL, with aluminum caps equipped with Teflon-lined silicone rubber septa.

Syringes. 10 µL and convenient sizes for making standards.

Macropipet. 1000 µL, with disposable plastic tips.

U-tube. Glass with at least one hose connection, approximately 75 mL internal volume.

Pump. Capable of drawing 200 mL/min through a front section of the sampling device.

Gas bag. 10 L volume for helium purge gas.

Test tubes. With close-fitting plastic caps.

Ring stand with clamps.

Reagents

Vinyl acetate, practical. Inhibited with hydroquinone and freshly distilled before use.
Hexane, UV grade.
Helium, Bureau of Mines grade A.
Hydrogen, prepurified.
Air, filtered and compressed.

Vinyl acetate

Procedure

Cleaning of equipment. All nondisposable glassware used for the laboratory analysis of vinyl acetate is washed with detergent and rinsed thoroughly with tap water and distilled water.

Collection and shipping of samples. Immediately before sampling remove the plastic caps from the inlet and outlet ends of the sampling tube. Connect the tube to the sampling pump using a short piece of flexible tubing. The backup section is positioned nearest the pump. The sampling tube is kept vertical during sampling to prevent channeling through the device. Air being sampled does not pass through any hose or tubing before entering the sampling device.

The temperature, pressure, and volume of air sampled is measured and reported. The volume sampled should not exceed 3 L sampled at a flow rate of 0.1 L/min or less. Record either the flow rate and sampling time or the initial and final stroke readings and the volumetric stroke factor. Immediately after sampling, disassemble the two sections, cap the sections with plastic caps, and label the section. Do not use rubber caps. For every ten samples taken, handle one sampling device in the same manner as the samples (uncap, disassemble, cap, label and transport) however do not sample any air through this device. Label this device as a blank. Samples received at the laboratory are logged in and analyzed as soon as possible. Samples stored at room temperature for 14 days exhibit an 11% loss of analyte.

Preparation of samples. Remove the caps from either a front or a back section. Wipe the outside of the tube with a clean laboratory wiper.

Thermal desorber conditions. Typical operating conditions for the thermal desorber are:

150° C. desorbing oven temperature
70 mL/min desorbing rate, helium gas
160° C. transfer line temperature
15 sec pressure equalization time

GC conditions. Typical operating conditions for the gas chromatograph are:

33 mL/min helium carrier gas flow rate
40 mL/min hydrogen flow to detector
435 mL/min air flow to detector
160° C. injector temperature
160° C. manifold (detector) temperature
60° C. oven temperature

Under these conditions, the capacity ratio for vinyl acetate was 4.4.

Thermal desorption of samples. Wipe off the tube and insert it in the desorbing oven. Desorb with helium at atmospheric pressure. The helium is stored in the 10 L gas bag. Desorbing with air chars the Chromosorb and renders it unsuitable for reuse.

Injection. Inject a 2 mL aliquot of the desorbed vapors into the GC column. Since the desorbed vapors are stored in a reservoir, as many as five replicate injections of each sample can be made.

Measurement of area. Measure the area of the sample peak with an electronic integrator or other suitable technique of area measurement. The results are read from a standard curve.

Preparation for next sample. After satisfactory analysis is obtained, purge the thermal desorber with helium for 2 min. Remove the tube from the desorbing oven, place it in a test tube and cap the test tube. When the tube is cool, remove it from the test tube and cap it with the plastic cups.

Calibration and Standards

It is convenient to express the concentration in terms of μg of vinyl acetate/sample tube.

Standard curves are prepared by loading clean sampling tubes (front sections) with known amounts of vinyl acetate. The density of vinyl acetate (0.932 mg/μL at 20° C.) is used to convert the volume taken to mass.

Preparation of standards. Pipet 1.00 mL aliquots of hexane into clean glass vials. Crimp the vials shut with an aluminum serum cap equipped with a Teflon lined silicone rubber septum. Inject either 25, 10, 5 or 1 μL of freshly distilled vinyl acetate into each vial. These standard solutions are freshly prepared for each analysis.

Loading of standard. Support a U-tube on a ring stand. Using a short length of tubing attach the outlet end of a clean front section of a sampling tube to a small pump. The inlet end of the clean front section is attached to the side of the U-tube that has the hose connection. Use the solvent flush technique to withdraw a 2 μL aliquot of a standard solution. Turn on the pump and inject this 2 μL aliquot into the end of the U-tube farthest from the sampling tube. Sweep enough air through the U-tube (2 min at 200 mL/min approximately 5 volume changes) to insure that all the vinyl acetate is loaded on the sample tube. Stop the pump, remove the sample tube, cap both ends, and label. This tube now contains a known amount of vinyl acetate.

Standardization. Analyze each of the above tubes as described earlier. The standard curve is obtained by plotting the amount of vinyl acetate loaded on a tube versus the peak area found. If conditions warrant, prepare standards at higher or lower concentrations.

Calculations

The sample weight in μg is read from the standard curve. Blank corrections are not expected. If the analysis shows a blank correction is needed, the correction is:

$$W_f = W_s - W_b$$

where:

W_f = corrected amount (μg) on the front section of the charcoal tube
W_s = amount (μg) found on the front section of the charcoal tube
W_b = amount (μg) found on the front section of the blank charcoal tube

The concentration, C, of vinyl acetate in the air sampled is expressed in mg/m³, which is numerically equal to μg/L:

$$C = (W_f + W_b) \div V$$

where:

W_f = amount of vinyl acetate found on front section in μg
W_b = amount of vinyl acetate found on backup section in μg
V = volume of air sampled in L

If desired the results may be expressed in ppm at 25° C. (298° K) and 760 torr:

$$\text{ppm} = (\mu g/L)(24.45)(760)(T + 273) \div (86.1)(P)(298)$$

where:

P = pressure of air sampled in torr
T = temperature of air sampled in °C.
24.45 = molar volume at 25° C. and 760 torr in L/mol
86.1 = molecular weight of vinyl acetate in g/mol

References

1. D.L. Foerst, A.W. Teass and M. Risholm-Sundman. A Sampling and Analytical Method for Vinyl Acetate in Air, (manuscript in preparation). Measurement Research Branch, NIOSH, Cincinnati, OH, 1978.

VINYL CHLORIDE

Synonyms: chloroethylene; chloroethene $\qquad$ CH$_2$=CHCl
ACGIH TLV: 5 ppm (10 mg/m^3) - Human carcinogen
OSHA Standard: 1 ppm
NIOSH Recommendation: minimum detectable level; 1 ppm ceiling (15 min)

Method	Sampling Duration	Sampling Location	Useful Range (ppm)	System Cost ($)	Test Cost ($)	Manufacturer
DT	Grab	Portable	0.25-3000	150	2.70	Nat'l Draeger
DT	Grab	Portable	0.25-10,000	165	2.30	Matheson
DT	Grab	Portable	0.1-20,000	180	2.25	Bendix
DT	8 hr	Personal	1-50	850	3.10	Nat'l Draeger
IR	Cont	Portable	0.6-4	4374	0	Foxboro
PI	Cont	Fixed	0.2-20	4950	0	HNU

NIOSH METHOD NO. P&CAM 178

Principle

A known volume of air is drawn through two small sorbent tubes in series containing activated carbon (made from coconut shells), which absorbs the vinyl chloride present in the air sample. The collected vinyl chloride is then desorbed with carbon disulfide, and the resulting solutions are analyzed by gas chromatography with a flame-ionization detector. The areas under the resulting peaks are compared with areas obtained from the injection of standards.

Range and Sensitivity

The minimum detectable amount of vinyl chloride was found to be 0.2 ng/injection at a 1 × 1 attenuation on a gas chromatograph. This corresponds to an estimated concentration of 0.008 mg/m^3 in a 5 L air sample analyzed by this method. However, the desorption efficiency from activated carbon of amounts of vinyl chloride as small as 40 ng (0.008 μg/L × 5 L) has not been determined. Therefore, the detection limit of the overall method may be somewhat higher than 0.008 mg/m^3.

At the recommended sampling flow rate of 50 mL/min, the total volume to be sampled should not exceed 5 L. This value is based upon data which indicated that more than 10 L of air containing 2.6 µg/L (1 ppm) of vinyl chloride could be sampled on activated carbon before 5% breakthrough was observed. This indicates that 5 L of air containing no more than 5.2 mg/m^3 may be sampled without significant breakthrough. (The sorbent tube consists of two sections of activated carbon separated by a section of urethane foam.) If a particular atmosphere is suspected of containing high concentration of contaminants or a high humidity is suspected, the sampling volume should be reduced by 50%. A safety factor has been included in the recommended 5 L volume and the capacity of the first tube should be within these limits except under the most extreme conditions.

Interferences

When the amount of water in the air is so great that condensation actually occurs in the tube, organic vapors will not be trapped effectively. Experiments indicate that high humidity severely decreases the capacity of activated carbon for organic vapors. When two or more substances are known or suspected to be present in the air, such information, including their suspected identities, should be transmitted with the sample since these compounds may interfere with the analysis for vinyl chloride.

Any compound that has the same retention time as vinyl chloride at the opening conditions described in this method is an interference. Hence, retention time data on a single column, or even on a number of columns, may not provide proof of chemical identity. Often, operating conditions can be modified to eliminate interferences. Samples should be analyzed by an independent method when overlapping as chromatographic peaks cannot be resolved.

Precision and Accuracy

The coefficient of variation resulting from the analysis of two sets of sorbent tubes, one set of 27 tubes exposed to a vinyl chloride concentration of 7.2 mg/m^3 in air and another set of 29 tubes exposed to a concentration of 71.3 mg/m^3, were 0.076 and 0.075, respectively. These values reflect total sampling and analytical error as well as desorption efficiency correction errors.

Experiments were performed to obtain some indication of the accuracy, although accuracy was difficult to evaluate in the absence of a primary standard. These experiments generally involved six sorbent tube samples exposed to a synthetic atmosphere. The calculated value was the concentration expected based on the measured amounts of vinyl chloride and air mixed to prepare the synthetic atmosphere. Therefore the calculated value was not the "true" value, since it was subject to experimental error. The value found from analysis of each sorbent tube, after correction for desorption efficiency, was also compared to that found by the direct injection of gas samples from the same synthetic atmosphere used in loading the tubes. The results of these experiments are shown in the table below. It should be noted that average concentrations determined by analysis of sorbent tubes were within 6% of the average concentrations determined by analysis of gas samples.

Advantages and Disadvantages

The sampling device is small, portable, and involves no liquids. Interferences are minimal, and most of those that do occur can be eliminated by altering chromatographic conditions. The tubes are analyzed by means of a rapid instrumental method. The method can also be used for the simultaneous determination of two or more components suspected to be present in the same sample by changing gas chromatographic conditions from isothermal to a temperature-programmed mode of operation.

One disadvantage of the method is that the amount of sample that can be taken is limited by the

Vinyl chloride

Experiment No.		Concentration, calculated, mg/m³	Concentration, experimental, mg/m³	Estimated error, %[a]
I	Gas samples	64	71.2 ± 0.7[b]	
	Sorbent tubes	64	69.8 ± 1.5	-2
II	Gas samples	13	14.5 ± 0.5	
	Sorbent tubes	13	13.6 ± 0.4	-6
III	Gas samples	2.6	2.88 ± 0.07	
	Sorbent tubes	2.6	2.91 ± 0.13	+1
IV	Gas samples	1.3	-	
	Sorbent tubes	1.3	1.27 ± 0.09	—

a. The estimated error is the average of concentrations determined from sorbent tubes minus the average of concentrations determined from gas samples, divided by the average of concentrations determined from gas samples, multiplied by 100.

b. The number given is the mean value plus or minus the 95% confidence level. The 95% confidence level is defined as the standard deviation multiplied by Student's t at the 0.05 significance level, divided by the square root of the number of samples.

amount of vinyl chloride that the tube will hold before it becomes overloaded. When the sample value obtained for the backup section of the sorbent tube exceeds 20% of that found on the front section, the possibility of sample loss exists. During storage, volatile compounds such as vinyl chloride will migrate throughout the tube until equilibrium is reached. At this time, 33% of these compounds will be found in the backup section. This may lead to some confusion as to whether sample loss has occurred. This migration effect can be considered decreased by shipping and storing the tubes at -20° C.

The precision of the method is limited by the reproducibility of the pressure drop and, therefore, the flow rates across the tubes. Because the pump is usually calibrated for one particular tube, differences in flow rates can occur when sampling through other tubes and can cause sample volumes to vary.

Apparatus

Personal sampling pump. The pump should be a properly calibrated personal sampling pump for personal and area samples. It should be calibrated with a representative sorbent tube in the sampling line. A dry or wet test meter or a glass rotameter that will determine the flow rate (50 mL/min) to within 5% may be used for the calibration.

Sorbent tubes. The glass tubes have both ends flame sealed. Each is 7 cm long, 6 mm O.D., 4 mm I.D., and contains two sections of 20/40 mesh activated carbon separated by a 2 mm portion of urethane foam. The activated carbon is prepared from coconut shells and is fired at 600° C. prior to packing to remove absorbed materials. The primary adsorbing section contains 100 mg of sorbent, the backup section 50 mg. A 3 mm portion of urethane foam is placed between the outlet end of the tube and the backup section. A plug of silanized glass wool is placed in front of the adsorbing section. The pressure drop across the tube must be less than 2 in of water at a flow rate of 0.2 L/min.

Gas chromatograph. Equipped with a flame-ionization detector.
Stainless steel column. 20 ft × 0.125 in, packed with 10% SE-30 on 80/100 mesh Chromosorb

W (acid washed, silanized with dimethyldichlorosilane). Other columns capable of performing the required separations may be used.

Electronic integrator. Or a recorder and some method for determining peak area.

Vials. 2 mL, capable of being sealed with caps containing Teflon-lined silicone rubber septa.

Syringes. 10 µL, and convenient sizes for making standards.

Gas-tight syringe. 1 mL, with a gas tight valve.

Pipets. 0.5 mL delivery pipets or 1.0 mL type graduated in 0.1 mL increments.

Volumetric flasks. 10 mL, or convenient sizes for making solutions. It is preferable to have plastic stoppers for the volumetric flasks.

Reagents

Carbon disulfide, spectroquality or better grade.
Vinyl chloride, lecture bottle, 99.9% minimum purity.
Toluene, chromatographic quality.
Purified helium.
Prepurified hydrogen.
Filtered compressed air.

Procedure

Cleaning of equipment. All glassware used for the laboratory analysis should be detergent washed and thoroughly rinsed with tap water and distilled water.

Collection and shipping of samples. Immediately before sampling, break the ends of the tube to provide an opening at least one-half the internal diameter of the tube (2 mm). The second sorbent tube is used as a backup and is positioned immediately downstream and in tandem with the first tube. The sorbent tubes are placed in a vertical position with the larger section of sorbent pointing up during sampling to minimize channeling of the vinyl chloride through the sorbent. Air being sampled is not to be passed through any hose or tubing before entering the sorbent tubes.

The flow rate and time, or volume, must be measured as accurately as possible. The sample is taken at a flow rate of 50 mL/min. The maximum volume to be sampled should not exceed 5 L. Relatively large volumes (10-20 L) of air also should be sampled through other sorbent tubes at the same time personal samples are taken. These bulk air samples will be used by the analyst to identify possible interferences before the personal samples are analyzed. If the temperature and pressure of the atmosphere being sampled are significantly different from 25° C. or 760 mm Hg, they should be measured and recorded.

The sorbent tubes are capped with the supplied plastic caps immediately after sampling. Under no circumstances should rubber caps be used. One tube is handled in the same manner as the sample tube (break, seal, and transport), except that no air is sampled through this tube. This tube is labeled as a blank. Capped tubes are packed tightly before they are shipped to minimize tube breakage during transport to the laboratory. The use of two tubes in series has eliminated the need for cooling during shipping. However, if two tubes are not used, i.e., only one tube is used, and if the samples will spend a day or more in transit, then cooling (e.g., with dry ice) is necessary to minimize migration of vinyl chloride to the backup section..

Samples received at the laboratory are logged in and immediately stored in a freezer (around -20° C.) until time for analysis. Samples may be stored in this manner for long periods of time with no appreciable loss of vinyl chloride (2 months). Even around -20° C., vinyl chloride will equilibrate between the two sections of activated carbon, i.e., it will migrate to the backup section. This phenomenon is observable after 2 weeks and may be confused with sample loss after 1-2 months.

Vinyl chloride

Preparation and desorption of samples. The two tubes used in the collection of a single sample are analyzed separately. Each tube is scored with a file and broken open at each end. The glass wool is discarded. Both sections of each tube are transferred to a small vial containing 1 mL of carbon disulfide. It is important to add the sorbent to the carbon disulfide and not the carbon disulfide to the sorbent. The vial is topped with a septum cap. The separating section in each tube is discarded. Tests indicate that desorption is complete in 30 min if the sample is agitated occasionally during this period. The samples should be analyzed within 60 min after addition to carbon disulfide. If only one tube is used for sampling, then each section of activated carbon should be analyzed separately.

GC conditions. The typical operating conditions for the gas chromatograph are:

40 mL/min (80 psig) helium carrier gas flow
65 mL/min (20 psig) hydrogen gas flow to detector
500 mL/min (50 psig) air flow to detector
230° C. injector temperature
230° C. detector temperature
60° C. column temperature

Injection. The first step in the analysis is the injection of the sample into the gas chromatograph. To eliminate difficulties arising from blow back or distillation within the syringe needle, one should employ the solvent flush injection technique. The 10 μL syringe is first flushed with solvent several times to wet the barrel and plunger. Two μL of solvent are drawn into the syringe to increase the accuracy and reproducibility of the injected sample volume. The needle is removed from the solvent, and the plunger is pulled back about 0.4 μL to separate the solvent flush from the sample with a pocket of air to be used as a marker. The needle is then immersed in the sample, and a 5 μL aliquot is withdrawn to the 7.4 μL mark (2 μL solvent + 0.4 μL air + 5 μL sample = 7.4 μL). After the needle is removed from the sample and prior to injection, the plunger is pulled back 1.2 μL to minimize evaporation of the sample from the tip of the needle. Duplicate injections of each sample and standard should be made. No more than a 3% difference in area is to be expected. Automatic sampling devices may also be used.

Measurement of area. The area of the sample peak is measured by an electronic integrator or some other suitable form of area measurement, and preliminary results are read from a standard curve prepared as discussed below.

Determination of desorption efficiency. The desorption efficiency of a particular compound can vary from one laboratory to another and also from one batch of sorbent to another. Thus, it is necessary to determine at least once the percentage of vinyl chloride that is removed in the desorption process. Desorption efficiency should be determined on the same batch of sorbent tubes used in sampling. Results indicate that desorption efficiency varies with loading (total vinyl chloride on the tube), particularly at lower values, e.g., 2.5 μg.

Sorbent tubes from the same batch as that used in obtaining samples are used in this determination. A measured volume of vinyl chloride gas is injected into a bag containing a measured volume of air. The bag is made of Tedlar (or a material that will retain the vinyl chloride and not absorb it) and should have a gas sampling valve and a septum injection port. The concentration in the bag may be calculated if room temperature and pressure are known. A measured volume is then sampled through a sorbent tube with a calibrated sampling pump. At least 5 tubes are prepared in this manner. These tubes are desorbed and analyzed in the same manner as the samples. Samples taken with a gas tight syringe from the bag are also injected into the gas chromatograph. The concentration in the bag is compared to the concentration obtained from the tubes.

The desorption efficiency equals the amount of vinyl chloride desorbed from the sorbent divided by the quantity of vinyl chloride contained in the volume of synthetic atmosphere sampled or:

D.E. = (amount of vinyl chloride from sorbent) ÷ (conc of vinyl chloride in bag)(vol of atmosphere sampled)

Calibration and Standards

CAUTION: Laboratory operations involving carcinogens. Vinyl chloride has been identified as a human carcinogen and appropriate precautions must be taken in handling this gas. The Occupational Safety and Health Administration has promulgated regulations for the use and handling of vinyl chloride. They may be found in 29 CFR 1910.93q (section 1910.93q in title 29 of the Code of Federal Regulations available in the Federal Register, Vol. 39, No. 194, Friday, October 4, 1974, pp. 35890-35898).

A series of standards, varying in concentration over the range of interest, is prepared and analyzed under the same gas chromatographic conditions and during the same time period as the unknown samples. Curves are established by plotting concentration in μg/mL versus peak area. There are two methods of preparing standards and, as long as highly purified vinyl chloride is used, both are comparable. If no internal standard is used in the method, standard solutions must be analyzed at the same time that the sample analysis is done. This will minimize the effect of day-to-day variations of the flame ionization response.

Gravimetric method. Vinyl chloride is slowly bubbled into a weighed 10 mL volumetric flask containing approximately 5 mL of toluene. After 3 min, the flask is again weighed. A weight change of 100-300 mg is usually observed. The solution is diluted to exactly 10 mL with carbon disulfide and is used to prepare other standards by removal of aliquots with different sized syringes. Subsequent dilution of these aliquots with carbon disulfide results in a series of values that are linear from the range of 0.2 ng/injection, the minimum detectable amount of vinyl chloride, to 1.5 μg/injection.

Volumetric method. A 1 mL gas sample of pure vinyl chloride is drawn into a gas tight syringe and the valve is closed. The tip of the needle is inserted into a 10 mL volumetric flask containing approximately 5 mL of CS_2. The valve is opened and the plunger is withdrawn slightly to allow the CS_2 to enter the syringe. The action of the vinyl chloride dissolving in the CS_2 creates a vacuum and the syringe becomes filled with the solvent. An air bubble (2%) is present and has been found to be due to the void volume in the needle of the syringe. The solution is returned to the flask and the syringe is rinsed with clean CS_2 and the washings added to the volumetric flask. The volumetric flask is then filled to the mark with CS_2. Other standards are then prepared from this stock solution.

Standards are stored in a freezer at -20° C. and have been found to be stable at this temperature for 3 days. Tight fitting plastic tops on the volumetric flasks seem to retain the vinyl chloride better than ground glass stoppers.

Calculations

The weight, in μg, corresponding to the area under each peak is read from the standard curve for vinyl chloride. No liquid volume corrections are needed because the standard curve is based on the number of μg in 1.0 mL of CS_2 and the volume of sample injected is identical to the volume of the standards injected. Corrections for the blank are made for each sample:

μg = μg sample − μg blank

Vinyl chloride

where:

μg sample = μg found on sample tube
μg blank = μg found in blank tube

A similar procedure is followed for the backup sections. The amounts present in the front and backup sections of the same sample tube are added to determine the total amount of vinyl chloride in the sample. The total amount is corrected for the desorption efficiency at the level of vinyl chloride measured:

Corrected amount (μg) = amount (μg) ÷ D.E.

The concentration of vinyl chloride in air may be expressed in mg/m³:

mg/m³ = corrected wt in μg ÷ volume of air sampled in L

The concentration may also be expressed in terms of ppm by volume:

ppm = (mg/m³)(24.45)(760)(T + 273) ÷ (M.W.)(P)(298)

where:

P = pressure (mm Hg) of air sampled
T = temperature (°C.) of air sampled
24.45 = molar volume (L/mol) at 25° C. and 760 mm Hg
M.W. = molecular weight
760 = standard pressure (mm Hg)
298 = standard temperature (°K)

References

1. Hill, R.H., C.S. McCammon, A.T. Saalwaechter, et al. Determination of vinyl chloride in air. Anal. Chem. 48: 1395, 1976.
2. White, L.D., D.G. Taylor, P.A. Mauer and R.E. Kupel. A convenient optimized method for the analysis of selected solvent vapors in the industrial atmosphere. Amer. Ind. Hyg. Assoc. J. 31: 225, 1970.

VINYLIDENE CHLORIDE

Synonym: 1,1-dichloroethene
ACGIH TLV: 5 ppm (20 mg/m^3)
OSHA Standard: 1 ppm
NIOSH Recommendation: as for vinyl chloride

$CCl_2=CH_2$

Method	Sampling Duration	Sampling Location	Useful Range (ppm)	System Cost ($)	Test Cost ($)	Manufacturer
DT	Grab	Portable	0.3-30	180	2.25	Bendix
IR	Cont	Portable	0.2-20	4374	0	Foxboro
PI	Cont	Fixed	0.2-20	4950	0	HNU

NIOSH METHOD NO. P&CAM 178

(Use as described for vinyl chloride, with 10 ft silanized glass column packed with 100/120 mesh Durapak OPN operated at 65° C.)

VINYLTOLUENE

Synonym: 60/40 mixture of m- and p-isomers
ACGIH TLV: 50 ppm (240 mg/m^3)
OSHA Standard: 100 ppm

$CH_3C_6H_4CH=CH_2$

Method	Sampling Duration	Sampling Location	Useful Range (ppm)	System Cost ($)	Test Cost ($)	Manufacturer
IR	Cont	Portable	2-200	4374	0	Foxboro

NIOSH METHOD NO. P&CAM 127

(Use as described for acetone, with 120° C. column temperature)

XYLENE

$CH_3C_6H_4CH_3$

Synonym: o-, m-, or p-dimethylbenzene
ACGIH TLV: 100 ppm (435 mg/m³) - Skin
OSHA Standard: 100 ppm
NIOSH Recommendation: 100 ppm TWA; 200 ppm ceiling (10 min)

Method	Sampling Duration	Sampling Location	Useful Range (ppm)	System Cost ($)	Test Cost ($)	Manufacturer
DT	Grab	Portable	25-1600	150	2.60	Nat'l Draeger
DT	Grab	Portable	5-1000	165	1.70	Matheson
DT	Grab	Portable	10-500	180	2.25	Bendix
IR	Cont	Portable	2-200	4374	0	Foxboro
PI	Cont	Fixed	2-200	4950	0	HNU
GC	Cont	Portable	$1-10^6$	10,000	0	Microsensor

NIOSH METHOD NO. P&CAM 127

(See acetone)

APPENDIX

Equipment Manufacturers

Airco Industrial Gases
575 Mountain Avenue
Murray Hill, NJ 07974
(201) 464-8100

Bacharach Instrument Company
625 Alpha Drive
Pittsburgh, PA 15238
(412) 782-3500

Bendix Corporation
12345 Starkey Road
Largo, FL 33543
(813) 536-6523

CEA Instruments, Inc.
15 Charles Street
Westwood, NJ 07675
(201) 664-2300

Dynamation, Inc.
168 Enterprise Drive
Ann Arbor, MI 48103
(313) 769-0573

Energetics Science
6 Skyline Drive
Hawthorne, NY 10532
(914) 592-3010

Foxboro Analytical
140 Water Street
South Norwalk, CT 06856
(203) 853-1616

GCA Corporation
213 Burlington Road
Bedford, MA 01730
(617) 275-5444

HNU Systems, Inc.
160 Charlemont Street
Newton Highlands, MA 02161
(617) 964-6690

Houston Atlas, Inc.
9441 Baythorne Drive
Houston, TX 77041
(713) 462-6116

Interscan Corporation
21700 Nordhoff Street
Chatsworth, CA 91311
(213) 882-2331

Jerome Instrument Corporation
P.O. Box 336
Jerome, AZ 86331
(602) 634-4263

Matheson
932 Paterson Plank Road
East Rutherford, NJ 07073
(201) 933-2400

MDA Scientific, Inc.
1815 Elmdale
Glenview, IL 60025
(312) 998-1600

Microsensor Technology, Inc.
47747 Warm Springs Blvd.
Fremont, CA 94539
(415) 490-0900

National Draeger, Inc.
101 Technology Drive
Pittsburgh, PA 15275
(412) 787-8383

Rexnord
45 Great Valley Parkway
Malvern, PA 19355
(215) 647-7200

Analytical Accessories Suppliers

Ace Glass Company
1430 N.W. Boulevard
Vineland, NJ 08360
(609) 692-2050
(bubbler tubes)

Applied Science Laboratories, Inc.
P.O. Box 440
State College, PA 16801
(800) 458-3769
(chromatographic supplies, sorbent materials)

Calibrated Instruments, Inc.
731 Saw Mill River Road
Ardsley, NY 10502
(914) 693-9232
(gas sampling bags)

SKC, Inc.
395 Valley View Road
Eighty-Four, PA 15330
(412) 941-9701
(pumps, impingers, filters, sorbent tubes)

Supelco, Inc.
Supelco Park
Bellefonte, PA 16823
(814) 359-2784
(chromatographic supplies, sorbent materials)

INDEX

acetaldehyde, 1
acetic acid, 6
acetic anhydride, 11
acetone, 14
acetonitrile, 20
acetylene dichloride, 155
acetylene tetrabromide, 25
acetylene tetrachloride, 393
acrolein, 30
acrylaldehyde, 30
acrylonitrile, 33
AGE, 43
allyl alcohol, 37
allyl chloride, 42
allyl 2,3-epoxypropyl ether, 43
allyl glycidyl ether, 43
aminoethane, 193
2-aminoethanol, 190
aminomethane, 302
ammonia, 49
n-amyl acetate, 51
aniline, 51
arsine, 57
aziridine, 206

benzene, 62
benzyl chloride, 62
BGE, 84
bis-β-chloroethyl ether, 156
bis(chloromethyl)ether, 63
bis-CME, 63
bromochloromethane, 107
bromoethane, 194
bromoform, 66
bromomethane, 303
1,3-butadiene, 67
1-butanethiol, 85
n-butanol, 73
sec-butanol, 78
tert-butanol, 79
2-butanone, 67
2-butenol, 136
1-n-butoxy-2,3-epoxypropane, 84
2-butoxyethanol, 68

n-butyl acetate, 73
n-butyl alcohol, 73
sec-butyl alcohol, 78
tert-butyl alcohol, 79
n-butylamine, 79
butyl cellosolve, 68
n-butyl glycidyl ether, 84
n-butyl mercaptan, 85
p-tert-butyltoluene, 90

carbolic acid, 346
carbon dioxide, 91
carbon disulfide, 94
carbon hexachloride, 249
carbon monoxide, 100
carbon tetrachloride, 103
carbonyl chloride, 353
cellosolve, 191
chlorine, 104
chlorobenzene, 107
chlorobromomethane, 107
2-chloro-1,3-butadiene, 114
1-chloro-2,3-epoxypropane, 190
chloroethane, 195
2-chloroethanol, 196
chloroethene, 411
chloroethylene, 411
chloroform, 108
chloroformyl chloride, 353
chloromethane, 304
1-chloro-1-nitropropane, 108
chloroprene, 114
3-chloropropene, 42
chromate, 120
chromic acid, 120
cinnamene, 377
cobalt dust or fume, 123
copper dust or fume, 127
cresol, 131
cristobalite, 373
crotonaldehyde, 136
cumene, 140
cyanide, 141
cyclohexane, 145

cyclohexanol, 146
cyclohexanone, 146
cyclohexene, 147
cyclohexylmethane, 305

DDVP, 157
DEAE, 163
diacetone alcohol, 147
dibromodifluoromethane, 169
1,2-dibromoethane, 201
o-dichlorobenzene, 148
p-dichlorobenzene, 148
dichlorodifluoromethane, 149
1,1-dichloroethane, 155
1,2-dichloroethane, 206
1,1-dichloroethene, 418
1,2-dichloroethylene, 155
dichloroethyl ether, 156
dichloromethane, 307
dichloromonofluoromethane, 156
1,2-dichloropropane, 363
dichlorotetrafluoroethane, 157
dichlorvos, 157
diethylamine, 163
2-diethylaminoethanol, 163
diethylenimide oxide, 315
diethyl ether, 217
difluorodibromomethane, 169
dihydroazirine, 206
diisobutyl ketone, 174
diisopropylamine, 174
diisopropyl ether, 277
diisopropylideneacetone, 275
dimethoxymethane, 296
dimethylacetamide, 175
N,N-dimethylacetamide, 175
dimethylbenzene, 419
dimethylformamide, 180
N,N-dimethylformamide, 180
2,6-dimethyl-4-heptanone, 174
dimethyl sulfate, 180
2,4-dinitrotoluene, 185
1,4-dioxane, 189
p-dioxane, 189
diphenyl ether, 350
divinyl, 67
DMF, 180

EDB, 201

EGDN, 335
epichlorohydrin, 190
epihydric alcohol, 243
1,2-epoxyethane, 211
1,2-epoxypropane, 363
2,3-epoxy-1-propanol, 243
1,2-ethanediamine, 196
ethanol, 193
ethanolamine, 190
ethenyl ethanoate, 406
ether, 217
ethinyl trichloride, 400
2-ethoxyethanol, 191
2-ethoxyethyl acetate, 191
ethyl acetate, 192
ethyl acrylate, 192
ethyl alcohol, 193
ethylamine, 193
ethyl amyl ketone, 308
ethylbenzene, 194
ethyl bromide, 194
ethyl butyl ketone, 195
ethyl chloride, 195
ethylene chlorohydrin, 196
ethylenediamine, 196
ethylene dibromide, 201
ethylene dichloride, 206
ethylene glycol dinitrate, 335
ethyleneimine, 206
ethylene oxide, 211
ethyl ether, 217
ethyl formate, 222
ethylidene chloride, 155
ethyl orthosilicate, 222
ethyl propenoate, 192
ethyl silicate, 222

F-11, 228
F-12, 149
F-13B1, 402
F-21, 156
F-112, 392
F-113, 401
F-114, 157
ferric oxide, 269
fluorotrichloromethane, 228
formaldehyde, 228
formic acid, 234
Freon 11, 228

Freon 12, 149
Freon 13B1, 402
Freon 21, 156
Freon 112, 392
Freon 113, 401
Freon 114, 157
2-furaldehyde, 238
2-furancarbinol, 243
2-furancarbonal, 238
furfural, 238
furfuraldehyde, 238
furfuryl alcohol, 243

glycidol, 243
glycol monoethyl ether, 191
glycol monomethyl ether, 303

n-heptane, 248
2-heptanone, 302
3-heptanone, 195
hexachloroethane, 249
hexahydrocresol, 306
Hexalin, 146
n-hexane, 249
2-hexanone, 250
hexone, 250
hydrazine, 251
hydrocarbons, 345, 376
hydrochloric acid, 254
hydrocyanic acid, 258
hydrofluoric acid, 259
hydrogen chloride, 254
hydrogen cyanide, 258
hydrogen fluoride, 259
hydrogen sulfide, 263
hydroxybenzene, 346
4-hydroxy-4-methyl-2-pentanone, 147

iodomethane, 308
iron oxide fume, 269
isoamyl acetate, 273
isoamyl alcohol, 274
isobutyl acetate, 274
isobutyl alcohol, 275
isobutylmethylcarbinol, 313
isophorone, 275
isopropanol, 276
isopropyl acetate, 276
isopropyl alcohol, 276

isopropylamine, 277
isopropylbenzene, 140
isopropyl ether, 277
isovalerone, 174

LPG, 278
liquified petroleum gas, 278

magnesium oxide fume, 279
MEK, 67
mercury vapor, 283
mesityl oxide, 291
methanoic acid, 234
methanol, 297
2-methoxyethanol, 303
2-methoxyethyl acetate, 304
methyl acetate, 291
methyl acetylene, 292
methyl acrylate, 296
methylal, 296
methyl alcohol, 297
methylamine, 302
methyl n-amyl ketone, 302
N-methylaniline, 315
o-methylaniline, 399
methylbenzene, 394
methyl bromide, 303
3-methyl-1-butanol, 274
3-methyl-1-butanol acetate, 273
methyl n-butyl ketone, 250
methyl cellosolve, 303
methyl cellosolve acetate, 304
methyl chloride, 304
methyl chloroform, 305
methylcyclohexane, 305
methylcyclohexanol, 306
2-methylcyclohexanone, 306
o-methylcyclohexanone, 306
1-methyl-2,4-dinitrobenzene, 185
methylene chloride, 307
methyl ethyl ketone, 67
methyl formate, 307
5-methyl-3-heptanone, 308
methyl iodide, 308
methylisobutylcarbinol, 313
methyl isobutyl ketone, 250
methyl methacrylate, 314
methyloxirane, 363
4-methyl-2-pentanol, 313

4-methyl-2-pentanone, 250
4-methyl-3-penten-2-one, 291
methylphenol, 131
2-methyl-1-propanol, 275
methyl propyl ketone, 345
α-methylstyrene, 314
methyl sulfate, 180
MIBK, 250
monochlorobenzene, 107
monomethylaniline, 315
morpholine, 315
muriatic acid, 254

nickel, 316
NIOSH Method No.
  P&CAM 106, 373
  P&CAM 112, 100
  P&CAM 127, 14
  P&CAM 141, 395
  P&CAM 168, 52
  P&CAM 175, 283
  P&CAM 178, 411
  P&CAM 187, 386
  P&CAM 202, 33
  P&CAM 203, 335
  P&CAM 205, 49
  P&CAM 209, 104
  P&CAM 211, 30
  P&CAM 213, 63
  P&CAM 219, 354
  P&CAM 221, 79
  P&CAM 231, 324
  P&CAM 232, 234
  P&CAM 235, 228
  P&CAM 247, 297
  P&CAM 260, 201
  P&CAM 265, 58
  P&CAM 270, 163
  P&CAM 276, 196
  P&CAM 278, 407
  P&CAM 285, 136
  P&CAM 290, 403
  P&CAM 298, 316
  P&CAM 295, 157
  P&CAM 300, 206
  P&CAM 301, 180
S4, 263
S8, 341
S17, 238
S52, 37
S66, 73
S70, 243
S76, 68
S80, 217
S84, 292
S93, 278
S98, 308
S107, 169
S111, 149
S112, 114
S117, 25
S160, 350
S161, 364
S165, 20
S167, 131
S169, 6
S170, 11
S176, 259
S188, 369
S203, 123
S211, 108
S215, 185
S217, 329
S237, 251
S244, 383
S245, 389
S246, 254
S248, 94
S249, 91
S250, 141
S254, 175
S264, 222
S286, 212
S308, 378
S317, 120
S319, 320
S330, 346
S332, 356
S345, 1
S346, 43
S350, 85
S354, 127
S366, 269
S369, 279
nitric acid, 319
nitric oxide, 324
nitrobenzene, 329
nitroethane, 334